Ti-Based Biomaterials

Ti-Based Biomaterials

Synthesis, Properties and Applications

Special Issue Editor
Jarosław Jakubowicz

MDPI • Basel • Beijing • Wuhan • Barcelona • Belgrade • Manchester • Tokyo • Cluj • Tianjin

Special Issue Editor
Jarosław Jakubowicz
Institute of Materials Engineering,
Poznan University of Technology
Poland

Editorial Office
MDPI
St. Alban-Anlage 66
4052 Basel, Switzerland

This is a reprint of articles from the Special Issue published online in the open access journal *Materials* (ISSN 1996-1944) (available at: https://www.mdpi.com/journal/materials/special_issues/Ti_Based_Biomaterials).

For citation purposes, cite each article independently as indicated on the article page online and as indicated below:

LastName, A.A.; LastName, B.B.; LastName, C.C. Article Title. *Journal Name* **Year**, *Article Number*, Page Range.

ISBN 978-3-03928-987-5 (Hbk)
ISBN 978-3-03928-988-2 (PDF)

Contents

About the Special Issue Editor

Jarosław Jakubowicz, he is a full professor working at Poznan University of Technology (Poland) in the field of materials science. His scientific career started in 1997 at Poznan University of Technology, when he completed his M.Sc. study. In 1999, at Warsaw University of Technology (Poland) he finished his Ph.D. on the topic of nanocrystalline hard magnetic materials. In 2001/2002, he held a postdoc position in Helmholtz-Zentrum Berlin (Germany). In 2009, at Military Technical Academy in Warsaw (Poland) he finished D.Sc. (habilitation) on the topic of silicon electrochemistry. He became a full professor in 2019. He is the author of over 160 papers, several books, and chapters related to nanomaterials and biomaterials. His H-index is 16. He has promoted 4 PhDs. He is a member of the Surface Engineering Team of the Materials Science Committee Polish Academy of Sciences. He was the head of 3 scientific grants in the field of biomaterials engineering financed by the Polish National Science Centre. He has previously held positions of Faculty Vice Dean and Institute Vice Head. Currently, he is the Head of the Division of Functional Nanomaterials in the Institute of Materials Engineering. Currently, his field of interest is related to biomaterials and their surface treatment, as well as nanocrystalline materials. Further information on his scientific achievements is available at https://orcid.org/0000-0001-6339-0796 and https://publons.com/researcher/2528908/jaroslaw-jakubowicz/.

Preface to "Ti-Based Biomaterials"

Dear Readers,

Recently, great attention has been paid to biomaterials that can be safely used in the human body to prepare parts that replace failed bone structures. Of all materials, Ti-based materials are the most desirable, because they provide an optimum combination of mechanical, chemical, and biological properties. Ti and its compounds are one of the most-investigated biomaterials, and, thus, this book has been proposed as a means to present recent developments in the field of Ti-based biomaterials and is a compilation of 20 articles included in this Special Issue of the MDPI journal Materials, entitled "Ti-Based Biomaterials: Synthesis, Properties and Applications" (https://www.mdpi.com/journal/materials/special_issues/Ti_Based_Biomaterials). The articles included in this Special Issue demonstrate the top results of scientists working in Australia, Brazil, China, Czech Republic, Egypt, Germany, Italy, Japan, Poland, Romania, Slovenia, South Africa, South Korea, Sweden, and Taiwan. Most of the articles are the results of international cooperation. The included articles mainly focus on β-type Ti alloys, shape memory alloys, nanostructure formation, additive manufacturing, and surface treatment. Microstructure, mechanical properties, and biocompatibility were reported.

The goal of this Special Issue is to provide comprehensive and broad knowledge about Ti-based biomaterials. The content of the book can be useful for students, engineers, researchers, academics, and other specialists whose interests are focused on Ti-based biomaterials development, manufacturing, and application.

I express my appreciation to all of the authors for their contribution.

Jarosław Jakubowicz
Special Issue Editor

materials

MDPI

Editorial

Special Issue: Ti-Based Biomaterials: Synthesis, Properties and Applications

Jarosław Jakubowicz

Poznan University of Technology, Institute of Materials Engineering, Jana Pawla II no 24, 61-138 Poznań, Poland; jaroslaw.jakubowicz@put.poznan.pl; Tel.: +48-061-665-3571

Received: 2 April 2020; Accepted: 3 April 2020; Published: 5 April 2020

Abstract: In the last half century, great attention has been paid to materials that can be used in the human body to prepare parts that replace failed bone structures. Of all materials, Ti-based materials are the most desirable, because they provide an optimum combination of mechanical, chemical and biological properties. The successful application of Ti biomaterials has been confirmed mainly in dentistry, orthopedics and traumatology. The Ti biomaterials provide high strength and a relatively low Young's modulus. Titanium biocompatibility is practically the highest of all metallic biomaterials, however new solutions are being sought to continuous improve their biocompatibility and osseointegration. Thus, the chemical modification of Ti results in the formation of new alloys or composites, which provide new perspectives for Ti biomaterials applications. Great attention has also been paid to the formation of nanostructures in Ti-based biomaterials, which has leads to extremely good mechanical properties and very good biocompatibility. Additionally, the surface treatment applied to Ti-based biomaterials provides faster osseointegration and improve in many cases mechanical properties. The special issue "Ti-Based Biomaterials: Synthesis, Properties and Applications" has been proposed as a means to present recent developments in the field. The articles included in the special issue cover broad aspects of Ti-based biomaterials formation with respect to design theirs structure, mechanical and biological properties, as highlighted in this editorial.

Keywords: titanium; biomaterials; biofunctionalization

Biomaterials, especially metal ones, are currently one of the most developed and used groups of materials. Titanium is the basic and one from the best biomaterial elements used in the pure form or as the main component in the biocompatible alloys or composites. Very good properties of titanium and its alloys are not able to cover up a few defects, which include for example chemical reactivity with atmospheric gases. Pure Ti, Ti alloys and Ti matrix composites are widely used in different fields. The titanium materials classification is based on structure at equilibrium state. Alloys of α, $\alpha + \beta$ and β structure are used in biomedical applications. The last one seems to be the most perspective, because theirs high biocompatibility and lowest among Ti alloys Young's modulus, close to that of human bone, causes reduce stress shielding. Careful choice of chemical composition and processing parameters allow to design titanium biomaterials for most biomedical applications. The properties of Ti-based biomaterials should be as close as possible into the properties of the body element that the implant replaces. Hence the Young's modulus is a key factor in biomaterials selection, preferring the titanium for hard tissue implants. Despite the very good properties of titanium and its alloys or composites, some properties are inadequate for biomedical device requirements. For example, wear resistance is poor and requires surface treatment. Deposition of more biocompatible surface layers provides safer used of implants. The nanocrystalline materials are a new group with interesting properties and a growing importance among of titanium biomaterials.

The special issue Ti-Based Biomaterials: Synthesis, Properties and Applications has been proposed to shows recent developments in the field of most extensively investigated and promising metallic biomaterials, and for this reason the nineteen articles included many different, however related aspects of Ti-based biomaterials. All the articles published in the special issue show the most important achievements in Ti-based biomaterials, useful for broader applications. A significant part of the articles included in the special issue focus on β-type Ti alloys, which are the most interesting among Ti alloys family. The β-type Ti alloys shows low modulus of elasticity and high biocompatibility, pointing the perspective and successful applications in hard tissue surgery. In the article by Bai et al. [1], the perspective β-type Ti-Mo alloys modified with Zr were presented. In the Ti-Mo-Zr system the diffusion coefficients were assessed to develop a self-consistent atomic mobility database of bcc phase. The calculated diffusion coefficients were compared with the experimental results showing good agreement. Sandu et al. [2] presents new β-type Ti-Mo alloys modified by Si, Zr and Ta, which shows very low Young's modulus with perspective dental and orthopedic implant applications. Kuroda et al. [3] focus on the β-type Ti-Ta-Zr ternary alloys. The article shows combination of different processing routes (hot-rolling, annealing and solution treatment) for design of alloys of different mechanical properties (hardness and Young's modulus). Fowler et al. [4] modified Ti-Nb-Ta-Zr alloy by addition of antibacterial Cu. As the effect of microstructural changes, the few % of Cu addition results in significant increase of alloys hardness. Consequently, in work [5], Fowler et al. continue study of Cu effect on antibacterial properties of Ti-Cu alloys. Generally, the Ti alloys had good antibacterial rates above 5 wt.% Cu content, however, the un-aged Ti-10 wt.% Cu alloy provide a good antibacterial effect, with superior hardness and corrosion protection. In addition to bulk β alloys, its porous form is very attractive for implant application. Formation of β-type Ti-Ta foams of 60%–76% porosity in the dealloying process was presented by Adamek et al. [6]. The nanocrystalline Ti-Ta-Mg powder mixture was prepared by mechanical alloying. Cold pressing followed by sintering at temperature above boiling point of Mg lead to its removal from the alloy matrix leaving the pores. The foams show mechanical and biological properties respective for hard tissue implant application.

Shape memory alloys are the second group of Ti-based materials with growing research interest in biomedical industry. Lee et al. [7] shows mechanical stability of the elastic nails made of the Nitinol shape memory alloy and compare them to titanium and stainless steel nails in the fixation of diaphyseal long bone fractures. The titanium nails have been recommended as C-shaped ones for diaphyseal long bone fractures when the end cap is not used. For Nitinol and stainless steel nails, use of the end cap prevent the nail from dropping out and provide bone fracture stabilization [7]. The Finite Element Method (FEM) model was compared with experimental results. The Nitinol was studied by Feng et al. [8], with respect to surface properties. The surface was modified using magnetic field-assisted Electrical Discharge Machining (EDM). Under the careful control of discharge conditions is generated a porous surface of different morphology and roughness, which finally effect on surface wettability. The main problem related with most commonly used Nitinols as shape memory alloys in medical industry is Ni as carcinogen. Nagoshi et al. [9] shows shape memory Ni-free β-type Ti-Nb alloys. The mechanical behavior of single crystalline micropillars was investigated in the microscale at various temperatures and crystal orientations to shows anisotropic deformation behavior and shape memory effect. The shape memory parts can work in biomedical microelectromechanical systems (Bio-MEMS).

Another group of Ti-based materials was presented by Miklaszewski and Jurczyk [10]. The Ti-TiB Metal Matrix Composites (MMC) were made using mechanical alloying of elemental Ti and B powders and pulse plasma sintering (PPS) for bulk sample generation. The Ti matrix is reinforced by TiB phase, which crystallizes during powders sintering in the PPS process. The processing conditions strongly effect on the structure and high mechanical properties of the ultrafine grains Ti-TiB composite.

New manufacturing methods are as often investigated as the Ti-based biomaterials themselves. In the work by Wei et al. [11], the Laser Additive Method (LAM) for the production of β-type Ti-Nb alloy has been shown. They used mixture of the elemental powders, which was directly melted under the laser beam. The additive technology in formation of alloy by melting of the powders mixture has many advantages over the melting prealloyed powders. The further annealing plays a key role for structure optimization and tailoring the mechanical properties. In the work by Bassoli and Denti [12], they focus on additive manufacturing (Laser Powder Bed Fusion—L-PBF) applied for synthesis Ti-6Al-4V for dental applications. The mechanical and structural properties were investigated, correlated and compared to those obtained for Co-Cr-Mo alloys. They found absence of secondary anisotropy in the L-PBF prepared specimens. Hsu et al. [13] applied Selective Electron Beam Additive Manufacturing (SEBAM) for the formation of Ti-6Al-4V under gradient energy density carried out on a specimen from the bottom to the top. This implies structural changes affecting the mechanical behavior.

According to Hall-Petch relation, the grain refinement leads to increase of materials strength. This relationship is used in Ti and its compounds. Different methods, based on severe plastic deformation (SPD) processes provide nano- or ultrafine structure of high strength and additionally good biocompatibility. Palán et al. [14] show properties of pure Ti after SPD processes (conform SPD and rotary swaging). They produced wires with ultrafine grains for medical implants. The process can be used in the large scale with low cost and the final products have ultimate strength around 1 GPa for Grade 2 cp-Ti.

Titanium shows good biocompatibility, however in some cases not sufficient for many applications. For the better osseointegration a different surface treatment is used for Ti parts surface modification. In the work by Elsayed et al. [15], the sphene (Titanite—$CaTiSiO_5$) bioactive coatings were deposited on Ti substrate by airbrush spray technology using preceramic polymer with nano-sized precursors. The room temperature process deposition with subsequent heating results in crack-free coatings. The sphene coatings have good chemical stability, support hADSCs attachment, proliferation and differentiation in vitro. Authors point a potential for use of sphene bioactive coatings on orthopedic and dental implants. The surface treatment was undertaken by Benčina et al. [16] to generate TiO_2 nanotubes on Ti substrate. The work describes crystallization process of the electrochemically made amorphous TiO_2 nanotubes under a low pressure non-thermal oxygen plasma conditions. The surface can be potentially used for blood contacting devices because no platelet adhesion or activation on surfaces was observed. The biocompatibility improvement of the titanium maxillofacial plates treated with amine plasma-polymerization has been presented by Jeong et al. [17]. The Ti surface treatment undergoes through formation of polyallylamine ultra-thin film, which improved the hydrophilicity and biocompatibility confirmed in both in vitro and in vivo tests. The process applied for Ti surfaces can lead to shorten the time required for osseointegration and bone regeneration. In addition to the Ti surface modification, the Ti coatings can be deposited on different substrates for improvement their biocompatibility. In the work [18], Wypych et al. shows the Titanium Plasma Spraying (TPS) applied to polymer substrates. The Ti coatings exhibit a typical for TPS multilayer morphology of low porosity and good adhesion to polyethylene and glass fiber-reinforced polyamide. The Ti coatings did not fall off the substrate after its significant bending deformation. The TPS coatings have significantly greater hardness and Young's modulus in comparison to the properties of the polymer substrates.

Finally, having materials and technology, it is possible to produce a biodevice. In this case, the numerical simulations including FEM can be helpful and are one of the steps for successful biodevice implementations. Gruenwald et al. [19] pointed that Ti electrical conductivity presents a challenge for the electromagnetic transmission of data and power. In article [19], the authors proposed a fast and practical method to determine the necessary transmission parameters for titanium encapsulated implants with inductive transmission, for example implanted infusion pump (with an external controlling device).

Summarizing, the special issue Ti-Based Biomaterials: Synthesis, Properties and Applications shows current achievements in Ti-based materials, pointing future research trends in this field of interest.

Conflicts of Interest: The author declares no conflict of interest.

References

1. Bai, W.; Xu, G.; Tan, M.; Yang, Z.; Zeng, L.; Wu, D.; Liu, L.; Zhang, L. Diffusivities and Atomic Mobilities in bcc Ti-Mo-Zr Alloys. *Materials* **2018**, *11*, 1909. [CrossRef] [PubMed]

2. Sandu, A.V.; Baltatu, M.S.; Nabialek, M.; Savin, A.; Vizureanu, P. Characterization and Mechanical Proprieties of New TiMo Alloys Used for Medical Applications. *Materials* **2019**, *12*, 2973. [CrossRef] [PubMed]
3. Kuroda, P.A.B.; Quadros, F.F.; Araújo, R.O.; Afonso, C.R.M.; Grandini, C.R. Effect of Thermomechanical Treatments on the Phases, Microstructure, Microhardness and Young's Modulus of Ti-25Ta-Zr Alloys. *Materials* **2019**, *12*, 3210. [CrossRef] [PubMed]
4. Fowler, L.; Janse Van Vuuren, A.; Goosen, W.; Engqvist, H.; Öhman-Mägi, C.; Norgren, S. Investigation of Copper Alloying in a TNTZ-Cu_x Alloy. *Materials* **2019**, *12*, 3691. [CrossRef] [PubMed]
5. Fowler, L.; Masia, N.; Cornish, L.A.; Chown, L.H.; Engqvist, H.; Norgren, S.; Öhman-Mägi, C. Development of Antibacterial Ti-Cu_x Alloys for Dental Applications: Effects of Ageing for Alloys with Up to 10 wt% Cu. *Materials* **2019**, *12*, 4017. [CrossRef] [PubMed]
6. Adamek, G.; Kozlowski, M.; Jurczyk, M.U.; Wirstlein, P.; Zurawski, J.; Jakubowicz, J. Formation and Properties of Biomedical Ti-Ta Foams Prepared from Nanoprecursors by Thermal Dealloying Process. *Materials* **2019**, *12*, 2668. [CrossRef] [PubMed]
7. Lee, P.-Y.; Chen, Y.-N.; Hu, J.-J.; Chang, C.-H. Comparison of Mechanical Stability of Elastic Titanium, Nickel-Titanium, and Stainless Steel Nails Used in the Fixation of Diaphyseal Long Bone Fractures. *Materials* **2018**, *11*, 2159. [CrossRef] [PubMed]
8. Feng, C.; Li, L.; Zhang, C.; Zheng, G.; Bai, X.; Niu, Z. Surface Characteristics and Hydrophobicity of Ni-Ti Alloy through Magnetic Mixed Electrical Discharge Machining. *Materials* **2019**, *12*, 388. [CrossRef] [PubMed]
9. Nagoshi, T.; Yasuda, T.; Otaki, N.; Tahara, M.; Hosoda, H.; Sone, M. Evaluation of the Shape Memory Effect by Micro-Compression Testing of Single Crystalline Ti-27Nb Ni-Free Alloy. *Materials* **2020**, *13*, 110. [CrossRef] [PubMed]
10. Miklaszewski, A.; Jurczyk, M. Mechanical Alloying and Electrical Current-Assisted Sintering Adopted for In Situ Ti-TiB Metal Matrix Composite Processing. *Materials* **2019**, *12*, 653. [CrossRef] [PubMed]
11. Wei, J.; Sun, H.; Zhang, D.; Gong, L.; Lin, J.; Wen, C. Influence of Heat Treatments on Microstructure and Mechanical Properties of Ti–26Nb Alloy Elaborated In Situ by Laser Additive Manufacturing with Ti and Nb Mixed Powder. *Materials* **2019**, *12*, 61. [CrossRef] [PubMed]
12. Bassoli, E.; Denti, L. Assay of Secondary Anisotropy in Additively Manufactured Alloys for Dental Applications. *Materials* **2018**, *11*, 1831. [CrossRef] [PubMed]
13. Hsu, T.-I.; Jhong, Y.-T.; Tsai, M.-H. Effect of Gradient Energy Density on the Microstructure and Mechanical Properties of Ti6Al4V Fabricated by Selective Electron Beam Additive Manufacture. *Materials* **2020**, *13*, 1509. [CrossRef] [PubMed]
14. Palán, J.; Procházka, R.; Džugan, J.; Nacházel, J.; Duchek, M.; Németh, G.; Máthis, K.; Minárik, P.; Horváth, K. Comprehensive Evaluation of the Properties of Ultrafine to Nanocrystalline Grade 2 Titanium Wires. *Materials* **2018**, *11*, 2522. [CrossRef] [PubMed]
15. Elsayed, H.; Brunello, G.; Gardin, C.; Ferroni, L.; Badocco, D.; Pastore, P.; Sivolella, S.; Zavan, B.; Biasetto, L. Bioactive Sphene-Based Ceramic Coatings on cpTi Substrates for Dental Implants: An In Vitro Study. *Materials* **2018**, *11*, 2234. [CrossRef] [PubMed]
16. Benčina, M.; Junkar, I.; Zaplotnik, R.; Valant, M.; Iglič, A.; Mozetič, M. Plasma-Induced Crystallization of TiO_2 Nanotubes. *Materials* **2019**, *12*, 626. [CrossRef] [PubMed]
17. Jeong, Y.-W.; Jung, S.; Han, J.J.; Park, H.-J.; Kim, R.Y.; Kim, B.-H.; Kook, M.-S. Effectiveness of Surface Treatment with Amine Plasma for Improving the Biocompatibility of Maxillofacial Plates. *Materials* **2019**, *12*, 2581. [CrossRef] [PubMed]
18. Wypych, A.; Siwak, P.; Andrzejewski, D.; Jakubowicz, J. Titanium Plasma-Sprayed Coatings on Polymers for Hard Tissue Applications. *Materials* **2018**, *11*, 2536. [CrossRef] [PubMed]
19. Gruenwald, W.; Bhattacharrya, M.; Jansen, D.; Reindl, L. Electromagnetic Analysis, Characterization and Discussion of Inductive Transmission Parameters for Titanium Based Housing Materials in Active Medical Implantable Devices. *Materials* **2018**, *11*, 2089. [CrossRef] [PubMed]

materials

MDPI

Article

Diffusivities and Atomic Mobilities in bcc Ti-Mo-Zr Alloys

Weimin Bai [1], Guanglong Xu [2], Mingyue Tan [1], Zhijie Yang [1], Lijun Zeng [1], Di Wu [1], Libin Liu [1,*] and Ligang Zhang [1,*]

1 School of Material Science and Engineering, Central South University, Changsha 410083, China; morrisbai@csu.edu.cn (W.B.); tmy_luna@163.com (M.T.); yangzhijiecsu@163.com (Z.Y.); ljzeng.mse@csu.edu.cn (L.Z.); wudi080133@163.com (D.W.)

2 Tech Institute for Advanced Materials & School of Materials Science and Engineering, Nanjing Tech University, Nanjing 211800, China; guanglongxu@njtech.edu.cn

* Correspondence: pdc@csu.edu.cn (L.L.); ligangzhang@csu.edu.cn (L.Z.); Tel.: +86-0731-8887-7732 (L.L. & L.Z.)

Received: 18 September 2018; Accepted: 5 October 2018; Published: 8 October 2018

Abstract: β-type (with bcc structure) titanium alloys have been widely used as artificial implants in the medical field due to their favorable properties. Among them, Ti-Mo alloy attracted numerous interests as metallic biomaterials. Understanding of kinetic characteristics of Ti alloys is critical to understand and manipulate the phase transformation and microstructure evolution during homogenization and precipitation. In this work, diffusion couple technique was employed to investigate the diffusion behaviors in bcc Ti-Mo-Zr alloys. The diffusion couples were prepared and annealed at 1373 K for 72 h and 1473 K for 48 h, respectively. The composition-distance profiles were obtained via electron probe micro-analysis (EPMA). The chemical diffusion coefficients and impurity diffusion coefficients were extracted via the Whittle-Green method and Hall method. The obtained diffusion coefficients were assessed to develop a self-consistent atomic mobility database of bcc phase in Ti-Mo-Zr system. The calculated diffusion coefficients were compared with the experimental results. They showed good agreement. Simulations were implemented by Dictra Module in Thermo-Calc software. The predicted composition-distance profiles, inter-diffusion flux, and diffusion paths are consistent with experimental data, confirming the accuracy of the database.

Keywords: bcc Ti-Mo-Zr alloys; Inter-diffusion coefficient; Impurity coefficient; Atomic mobility; CALPHAD modeling

1. Introduction

Due to their biocompatibility, low modulus, high specific strength, and high corrosion resistance, β-type (with bcc structure) and near β titanium alloys have been widely used as surgery implants [1–4]. Previous studies have verified that Nb, Zr, Mo, Hf, and Ta are effective alloying elements leading to low elastic modulus, nontoxic, and non-allergic β-Ti alloys [4–6]. Moreover, Zr can increase hardenability and corrosion resistance for the alloy [7,8]. Mo is a strong β-phase stabilizing element for titanium alloys and the Ti-Mo based alloys exhibit adequate mechanical compatibility and good cyto-compatibility [9–12]. Therefore, the Ti-Mo-Zr system, which showed good performances and magnificent prospects in biomaterial application, has been intensively studied [3,6,10–13]. Several alloys related to the Ti-Mo-Zr system have been investigated recently, for example, Ti-12Mo-6Zr-2Fe [14], Ti-15Mo-5Zr-3Al [15], and Ti-8Mo-4Nb-2Zr [6].

Inter-diffusion in alloys plays an important role in materials processing. Understanding diffusion kinetics of Ti alloys is critical to manipulate phase transformation and microstructure evolution during homogenization and precipitation [16–18]. The reliable inter-diffusion coefficients are the essential input parameters in various quantitative modeling of phase transformation [17,19,20]. It is also valuable to determine thermodynamic stability under long-term service conditions.

Inspired by the CALPHAD (CALculation of PHAse Diagram) method, Andersson and Ågren [21] developed a phenomenological kinetic technique which present the atomic mobilities of individual species using Redlick-Kister polynomials. This technique assesses the discrete diffusion coefficients that were extracted experimentally and then employs a set of parameters to describe composition dependent mobilities and diffusivities. Combined with thermodynamic information, diffusion simulation can be implemented in DICTRA module embedded in Thermo_Calc software [22,23].

Various of measurements of self- and impurity diffusion coefficients in bcc Ti, Zr, and Mo have been conducted [24] as well as inter-diffusion coefficients in bcc Ti-Mo [25–27], Ti-Zr [28–31] and Mo-Zr alloys [32]. Based these diffusion coefficients, Liu et al. [33,34] assessed the atomic mobilities in bcc Ti-Mo, Ti-Zr, and Mo-Zr alloys with the CALPHAD method. Simulations of concentration profiles and Kirkendall movement using these mobility parameters show good agreement with the experimental results from literatures. However, the ternary diffusion coefficients and mobilities in bcc Ti-Mo-Zr alloys is still open to be evaluated and barely reported. The purpose of this work is to study the diffusion behavior of Mo and Zr in β-Ti alloys and establish a CALPHAD-type atomic mobility database of bcc phase in Ti-Mo-Zr system.

2. Model

2.1. Extraction of Inter-Diffusion Coefficients and Impurity Diffusion Coefficients

According to Kirkaldy [35], in one dimension, the inter-diffusion flux of component i with concentration c_i in a ternary system can be expressed as:

$$\widetilde{J}_i = \frac{1}{2t}\int_{c_i^-}^{c_i} (z_i - z_0)dc_i = -\sum_{j=1}^{2} \widetilde{D}_{ij}^k \frac{\partial c_j}{\partial z}, \tag{1}$$

where $\widetilde{D}_{ij}^k$ are the inter-diffusion coefficients;c_i, c_j are the concentration of species i and j; c_i^- is the concentration of i at the left end of the diffusion couple; z is the distance; z_0 is the Matano plane for the diffusion couple; and, t is the diffusion time.

In a ternary system, it involves four independent diffusion coefficients, $\widetilde{D}_{11}^3$, $\widetilde{D}_{22}^3$, $\widetilde{D}_{12}^3$ and $\widetilde{D}_{21}^3$. However only two composition-distance profiles for independent elements 1 and 2 can be derived from one diffusion couple. Hence, two diffusion couples result in two diffusion paths which intersect at one point, are needed to obtain four profiles. Then, the diffusion coefficients at this point can be solved mathematically. The old-fashioned Boltzmann-Matano had to position the Matano planes of four profiles, which is time-consuming and might lead to unnecessary inaccuracy. To avoid this drawback, Whittle and Green [36] introduced a normalized concentration variable:

$$Y_i = \frac{c_i - c_i^-}{c_i^+ - c_i^-}, \tag{2}$$

where c_i^- and c_i^+ are concentrations of two elements in the left and right end of the diffusion couple.

With the variable Y, the inter-diffusion flux of element i is no longer referred to a fixed laboratory coordinate and can be transformed into:

$$\widetilde{J}_i = \frac{(c_i^+ - c_i^-)}{2t}\left[(1 - Y_i)\int_{-\infty}^{z} Y_i dz + Y_i \int_{-\infty}^{z} (1 - Y_i)dz\right]. \tag{3}$$

Reformulating the right-hand side of Equations (1) and (3) lead to:

$$\frac{1}{2t}\left(\frac{dz}{dY_i}\right)\left[(1-Y_i)\int_{-\infty}^{z} Y_i dz + Y_i\int_{-\infty}^{z}(1-Y_i)dz\right] = \widetilde{D}_{ii}^{k} + \widetilde{D}_{ij}^{k}\frac{dc_j}{dc_i}(i=1,2). \tag{4}$$

By solving a set of four equations from one pair of diffusion couples, the diffusion coefficients $\widetilde{D}_{MoMo}^{Ti}$, $\widetilde{D}_{ZrZr}^{Ti}$, $\widetilde{D}_{MoZr}^{Ti}$ and $\widetilde{D}_{ZrMo}^{Ti}$ at the intersecting composition can be extracted. Note that the molar volume was taken to be constant for the lack of reliable data on the composition-dependent molar volume in bcc Ti-Mo-Zr alloys. Errors that are introduced by this approximate treatment are believed to be within the accuracy of the results obtained via the Whittle-Green method [36].

By applying a generalized Hall method [37,38], the impurity diffusion coefficients of Zr in Ti-Mo alloys and Mo in Ti-Zr at the terminal compositions of the diffusion couples can be obtained. The profiles were transformed to a plot of μ *vs* λ, in which $\mathrm{erf}\mu = 2Y - 1$ and $\lambda = z/\sqrt{t}$. By fitting the plot with a linear equation $\mu = h\lambda + k$, the h_1, k_1 for the left terminal, and h_2, k_2 for the right terminal of the diffusion couples can be obtained and the impurity diffusion coefficients can be derived via the following formulas:

$$\widetilde{D}(z') = \frac{1}{4h_1^2}\left[1 + \frac{2k_1}{\sqrt{\pi}}\exp\left(\mu^2\right) \times Y(z')\right], \tag{5}$$

$$\widetilde{D}(z') = \frac{1}{4h_2^2}\left[1 - \frac{2k_2}{\sqrt{\pi}}\exp\left(\mu^2\right) \times \left[1 - Y(z')\right]\right], \tag{6}$$

where c_i represents the content of i in binary alloys and c_0 is the content of i at the other end of the diffusion couple.

2.2. Atomic Mobility and Diffusivity

As initiated by Andersson and Ågren [21] and later modified by Jönsson [39], the atomic mobility M_i of species i can be expressed as:

$$M_i = M_i^0 \exp\left(\frac{-Q_i}{RT}\right)\frac{1}{RT} \times^{mg}\Omega = \exp\left(\frac{RT\ln M_i^0 - Q_i}{RT}\right)\frac{1}{RT} \times^{mg}\Omega, \tag{7}$$

where R is the gas constant; T is the temperature; M_i^0 is the frequency factor; Q_i is the activation energy; and, ${}^{mg}\Omega$ is the ferromagnetic contribution. Since there is no ferromagnetic transition reported in Ti-Mo-Zr alloys, ${}^{mg}\Omega$ can be taken as 1. Provided $\Delta G_i^{\phi} = RT\ln M_i^0 - Q_i$, Equation (7) can be written as [21]:

$$M_i = \exp\left(\frac{\Delta G_i^{\phi}}{RT}\right)\frac{1}{RT}, \tag{8}$$

the parameter ΔG_i^{ϕ} is composition dependent and is expressed in the Redlich-Kister polynomial form as:

$$\begin{aligned}\Delta G_i^{\phi} &= \sum_p x_p \Delta G_i^p + \sum_p \sum_{q>p} x_p x_q \left[\sum_{r=0,1,2,\ldots} \Delta^{(r)} G_i^{p,q} \left(x_p - x_q\right)^r\right] \\ &+ \sum_p \sum_{q>p} \sum_{v>q} x_p x_q x_v \left\{\sum_{s=p,q,v} \left[x_s + \left(1 - x_p - x_q - x_v\right)/3\right] \Delta^{(s)} G_i^{p,q,v}\right\}\end{aligned} \tag{9}$$

where ϕ denotes the solid solution phase; x_p is the mole fraction of species p; ΔG_i^p is the value ΔG_i of species i in pure species p; and, $\Delta^{(r)} G_i^{p,q}$ and $\Delta^{(s)} G_i^{p,q,v}$ are the binary and ternary interaction parameters.

Assuming the mono-vacancy exchange is the dominant diffusion mechanism, and neglecting correlation factors, the tracer diffusion coefficients D_i^* of species i is directly related to the mobility M_i by means of the Einstein relation:

$$D_i^* = RTM_i, \tag{10}$$

For a substitutional solution phase, the inter-diffusion coefficients in terms of the volume-fixed reference frame is given by the following general expression:

$$\widetilde{D}_{ij}^n = \sum_k^n (\delta_{ki} - x_i) x_k M_k \left(\frac{\partial \mu_k}{\partial x_j} - \frac{\partial \mu_k}{\partial x_n} \right), \tag{11}$$

where δ_{ki} is the Kronecker delta ($\delta_{ki} = 1$ when i=k, and 0 otherwise).

Then, the inter-diffusion flux of species i, $\widetilde{J}_i$ can be calculated via Equation (1), and the concentration evolution of component i in ternary systems can be solved by the mass conservation law:

$$\frac{\partial c_i}{\partial t} + \frac{\partial \widetilde{J}_i}{\partial z} = 0. \tag{12}$$

What calls for special attention in the formula derivation is that the transformation of concentration of species i, c_i, and the mole fraction x_i:

$$c_i = \frac{x_i}{V_m}, \tag{13}$$

where V_m is the molar volume of a phase that was taken to be constant in this work.

3. Experiment

Pure Ti, binary Ti-Mo and Ti-Zr alloys, and ternary Ti-Mo-Zr alloys were prepared from pure Ti (99.99 wt %), Mo (99.99 wt %), and Zr (99.99 wt %) by arc melting in electric arc furnace under an argon atmosphere. All of the alloys compositions were designed in the bcc phase region according the binary phase diagram of Ti-Mo [40], Ti-Zr [41], and Zr-Mo [42] systems and ternary phase diagram of Ti-Mo-Zr system at 1473 K [43] (shown in Figure 1). The compositions of all alloys were listed in Table 1. The ingots were re-melted for six times to attain homogeneity and then annealed at 1473 K for 24 h to obtain microstructure with large grain size above millimeters, such that the effect of grain boundary diffusion can be ignored.

The annealed ingots were cut into blocks with a size of 10 × 10 × 5 mm using wire-electrode cutting. After one surface of the blocks polished to mirror-like quality, the well-contacted diffusion couples were assembled with appropriate pairs of blocks in Table 1 under vacuum at 1173 K for 4 h. The diffusion couples were sealed into quartz capsules that were filled with pure argon. The M1-M8 diffusion couples were annealed at 1373 K for 72 h and N1-N8 diffusion couples were annealed 1473 K for 48 h, followed by quenching in ice water.

After the annealing process, the diffusion couples were then cut into halves parallel to the ends using wire-electrode cutting, and then mounted, ground, and polished by standard metallographic techniques. The composition-distance profiles of 16 diffusion couples were determined using electron probe micro-analysis (EPMA, JEOL, JXA-8230, Tokyo, Japan, 15 kV, 20 nA) with 15 kV voltage, 20 nA current, and a 40° take-off angle. The accuracy of the EPMA test is >98% (mass percent).

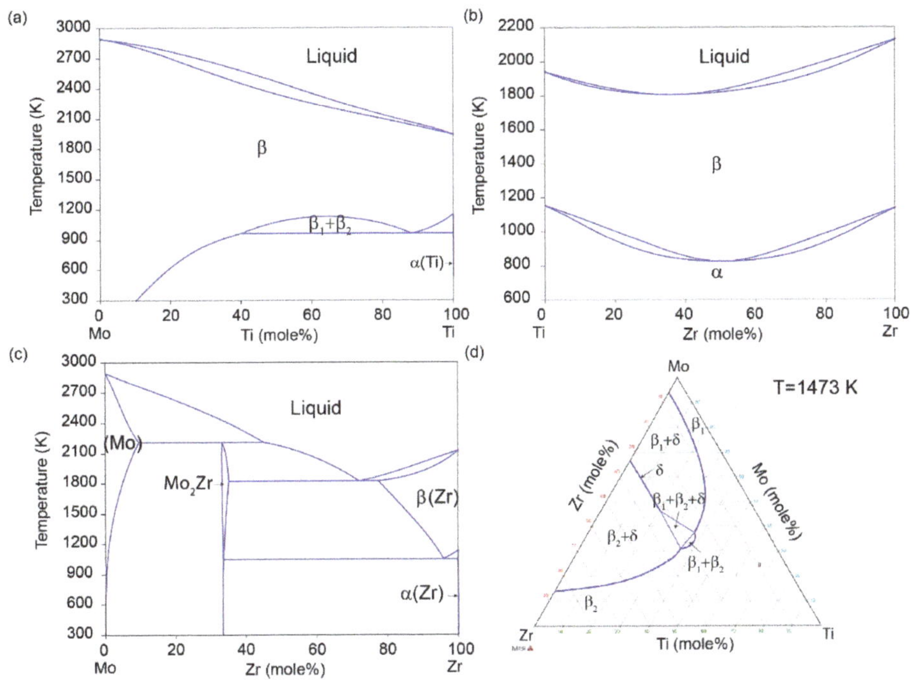

Figure 1. Phase diagram of (**a**) Ti-Mo [40], (**b**) Ti-Zr [41], (c) Zr-Mo [42], and (**d**) Ti-Mo-Zr [43] systems.

Table 1. Compositions of diffusion couples.

Temperature (K)	Diffusion Couples	Composition (mole %)
1373	M1	Ti-2.95Mo/Ti-4.95Zr
	M2	Ti-4.90Mo/Ti-7.81Zr
	M3	Ti-7.83Mo/Ti-11.17Zr
	M4	Ti-9.74Mo/Ti-13.64Zr
	M5	Ti-11.61Mo/Ti-16.57Zr
	M6	Pure Ti/Ti-17.18Mo-5.48Zr
	M7	Pure Ti/Ti-11.63Mo-11.51Zr
	M8	Pure Ti/Ti-5.91Mo-16.90Zr
1473	N1	Ti-2.85Mo/Ti-4.85Zr
	N2	Ti-4.92Mo/Ti-7.39Zr
	N3	Ti-7.90Mo/Ti-11.02Zr
	N4	Ti-9.90Mo/Ti-13.27Zr
	N5	Ti-13.11Mo/Ti-15.15Zr
	N6	Pure Ti/Ti-17.14Mo-5.81Zr
	N7	Pure Ti/Ti-11.67Mo-11.61Zr
	N8	Pure Ti/Ti-5.91Mo-17.32Zr

4. Results and Discussions

4.1. Inter-Diffusion and Impurity Diffusion Coefficients At 1373 K and 1473 K

All of the diffusion couples show single bcc phase microstructure. The SEM backscattered electron image (BSE) of diffusion zone of the diffusion couple M1 is taken as an example and is presented in Figure 2a. There is no Kirkendal void existing in the diffusion zone and no microstructure is evident, revealing that the alloys were annealed in single bcc phase. After being annealed at high temperature and quenched, the martensite transformation may present in the samples. However, because martensite transformation is a diffusionless phase transformation, the composition profile will not change. We can still extract the diffusion coefficients from the composition profile that was obtained in this sample.

The composition-distance profiles that were determined from couple M2 is shown in Figure 2b. It is worth mentioning that, to avoid the errors from fitting or smoothing, a robust error function expansion (ERFEX) was put forward to represent the experimental profiles in an accurate analytical form [37,44]:

$$X(r) = \sum_i [a_i \mathrm{erf}(b_i z + c_i) + d_i], \tag{14}$$

where $X(r)$ is the effective alloying element content at location z; a, b, c, and d are the fitting parameters.

It could be evidenced by practice that the profiles smoothed using ERFEX method exhibits higher accuracy than using traditional method, such as moving average smoothing, Savitzky-Golay smoothing method, and Piecewise Cubic Hermite Interpolating Polynomial (PCHIP) interpolation, and it can present the details of the composition profiles. For example, the Mo pill-up at the right end of couple M6 (shown in Figure 2c). The ERFEX method also has its advantages in fitting asymmetrical curves when the diffusivities of the elements in one end of the diffusion couple are much larger than that of the other end. The couple M6 is compared with M8, for instance. The left-hand side of M6 is pure titanium and the other end has a high Mo concentration. The slop of profile M6 is steep at the right end of diffusion (Figure 2c). It is because the diffusivity of Mo in Ti alloys is much lower than that of Ti. In comparison, it is appropriately smooth and it shows the conventional S-shape (Figure 2d) in M8. The steep slope will introduce a larger error when calculating the diffusion coefficients at the concentration on it. That means that the accuracy of the calculated diffusion coefficients at the component close to Ti-Mo boundary will be lower.

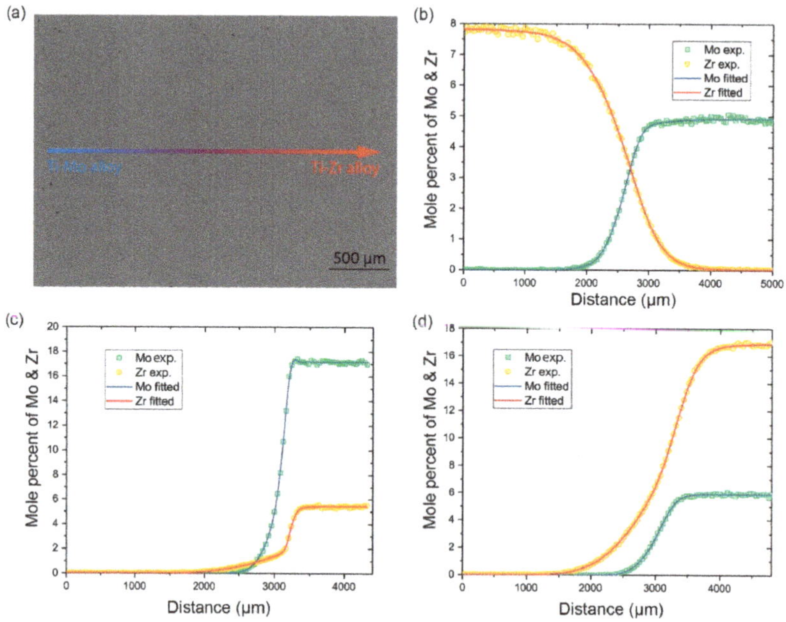

Figure 2. (**a**) Microstructure (BSE) of the diffusion couple M1 annealed at 1373 K for 72 h and the robust error function expansion (ERFEX) representation of the composition-distance profiles of the different couples annealed in 1373 K for 72 h: (**b**) for M2; (**c**) for M6; (**d**) for M8.

The diffusion paths of all diffusion couples after annealing process are presented in Figure 3. According to the Whittle-Green method, the inter-diffusion coefficients at the intersecting compositions can be extracted. The impurity diffusion of Zr in Ti-Mo alloys and Mo in Ti-Zr alloys can be obtained via the Hall method at the ends of couple M1-M5 and N1-N5.

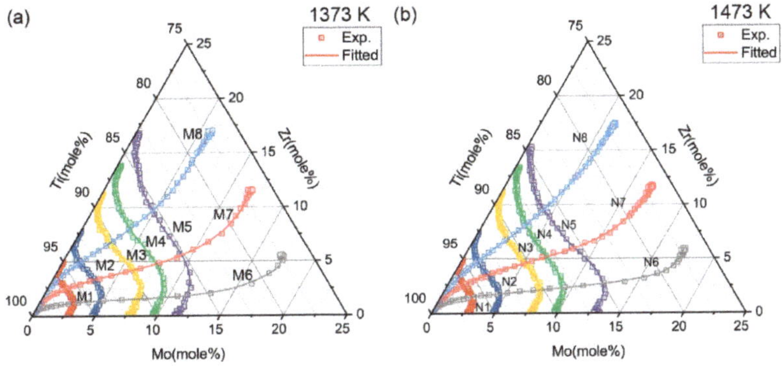

Figure 3. Diffusion paths of diffusion couples determined using electron probe micro-analysis (EPMA) after annealing process.

The determined inter-diffusion coefficients are summarized in Table 2 and graphically presented with three-dimensional plots in Figure 4. The standard deviations were determined from four independent calculations upon two independent measurements. All of the results can satisfactorily fulfill the thermodynamic constrains derived by Kirkaldy [45]:

$$\widetilde{D}^{Ti}_{ZrZr} + \widetilde{D}^{Ti}_{NbNb} > 0, \tag{15}$$

$$\widetilde{D}^{Ti}_{ZrZr} \times \widetilde{D}^{Ti}_{NbNb} - \widetilde{D}^{Ti}_{ZrNb} \times \widetilde{D}^{Ti}_{NbZr} > 0, \tag{16}$$

$$\left(\widetilde{D}^{Ti}_{ZrZr} + \widetilde{D}^{Ti}_{NbNb}\right)^2 - 4\left(\widetilde{D}^{Ti}_{ZrZr} \times \widetilde{D}^{Ti}_{NbNb} - \widetilde{D}^{Ti}_{ZrNb} \times \widetilde{D}^{Ti}_{NbZr}\right) \geq 0. \tag{17}$$

Table 2. Inter-diffusion coefficients in the bcc Ti-Mo-Zr alloys at 1373 K and 1473 K.

Temp. (K)	Diffusion Couple	Intersection Composition (mole %)		Inter-Diffusion Coefficients ($m^2 \cdot s^{-1}$)			
		Mo	Zr	$\widetilde{D}^{Ti}_{MoMo}$	$\widetilde{D}^{Ti}_{MoZr}$	$\widetilde{D}^{Ti}_{ZrMo}$	$\widetilde{D}^{Ti}_{ZrZr}$
				$\times 10^{-14}$	$\times 10^{-14}$	$\times 10^{-14}$	$\times 10^{-13}$
1373	M1-M6	2.71	1.18	9.96 ± 1.12	−7.26 ± 5.24	0.96 ± 0.38	6.00 ± 0.42
	M1-M7	1.46	2.54	13.84 ± 0.56	−2.89 ± 1.72	3.57 ± 2.24	6.71 ± 0.37
	M1-M8	0.45	3.45	17.08 ± 1.81	0.31 ± 0.25	12.63 ± 5.54	8.29 ± 0.31
	M2-M6	4.73	1.38	6.67 ± 0.60	−2.95 ± 1.93	1.55 ± 0.20	4.52 ± 0.11
	M2-M7	3.32	3.33	9.79 ± 0.25	−2.93 ± 0.27	2.52 ± 1.47	5.24 ± 0.10
	M2-M8	1.38	4.93	15.41 ± 0.56	−0.49 ± 0.33	11.30 ± 1.09	7.14 ± 0.10
	M3-M6	7.72	1.56	3.56 ± 0.41	−1.18 ± 0.41	1.07 ± 0.13	2.89 ± 0.10
	M3-M7	5.99	4.24	6.45 ± 0.11	−2.50 ± 0.36	1.25 ± 0.60	3.44 ± 0.15
	M3-M8	2.90	6.74	14.48 ± 0.28	−1.25 ± 0.48	7.11 ± 0.99	5.49 ± 0.15
	M4-M6	9.49	1.68	2.93 ± 0.17	−2.94 ± 1.04	0.50 ± 0.15	2.36 ± 0.06
	M4-M7	7.58	4.76	5.92 ± 0.28	−3.06 ± 0.58	0.34 ± 0.71	2.75 ± 0.14
	M4-M8	3.88	8.07	14.69 ± 0.93	−2.30 ± 0.84	5.98 ± 1.07	4.52 ± 0.04
	M5-M6	11.45	1.83	2.31 ± 0.29	−1.40 ± 1.31	0.16 ± 0.46	1.62 ± 0.06
	M5-M7	9.24	5.50	4.84 ± 0.60	−1.86 ± 0.68	0.72 ± 0.78	1.97 ± 0.02
	M5-M8	4.80	9.73	14.49 ± 1.04	−2.08 ± 0.50	4.76 ± 3.54	3.82 ± 0.16

Table 2. *Cont.*

Temp. (K)	Diffusion Couple	Intersection Composition (mole %)		Inter-Diffusion Coefficients ($m^2{\cdot}s^{-1}$)			
		Mo	Zr	$\widetilde{D}^{Ti}_{MoMo}$	$\widetilde{D}^{Ti}_{MoZr}$	$\widetilde{D}^{Ti}_{ZrMo}$	$\widetilde{D}^{Ti}_{ZrZr}$
				$\times 10^{-13}$	$\times 10^{-14}$	$\times 10^{-13}$	$\times 10^{-12}$
1473	N1-N6	1.524	3.605	3.37 ± 0.18	−4.00 ± 2.19	0.45 ± 0.22	1.59 ± 0.06
	N1-N7	2.597	1.898	4.43 ± 0.01	−2.46 ± 1.08	0.68 ± 0.40	1.69 ± 0.02
	N1-N8	3.365	0.81	5.42 ± 0.13	−1.39 ± 1.30	1.97 ± 1.16	1.83 ± 0.05
	N2-N6	1.707	4.399	2.39 ± 0.19	−13.24 ± 4.86	0.55 ± 0.06	1.23 ± 0.01
	N2-N7	3.589	3.386	3.95 ± 0.12	−3.33 ± 2.17	1.20 ± 0.23	1.43 ± 0.03
	N2-N8	5.701	1.951	5.19 ± 0.12	−2.16 ± 1.40	2.38 ± 0.21	1.62 ± 0.02
	N3-N6	2.43	8.144	1.68 ± 0.04	−2.24 ± 2.12	0.52 ± 0.09	0.82 ± 0.02
	N3-N7	4.724	5.243	2.80 ± 0.05	−3.75 ± 0.51	1.16 ± 0.15	1.06 ± 0.02
	N3-N8	6.881	2.508	4.91 ± 0.08	−2.11 ± 0.76	2.04 ± 0.51	1.36 ± 0.03
	N4-N6	2.552	8.725	1.50 ± 0.09	−13.40 ± 4.90	0.36 ± 0.10	0.71 ± 0.01
	N4-N7	5.489	6.461	2.72 ± 0.17	−11.73 ± 2.86	0.84 ± 0.06	0.93 ± 0.01
	N4-N8	8.913	3.466	6.04 ± 0.46	−8.67 ± 2.82	1.18 ± 0.19	1.27 ± 0.02
	N5-N6	3.438	12.795	0.85 ± 0.03	−3.56 ± 1.06	0.36 ± 0.06	0.40 ± 0.01
	N5-N7	6.723	8.091	2.03 ± 0.07	−1.85 ± 0.97	1.26 ± 0.26	0.68 ± 0.01
	N5-N8	9.993	3.932	5.20 ± 0.44	−3.24 ± 2.66	1.88 ± 0.66	1.12 ± 0.03

The impurity diffusion coefficients that were obtained by Hall methods are listed in Table 3, and also illustrated in Figure 4, along with the inter-diffusion coefficients. In Figure 4a,b, the main inter-diffusion coefficients coincide with the impurity diffusion coefficients when the concentration of diffusion elements tends to 0.

Scrutinizing the map of the diffusion coefficients in Figure 4, it can also be observed that, when the concentration of the gradient elements j tends to 0, the limit of main inter-diffusion coefficients $\widetilde{D}^k_{ii}$ will match the binary diffusion coefficients from literature [33] well, and when the concentration of the diffusion elements i tend to 0, the limits of cross inter-diffusion coefficients $\widetilde{D}^k_{ij}$ will be 0. The main diffusion coefficients are larger than the cross ones and they showed less fluctuation than cross ones. Values of $\widetilde{D}^{Ti}_{MoZr}$ are negative, which means that the element Zr has a negative effect on the diffusion of Mo in Ti-based alloys.

The variation of $\widetilde{D}^{Ti}_{MoMo}$ and $\widetilde{D}^{Ti}_{ZrZr}$ with the composition of Mo and Zr at 1373 K are presented in Figure 5. It can be found that $\widetilde{D}^{Ti}_{MoMo}$ decrease with the increase concentration of Mo in Figure 5a. In Figure 5b, $\widetilde{D}^{Ti}_{MoMo}$ also decrease with the increase concentration of Zr. However, the decreasing rate varies with different diffusion Mo contents (the three colors correspond to inter-diffusion coefficients extracted from three different diffusion couples, M6, M7, and M8). In general, a great ratio of x(Zr)/x(Mo) slows down the process of decrease. As to $\widetilde{D}^{Ti}_{ZrZr}$, as is shown in Figure 5c,d, show similar trends, like $\widetilde{D}^{Ti}_{MoMo}$, when it varies with the concentration of Mo and Zr. From these phenomena, we can infer that the influence of element Mo on the variations of $\widetilde{D}^{Ti}_{MoMo}$ and $\widetilde{D}^{Ti}_{ZrZr}$ is much larger than Zr.

When extent to boundary, the impurity diffusion coefficients of Mo in Ti-Zr alloys, $D^*_{Mo(Ti-Zr)}$ show an increasing trend with the increase of Zr concentration. The impurity diffusion coefficients of Zr in Ti-Mo alloys $D^*_{Zr(Ti-Mo)}$ show a decreasing trend with the increase of Mo concentration.

Figure 6 presents the three-dimensional (3D) plot of diffusion coefficients in the bcc Ti-Mo-Zr alloys and the variations of $\widetilde{D}^{Ti}_{MoMo}$ and $\widetilde{D}^{Ti}_{ZrZr}$ with of the concentration of Mo and Zr at 1473 K. They showed similar trends as those at 1373 K.

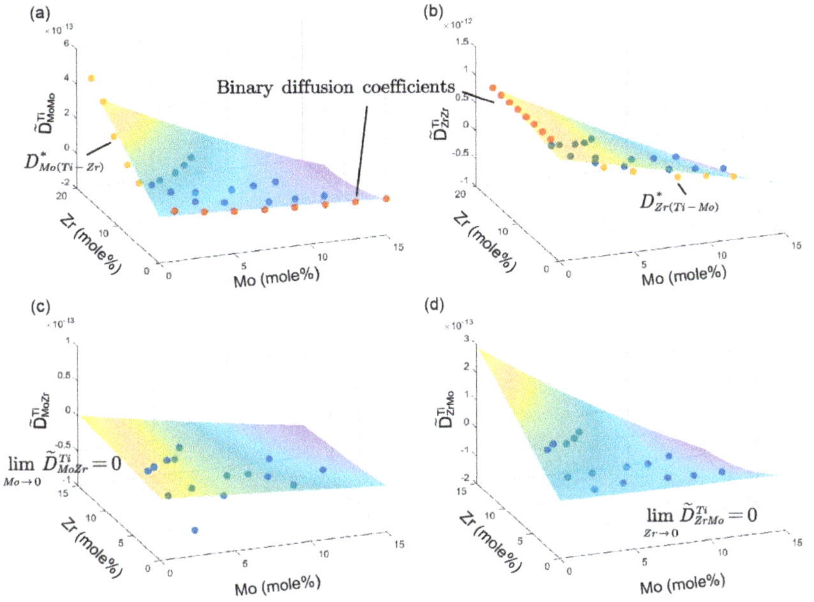

Figure 4. Three-dimensional (3D) plot of the ternary inter-diffusion coefficients (**a**) $\widetilde{D}_{MoMo}^{Ti}$, (**b**) $\widetilde{D}_{ZrZr}^{Ti}$, (**c**) $\widetilde{D}_{MoZr}^{Ti}$ and (**d**) $\widetilde{D}_{ZrMo}^{Ti}$ in the bcc Ti-Mo-Zr alloys at 1373 K, together with the impurity diffusion coefficients $D_{Mo(Ti-Zr)}^{*}$ and $D_{Zr(Ti-Mo)}^{*}$, and binary diffusion coefficients obtained from the literature [33].

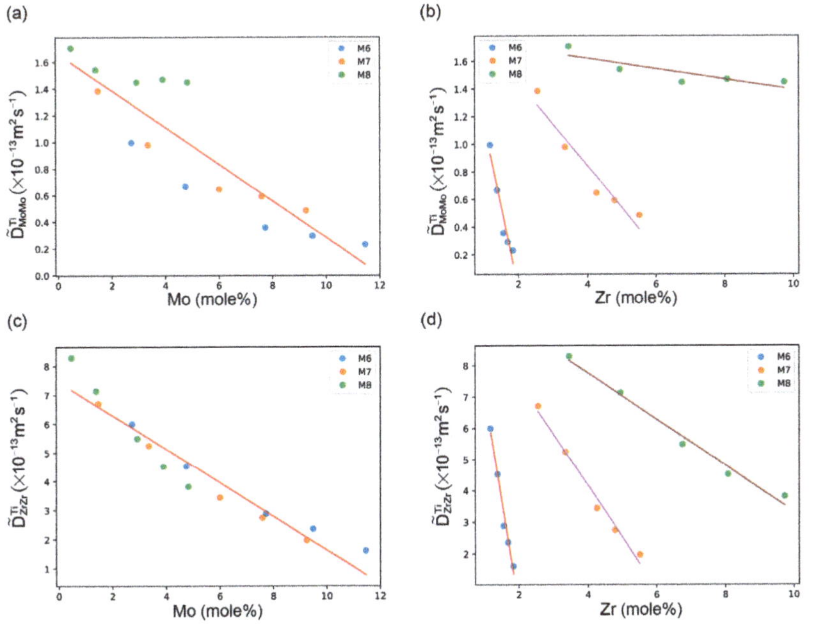

Figure 5. The variation of ternary inter-diffusion coefficients with the compositions: (**a**) $\widetilde{D}_{MoMo}^{Ti}$ with Mo, (**b**) $\widetilde{D}_{MoMo}^{Ti}$ with Zr, (**c**) $\widetilde{D}_{ZrZr}^{Ti}$ with Mo and (**d**) $\widetilde{D}_{ZrZr}^{Ti}$ with Zr at 1373 K.

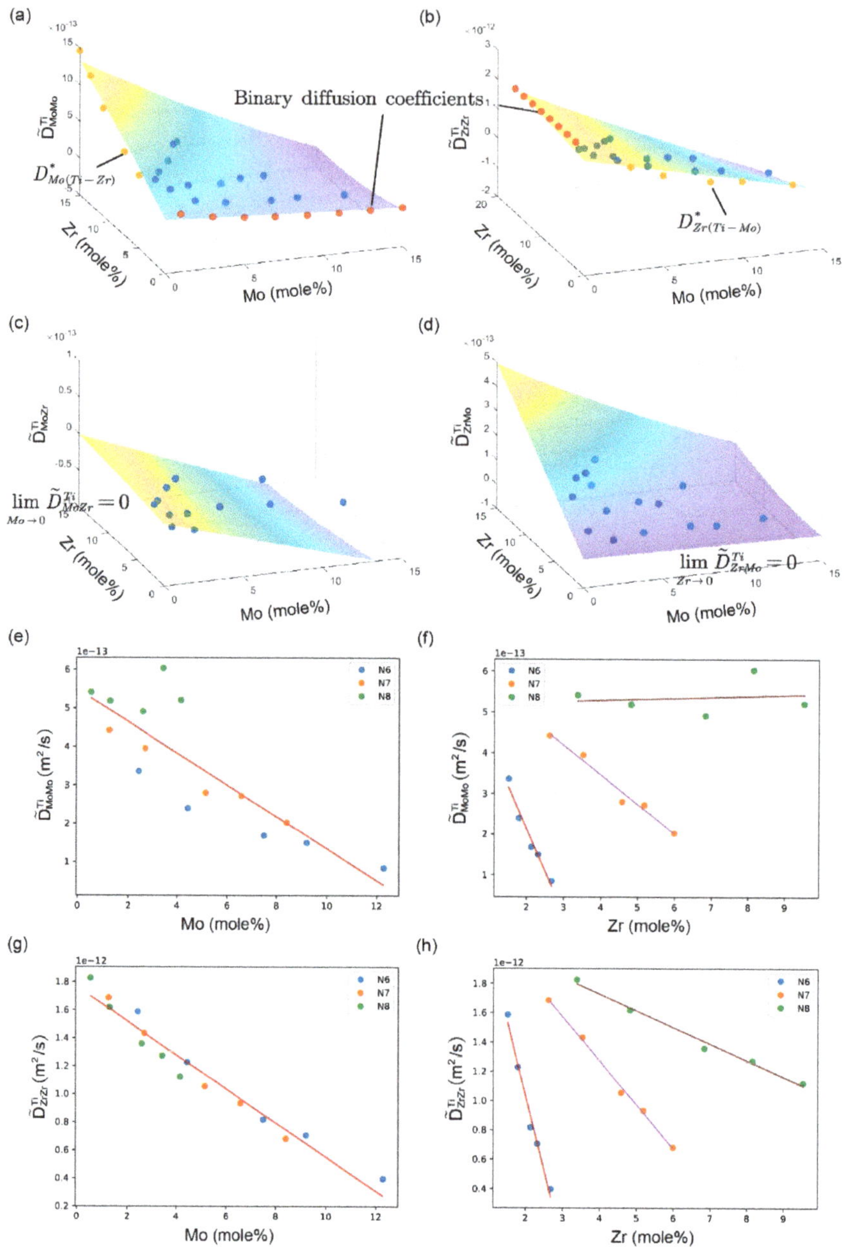

Figure 6. The 3D plot of the ternary inter-diffusion coefficients (**a**) $\widetilde{D}^{Ti}_{MoMo}$, (**b**) $\widetilde{D}^{Ti}_{ZrZr}$, (**c**) $\widetilde{D}^{Ti}_{MoZr}$, and (**d**) $\widetilde{D}^{Ti}_{ZrMo}$ in the bcc Ti-Mo-Zr alloys and the variation of ternary inter-diffusion coefficients with the compositions: (**e**) $\widetilde{D}^{Ti}_{MoMo}$ with Mo, (**f**) $\widetilde{D}^{Ti}_{MoMo}$ with Zr, (**g**) $\widetilde{D}^{Ti}_{ZrZr}$ with Mo, and (**h**) $\widetilde{D}^{Ti}_{ZrZr}$ with Zr at 1473 K.

Table 3. Impurity diffusion coefficients of Zr in Ti-Mo and Mo in Ti-Zr alloys at 1373 K and 1473 K.

Temperature/K	Composition	Impurity Diffusion Coefficients ($\times 10^{-13}$ m^2·s^{-1})	Composition	Impurity Diffusion Coefficients ($\times 10^{-13}$ m^2·s^{-1})
1373	$D^*_{Zr(Ti-2.95Mo)}$	5.85 ± 0.58	$D^*_{Mo(Ti-4.95Zr)}$	1.54 ± 0.09
	$D^*_{Zr(Ti-4.90Mo)}$	4.51 ± 0.88	$D^*_{Mo(Ti-7.81Zr)}$	1.96 ± 0.62
	$D^*_{Zr(Ti-7.83Mo)}$	2.55 ± 0.05	$D^*_{Mo(Ti-11.17Zr)}$	2.86 ± 0.71
	$D^*_{Zr(Ti-9.74Mo)}$	2.13 ± 0.14	$D^*_{Mo(Ti-13.64Zr)}$	4.37 ± 1.68
	$D^*_{Zr(Ti-11.61Mo)}$	1.39 ± 0.15	$D^*_{Mo(Ti-16.57Zr)}$	5.07 ± 1.45
1473	$D^*_{Zr(Ti-2.89Mo)}$	15.45 ± 1.22	$D^*_{Mo(Ti-4.85Zr)}$	5.04 ± 1.24
	$D^*_{Zr(Ti-4.92Mo)}$	11.31 ± 0.74	$D^*_{Mo(Ti-7.39Zr)}$	6.42 ± 2.38
	$D^*_{Zr(Ti-7.90Mo)}$	7.24 ± 0.07	$D^*_{Mo(Ti-11.02Zr)}$	9.78 ± 1.82
	$D^*_{Zr(Ti-9.90Mo)}$	6.08 ± 0.26	$D^*_{Mo(Ti-13.27Zr)}$	12.61 ± 3.46
	$D^*_{Zr(Ti-13.11Mo)}$	3.04 ± 0.17	$D^*_{Mo(Ti-15.15Zr)}$	14.66 ± 2.39

4.2. Atomic Mobilities in bcc Ti-Zr-Mo System

The DICTRA module that was embedded in ThermoCalc software [21,22] was employed to obtain the atomic mobility parameters for the Ti-Mo-Zr ternary bcc alloys. Mobility parameters of boundary binary systems were taken from Liu's work [33,34]. Thermodynamic description of boundary binary systems were also the same as Liu's work [40–42]. Thermodynamic modeling of bcc phase in the ternary system can be carried out by simply extrapolation based on three binary systems. By assessing the experimental diffusion coefficients in this work, a set of self-consistent atomic mobility parameters were obtained, as listed in Table 4.

Using these parameters and DICTRA software, the inter-diffusion coefficients at the intersection composition at 1373 K and 1473 K were calculated and compared with the experimental results, as shown in Figure 7. The calculated results agree with the main inter-diffusion coefficients well, within the error allowed. In contrast, the cross coefficients show certain fluctuation with a large error range. Since the calculated and experimental cross coefficients show values in the same order of magnitude, and a same variation tendency with compositions, the accessed mobilities are feasible to represent the cross coefficients.

Figure 8 shows the comparison of the calculated impurity diffusion coefficients and the experimental results. The calculated impurity diffusion coefficients also agree well with the experimental results.

In order to further verify the atomic mobility parameters that were obtained in the present work, the simulation of several diffusion couples were implemented. The diffusion simulation was set up to model the semi-infinite ternary diffusion couples in experiment. It was conducted using the same initial conditions and heat treatment processes in the experiment. The predict composition-distance profiles show good agreement with the experimental profiles. For instance, the predicted composition-distance profiles and fluxes of couple M3 and N8 compared with experimental data are exhibited in Figures 9 and 10 presented the simulated diffusion paths of 16 diffusion couples annealed at 1373 K and 1473 K. The calculated results show good agreement with the experimental results.

Table 4. Atomic mobility parameters for the bcc phase of the Ti-Mo-Zr ternary system.

Mobility	Parameter, J/mole	Reference
Mobility of Mo		
Q_{Mo}^{Mo}	−479740.87 − 63.98·T	[33]
Q_{Mo}^{Ti}	−196255.40 − 105.21·T	[33]
Q_{Mo}^{Zr}	−154895.63 − 140.91·T (T ≤ 1450 K) −214913.77 − 114.17·T (T ≥ 1450 K)	[34]
${}^{0}Q_{Mo}^{Mo,Ti}$	−24153.22 − 45.32·T	[33]
${}^{1}Q_{Mo}^{Mo,Ti}$	−61804.04	[33]
$Q_{Mo}^{Mo,Zr}$	150325.48 + 10.03·T	[34]
$Q_{Mo}^{Ti,Zr}$	−268357.13 + 283.18·T	This work
${}^{0}Q_{Mo}^{Mo,Ti,Zr}$	53485.75	This work
${}^{1}Q_{Mo}^{Mo,Ti,Zr}$	2707991.87	This work
${}^{2}Q_{Mo}^{Mo,Ti,Zr}$	−719018.41	This work
Mobility of Ti		
Q_{Ti}^{Mo}	−435701.23 − 72.67·T	[33]
Q_{Ti}^{Ti}	−151989.95 − 127.37·T	[33]
Q_{Ti}^{Zr}	−140356.54 − 138.12·T	[33]
${}^{0}Q_{Ti}^{Mo,Ti}$	−91728.48 + 64.56·T	[33]
${}^{1}Q_{Ti}^{Mo,Ti}$	−96300.05	[33]
${}^{0}Q_{Ti}^{Ti,Zr}$	−15826.04 + 62.55·T	[33]
${}^{1}Q_{Ti}^{Ti,Zr}$	8243.54	[33]
${}^{0}Q_{Ti}^{Mo,Ti,Zr}$	−1304851.67	This work
${}^{1}Q_{Ti}^{Mo,Ti,Zr}$	−6019149.28	This work
${}^{2}Q_{Ti}^{Mo,Ti,Zr}$	−1128202.89	This work
Mobility of Zr		
Q_{Zr}^{Mo}	−464587.32 − 64.72·T	[34]
Q_{Zr}^{Ti}	−131670.56 − 133.36·T	[33]
Q_{Zr}^{Zr}	−104624.81 − 163.15·T (T≤1573 K) −161543.53 − 126.10·T (T≥1573 K)	[33]
$Q_{Zr}^{Mo,Ti}$	−81189.65 + 88.51·T	This work
$Q_{Zr}^{Mo,Zr}$	210325.67 + 15.19·T	[34]
${}^{0}Q_{Zr}^{Ti,Zr}$	−12581.03 + 33.38·T	[33]
${}^{1}Q_{Zr}^{Ti,Zr}$	2898.60	[33]
${}^{0}Q_{Zr}^{Mo,Ti,Zr}$	2864755.37	This work
${}^{1}Q_{Zr}^{Mo,Ti,Zr}$	−340701.76	This work
${}^{2}Q_{Zr}^{Mo,Ti,Zr}$	−609499.74	This work

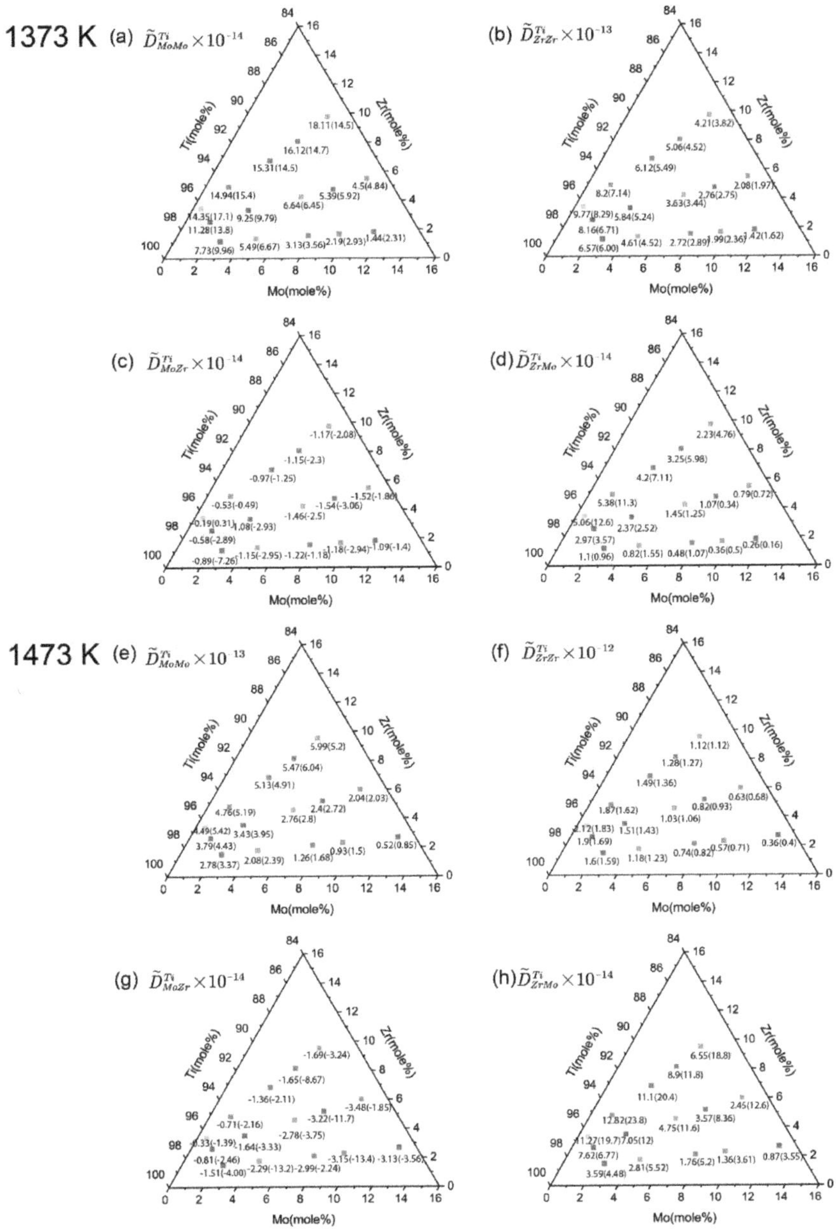

Figure 7. Calculated inter-diffusion coefficients of the ternary Ti-Mo-Zr system compared with the experimental measurements (in brackets) in this work: **(a)** $\widetilde{D}^{Ti}_{MoMo}$, **(b)** $\widetilde{D}^{Ti}_{ZrZr}$, **(c)** $\widetilde{D}^{Ti}_{MoZr}$ and **(d)** $\widetilde{D}^{Ti}_{ZrMo}$ at 1373 K and **(e)** $\widetilde{D}^{Ti}_{MoMo}$, **(f)** $\widetilde{D}^{Ti}_{ZrZr}$, **(g)** $\widetilde{D}^{Ti}_{MoZr}$ and **(h)** $\widetilde{D}^{Ti}_{ZrMo}$ at 1473 K.

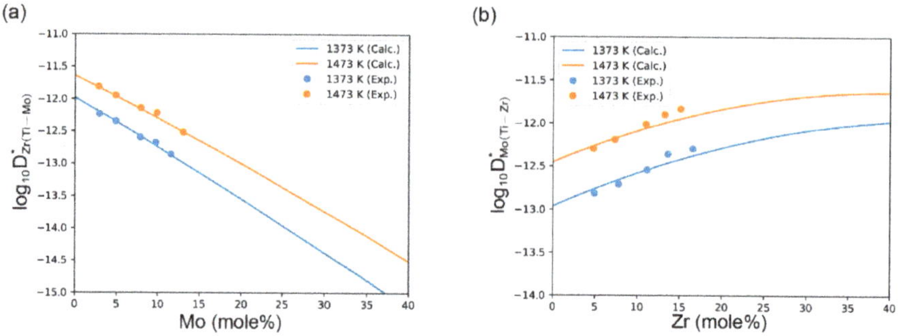

Figure 8. Calculated impurity diffusion coefficients (**a**) $D^*_{Zr(Ti-Mo)}$ and (**b**) $D^*_{Mo(Ti-Zr)}$ as compared with the experimental data.

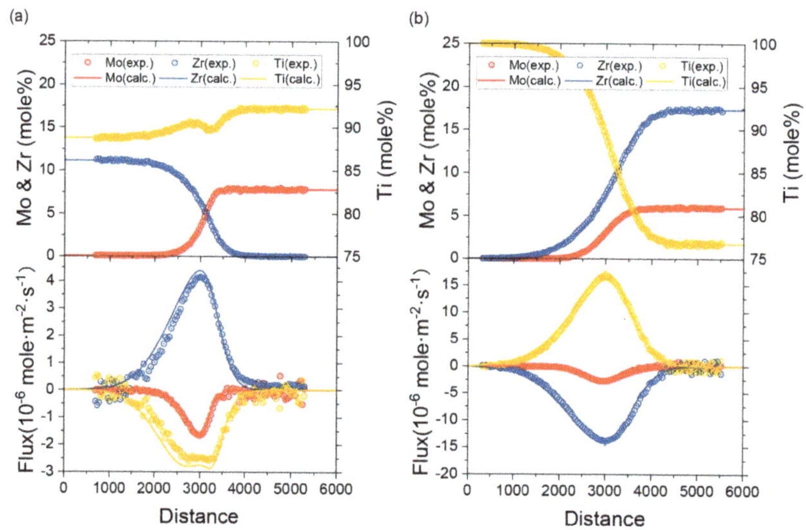

Figure 9. Predicted composition-distance profiles and inter-diffusion fluxes of M3 and N8 as compared with experimental data.

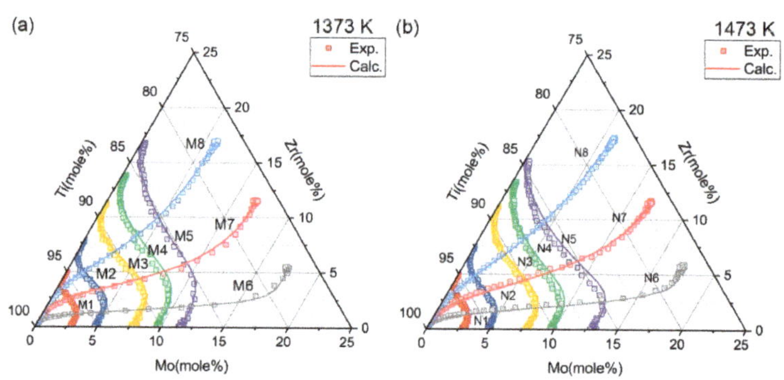

Figure 10. Simulated diffusion paths for diffusion couples compared with the experimental measurements: (**a**) 1373 K and (**b**) 1473 K.

5. Conclusions

In this work, two set of diffusion couples of bcc Ti-Mo-Zr alloys were made and annealed at 1373 K for 72 h and 1473 K for 48 h, respectively. The composition-distance profiles of these diffusion couples were determined using EPMA. The ternary inter-diffusion coefficients and impurity diffusion coefficients were extracted using the Whittle-Green and Hall method. Based on the experimental results and thermodynamic descriptions, as well as atomic mobility parameters of the sub-binary systems of Ti-Mo-Zr system, an atomic mobility database for the bcc phase in the Ti-Mo-Zr ternary system was developed. Inter-diffusion coefficients at the intersection points of diffusion couples and impurity diffusion coefficients were calculated using the database and compared with the data extracted directly from the composition-distance profiles. In addition, simulations of the diffusion couples with the same initial conditions and heat treatment processes of the experiment were proceeded. The composition-distance profiles and diffusion paths were compared with the experimental results. All of the calculated results show good agreement with experimental data.

Author Contributions: W.B., G.X., L.L. and L.Z. (Ligang Zhang) designed the study. W.B., M.T., L.Z. (Lijun Zeng), D.W and Z.Y. conducted the experiments; W.B., L.Z. (Lijun Zeng) and G.X. analyzed the results. W.B., G.X., L.L, M.T. and L.Z. (Ligang Zhang) contributed to preparation of the manuscript.

Funding: The work was funded by Aof the National Nature Science Foundation of China (Grant No. 51671218), the National Key Research and Development Program of China (Grant No. 2016YFB0701301) and the National Key Basic Research Program of China (973Program) (Grant No. 2014CB644000).

Conflicts of Interest: The authors declare no conflict of interest.

References

1. Gogia, A.K. High-temperature Titanium Alloys. *Defence Sci. J.* **2005**, *55*, 149–173. [CrossRef]
2. Geetha, M.; Singh, A.K.; Asokamani, R.; Gogia, A.K. Ti based biomaterials, the ultimate choice for orthopaedic implants—A review. *Prog. Mater. Sci.* **2009**, *54*, 397–425. [CrossRef]
3. Correa, D.R.N.; Kuroda, P.A.B.; Grandini, C.R. Structure, Microstructure, and Selected Mechanical Properties of Ti-Zr-Mo Alloys for Biomedical Applications. *Adv. Mater.* **2014**, *922*, 75–80. [CrossRef]
4. Niinomi, M. Recent metallic materials for biomedical applications. *Metall. Mater. Trans. A* **2002**, *33*, 477. [CrossRef]
5. Song, Y.; Xu, D.S.; Yang, R.; Li, D.; Wu, W.T.; Guo, Z.X. Theoretical study of the effects of alloying elements on the strength and modulus of β-type bio-titanium alloys. *Mater. Sci. Eng. A* **1999**, *260*, 269–274. [CrossRef]
6. Nnamchi, P.S.; Obayi, C.S.; Todd, I.; Rainforth, M.W. Mechanical and electrochemical characterisation of new Ti-Mo-Nb-Zr alloys for biomedical applications. *J. Mech. Behav. Biomed. Mater.* **2016**, *60*, 68–77. [CrossRef] [PubMed]
7. Eisenbarth, E.; Velten, D.; Müller, M.; Thull, R.; Breme, J. Biocompatibility of β-stabilizing elements of titanium alloys. *Biomaterials* **2004**, *25*, 5705–5713. [CrossRef] [PubMed]
8. Han, M.-K.; Hwang, M.-J.; Yang, M.-S.; Yang, H.-S.; Song, H.-J.; Park, Y.-J. Effect of zirconium content on the microstructure, physical properties and corrosion behavior of Ti alloys. *Mater. Sci. Eng. A* **2014**, *616*, 268–274. [CrossRef]
9. Ho, W.F.; Ju, C.P.; Lin, J.C. Structure and properties of cast binary Ti-Mo alloys. *Biomaterials* **1999**, *20*, 2115–2122. [CrossRef]
10. Trentani, L.; Pelillo, F.; Pavesi, F.C.; Ceciliani, L.; Cetta, G.; Forlino, A. Evaluation of the $TiMo_{12}Zr_6Fe_2$ alloy for orthopaedic implants: In vitro biocompatibility study by using primary human fibroblasts and osteoblasts. *Biomaterials* **2002**, *23*, 2863–2869. [CrossRef]
11. Gordin, D.M.; Gloriant, T.; Texier, G.; Thibon, I.; Ansel, D.; Duval, J.L.; Nagel, M.D. Development of a β-type Ti-12Mo-5Ta alloy for biomedical applications: Cytocompatibility and metallurgical aspects. *J. Mater. Sci. Mater. Med.* **2004**, *15*, 885–891. [CrossRef] [PubMed]
12. Nag, S.; Banerjee, R.; Fraser, H.L. Microstructural evolution and strengthening mechanisms in Ti-Nb-Zr-Ta, Ti-Mo-Zr-Fe and Ti-15Mo biocompatible alloys. *Mater. Sci. Eng. C* **2005**, *25*, 357–362. [CrossRef]
13. Yu, Z.T.; Wang, G.; Ma, X.Q.; Dargusch, M.S.; Han, J.Y.; Yu, S. Development of Biomedical Near β Titanium Alloys. *Mater. Sci. Forum* **2009**, *618*, 303–306. [CrossRef]

14. Wang, K.; Gustavson, L.; Dumbleton, J. The characterization of Ti-12Mo-6Zr-2Fe a new biocompatible titanium alloy developed for surgical implants. In *Beta Titanium in the 1990's*; The Mineral, Metals and Materials Society: Warrendale, PA, USA, 1993; pp. 2697–2704.
15. Steinemann, S.G.; Mäusli, P.A.; Szmukler-Moncler, P.; Semlitsch, M.; Pohler, O.; Hintermann, H.E.; Perren, S.M. Beta-titanium alloy for surgical implants. In *Beta Titanium in the 1990'*; The Mineral, Metals and Materials Society: Warrendale, PA, USA, 1993; pp. 2689–2696.
16. Helander, T.; Ågren, J. Diffusion in the B2-b.c.c. phase of the Al-Fe-Ni system—Application of a phenomenological model. *Acta Mater.* **1999**, *47*, 3291–3300. [CrossRef]
17. Chen, Q.; Ma, N.; Wu, K.; Wang, Y. Quantitative phase field modeling of diffusion-controlled precipitate growth and dissolution in Ti-Al-V. *Scr. Mater.* **2004**, *50*, 471–476. [CrossRef]
18. Mao, C.; Tan, M.; Zhang, L.; Wu, D.; Bai, W.; Liu, L. Experimental reinvestigation and thermodynamic description of Bi-Te binary system. *Calphad* **2018**, *60*, 81–89. [CrossRef]
19. Chen, F.; Xu, G.; Zhang, X.; Zhou, K. Exploring the Phase Transformation in β-Quenched Ti-55531 Alloy During Continuous Heating via Dilatometric Measurement, Microstructure Characterization, and Diffusion Analysis. *Metall. Mater. Trans. A* **2016**, 1–12. [CrossRef]
20. Zhang, L.; Stratmann, M.; Du, Y.; Sundman, B.; Steinbach, I. Incorporating the CALPHAD sublattice approach of ordering into the phase-field model with finite interface dissipation. *Acta Mater.* **2015**, *88*, 156–169. [CrossRef]
21. Andersson, J.-O.; Ågren, J. Models for numerical treatment of multicomponent diffusion in simple phases. *J. Appl. Phys.* **1992**, *72*, 1350–1355. [CrossRef]
22. Borgenstam, A.; Höglund, L.; Ågren, J.; Engström, A. DICTRA, a tool for simulation of diffusional transformations in alloys. *J. Phase Equilib.* **2000**, *21*, 269. [CrossRef]
23. Andersson, J.-O.; Helander, T.; Höglund, L.; Shi, P.; Sundman, B. Thermo-Calc & DICTRA, computational tools for materials science. *Calphad* **2002**, *26*, 273–312. [CrossRef]
24. Neumann, G.; Tuijn, C. *Self-Diffusion and Impurity Diffusion in Pure Metals: Handbook of Experimental Data*, 1st ed.; Pergamon: Oxford, UK, 2009.
25. Heumann, T.; Imm, R. *Untersuchungen über den Kirkendall-Effekt in kub. raumzentrierten Titan-Molybdän Legierungen*; Forschungsberichte des Landes Nordrhein-Westfalen; VS Verlag für Sozialwissenschaften: Wiesbaden, Germany, 1978.
26. Thibon, I.; Ansel, D.; Baliveau, M.; Debuigne, J. Interdiffusion in the β Mo-Ti solid solution at high temperatures. *Z. Metallkd.* **1998**, *89*, 187–191.
27. Sprengel, W.; Yamada, T.; Nakajima, H. Interdiffusion in Binary β-Titanium Alloys. *Defect Diffusion Forum* **1997**, *143*, 431–436. [CrossRef]
28. Thibon, I.; Ansel, D.; Gloriant, T. Interdiffusion in β-Ti-Zr binary alloys. *J. Alloys Compd.* **2009**, *470*, 127–133. [CrossRef]
29. Brunsch, A.; Steeb, S. Diffusionsuntersuchungen im System Ti-Zr mittels Mikrosonde / Interdiffusion in the System Ti-Zr. *Z. Naturforsch. A* **1974**, *29*, 1319–1324. [CrossRef]
30. Raghunathan, V.S.; Tiwari, G.P.; Sharma, B.D. Chemical diffusion in the β phase of the Zr-Ti alloy system. *Metallogr. Trans.* **1972**, *3*, 783–788. [CrossRef]
31. Bhanumurthy, K.; Laik, A.; Kale, G.B. Novel Method of Evaluation of Diffusion Coefficients in Ti-Zr System. In *Phase Transformation and Diffusion*; Defect and Diffusion Forum; Trans Tech Publications: Zurich, Switzerland, 2008; pp. 53–62.
32. Bhatt, Y.J.; Kumar, L.; Patil, R.V.; Kale, G.B.; Garg, S.P. Diffusion studies in Hf-Mo, Zr-Mo, Cr-Nb, Cr-Ta and Th-Re systems above 1900 K. *J. Alloys Compd.* **2000**, *302*, 177–186. [CrossRef]
33. Liu, Y.; Zhang, L.; Yu, D. Computational Study of Mobilities and Diffusivities in bcc Ti-Zr and bcc Ti-Mo Alloys. *J. Phase Equilib. Diffus.* **2009**, *30*, 334–344. [CrossRef]
34. Liu, Y.; Wang, G.; Wang, J.; Kang, Z. Mobilities and diffusivities for bcc Nb-W, Nb-Ta, Zr-Mo and Zr-Hf alloys. *J. Alloys Compd.* **2013**, *555*, 381–389. [CrossRef]
35. Kirkaldy, J.S. Diffusion in Multicomponent Metallic Systems. *Can. J. Phys.* **1957**, *35*, 435–440. [CrossRef]
36. Whittle, D.P.; Green, A. The measurement of diffusion coefficients in ternary systems. *Scr. Metall.* **1974**, *8*, 883–884. [CrossRef]
37. Chen, Y.; Tang, B.; Xu, G.; Wang, C.; Kou, H.; Li, J.; Cui, Y. Diffusion Research in BCC Ti-Al-Mo Ternary Alloys. *Metall. Mater. Trans. A* **2014**, *45*, 1647–1652. [CrossRef]

38. Hall, L.D. An Analytical Method of Calculating Variable Diffusion Coefficients. *J. Chem. Phys.* **1953**, *21*, 87–89. [CrossRef]
39. Jönsson, B. Assessment of the mobility of carbon in fcc C-Cr-Fe-Ni alloys. *Z. Metallkd.* **1994**, *85*, 502–509.
40. Shim, J.-H.; Oh, C.-S.; Lee, D.N. A thermodynamic evaluation of the Ti-Mo-C system. *Metall. Mater. Trans. B* **1996**, *27*, 955–966. [CrossRef]
41. Turchanin, M.A.; Agraval, P.G.; Abdulov, A.R. Thermodynamic assessment of the Cu-Ti-Zr system. II. Cu-Zr and Ti-Zr systems. *Powder. Metall. Metal Cream.* **2008**, *47*, 428. [CrossRef]
42. Jerlerud Pérez, R.; Sundman, B. Thermodynamic assessment of the Mo–Zr binary phase diagram. *Calphad* **2003**, *27*, 253–262. [CrossRef]
43. Watson, A. Molybdenum-Titanium-Zirconium. In *Refractory metal systems*; Springer: Berlin/Heidelberg, Germany, 2010; pp. 441–454.
44. Wang, C.; Xu, G.; Cui, Y. Mapping of Diffusion and Nanohardness Properties of Fcc Co-Al-V Alloys Using Ternary Diffusion Couples. *Metall. Mater. Trans. A* **2017**, *48*, 4286–4296. [CrossRef]
45. Kirkaldy, J.S.; Weichert, D.; Haq, Z.U. Diffusion in Multicomponent Metallic Systems: Vi. Some Thermodynamic Properties of the D Matrix and the Corresponding Solutions of the Diffusion Equations. *Can. J. Phys.* **1963**, *41*, 2166–2173. [CrossRef]

materials

MDPI

Article

Characterization and Mechanical Proprieties of New TiMo Alloys Used for Medical Applications

Andrei Victor Sandu [1,2,*], Madalina Simona Baltatu [1], Marcin Nabialek [3], Adriana Savin [4] and Petrica Vizureanu [1,*]

1 Faculty of Materials Science and Engineering, "Gheorghe Asachi" Technical University of Iasi, 700050 Iasi, Romania; simona.baltatu@tuiasi.ro
2 Romanian Inventors Forum, Iasi 700089, Romania
3 Institute of Physics, Faculty of Production Engineering and Materials Technology, Czestochowa University of Technology, 42-200 Częstochowa, Poland; nmarcell@wp.pl
4 National Institute of Research and Development for Technical Physics 47 Mangeron Boulevard, Iasi 700050, Romania; asavin@phys-iasi.ro
* Correspondence: sav@tuiasi.ro (A.V.S.); peviz@tuiasi.ro (P.V.); Tel.: +40-745-438604 (A.V.S.)

Received: 27 August 2019; Accepted: 11 September 2019; Published: 13 September 2019

check for updates

Abstract: Ti-based alloys are accessible for use in the human body due to their good mechanical properties, corrosion resistance, and biocompatibilities. These main properties of alloys are important criteria for choosing biomedical implants for human bones or for other kinds of applications in general medicine. This paper presents a comparison of four new Ti-based alloys desired to satisfy various requirements for biomedical implants. The materials were prepared with recipes for two new system alloys, TiMoZrTa (TMZT) and TiMoSi (TMS), alloys with nontoxic elements. The presented research contains microstructure images, indentation tests, Vickers hardness, XRD, and corrosion resistance, showing better characteristics than most commercial products used as implants (Young's modulus closer to the human bone).

Keywords: titanium alloy; biomaterial; TiMoZrTa; TiMoSi; mechanical properties; low elasticity modulus; corrosion

1. Introduction

At present, the market request for biomaterials is very high. The uses of biomaterials in medical applications includes several areas: Orthopedics, cardiovascular surgery, ophthalmology, dentistry, urology, aesthetic surgery, neurology, suture material for wound healing, and controlled release drug delivery systems. Due to this, it is very important to develop new materials with enhanced properties, biocompatibility, and long-term viability as an implant material. Fundamental requirements are that an implant should have biomechanical properties (stiffness, strength, fracture toughness, wear resistance, fatigue strength, corrosion resistance) and biomedical properties (toxicity, surface state, osseointegration) [1–3].

The usual materials currently found in medical applications are classical metallic alloys: Titanium alloys (TiAlV), cobalt alloys (CoCrMo), and stainless steels.

The microstructure, as well as the properties of the titanium alloys, differs depending on the amount of α or β stabilizing elements added to the titanium alloys. Ti-based are grouped into α-type, ($\alpha + \beta$)-type, and β-type alloys. Alloying the pure titanium with β-elements leads to the widening of the β-phase domain, as well as the improvement of the mechanical properties proven in the literature [4–7].

Mo, V, Nb, and Ta are β-stabilizers elements that decrease the temperature of allotropic transformation and neutrals, such as Zr. Commercial pure titanium and Ti_6Al_4V alloy were the first titanium-based materials for commercial use. In recent years, further studies have indicated that vanadium, used to stabilize the α-phase, may produce harmful oxides for the human body [8,9].

The authors of [10,11] concluded that Ti, B, Mg, Si, P, Ca, Sr, Zr, Nb, Mo, Pd, In, Sn, Ta, Pt, and Au are biocompatible elements, while harmful elements include Be, Al, V, Cr, Mn, Fe, Co, Ni, Cu, Zn, and Ag [10,11].

One of the most important features of an implant is that it comes in contact with the living tissues of the body, thus creating an interface between them. The phenomena that occur at this interface are of great interest because they ultimately determine the success or failure of the implant, both in terms of immediate reaction and long-term response. The biological response between the implant and the host tissue is largely dependent on the place of implantation and the surface properties of the implant. The role of biomaterials is to come into contact with a biological system. When a biomaterial is placed in the human body in the form of a medical device, human tissues react to its implantation in different ways, depending on the type of biomaterial used. The mechanism of cell attachment to the implant depends on the tissue response to the surface of the alloy.

Most recently, researchers have aimed to improve properties by introducing nontoxic elements, for enhancing surface characteristics, mechanical properties, corrosion resistance, biocompatibility and developing alloys with low modulus β or near β Ti alloys, such as Ti–Mo systems, Ti–36Nb–2Ta–3Zr, Ti–24Nb–4Zr–7.8Sn [2,12], etc.

Titanium is a nontoxic element, even in larger quantities; some studies have demonstrated the influence of ingestion by humans of up to 0.8 mg of titanium daily, proving that titanium was eliminated without being digested/assimilated. Its uses in medical applications were due to good interaction with the host bone, the titanium implants not being rejected by the body, and having a high corrosion resistance [13–15].

Molybdenum is an element with a lower degree of toxicity compared to Co, Ni, and Cr, and moreover, is a β-stabilizing element. Field studies have highlighted that titanium alloyed with molybdenum in varying percentages between 15%–20% can decrease the modulus of elasticity leading to adequate mechanical properties.

Silicon reduces ductility, improves creep and high-temperature resistance, and increases corrosion resistance.

Zirconium is used in medical applications due to its low modulus of elasticity, high corrosion resistance, and high biocompatibility with human tissue [13–15].

Tantalum is considered biocompatible, being a β-stabilizing element that influences the value of the modulus of elasticity [13–15].

Silicon is an element found in the human bone and is considered biocompatible. It is also a β-stabilizing element that influences the decrease of the modulus of elasticity.

Despite these facts, Mo, Zr, Ta, and Si were selected for designing a new ß-type of Ti-based alloys (Table 1), systems like TMZT (TiMoZrTa) and TMS (TiMoSi) with lower elastic modulus, corrosion resistance, and good biocompatibility.

Table 1. Chemical composition of alloys after melting obtained by Energy Dispersive X-ray analysis.

Chemical Element	Mo	Si	Zr	Ta	Ti
wt.(%)	13.56	-	6.34	4.64	Balance
	12.56	-	6.99	14.45	Balance
	15.93	0.28	-	-	Balance
	15.19	1.02	-	-	Balance

This article presents new recipes for Ti-Mo system alloys, with good characteristics and using nontoxic elements, which can be used in medical applications.

2. Experimental

Materials and Preparation

In order to obtain titanium alloys, it was chosen to use a vacuum arc melting plant, produced in house by Institute of Physics, Faculty of Production Engineering and Materials Technology, Czestochowa University of Technology, Poland, and we used raw materials with a high purity: Ti—99.8%, Mo—99.7%, Zr—99.2%, Ta—99.5%, and Si—99.2%.

The advantages of this equipment are as follows: High melting temperatures, the use of vacuum or protective environment, uniform composition by repetitive melting, and the possibility of melting elements at different melting points (Figure 1).

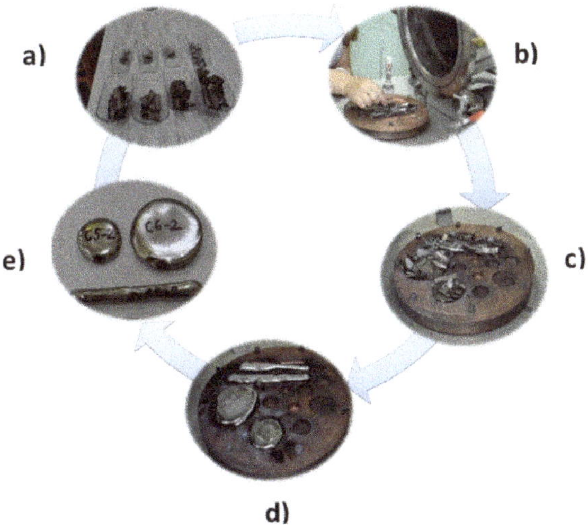

Figure 1. Stages of development of the alloys: (**a**) weighing of raw materials and gravimetric dosing; (**b**,**c**) loading of the raw material; (**d**,**e**) TiMoZrTa (TMZT) semi-products obtained after solidification.

The chemical composition of the alloys was analyzed using a scanning electron microscope (SEM) VEGA II LSH manufactured by the TESCAN Co., Brno, the Czech Republic, coupled with an EDX QUANTAX QX2 detector manufactured by the BRUKER/ROENTEC Co., Berlin, Germany. The chemical analyses (EDX) of the alloys studied were performed at many different points for a precise determination, with a uniform distribution of elements and a homogeneous alloy. The chemical composition is shown in Table 1.

The phase analysis was carried out using a Panalytical X'Pert Pro MPD diffractometer, by Malvern Panalytical a Spectris company, Almelo, The Netherlands. The parameters used for sample analysis were: Range of angle θ–2θ between 20° and 80°; continuous scanning. In order to determine the constituent phases, by X-ray diffraction, from the Ti-based alloys developed, samples with size of 10 mm × 10 mm × 5 mm were cut and used after they had been polished.

A Wilson Wolpert universal hardness tester 751 N model, by Wilson Instruments an Instron Company, Heerlen, The Netherlands, was used for the Vickers hardness measurements, with a load of 9807 N during 12 s. The hardness measurements were performed on samples obtained from Ti-based alloys with dimensions 10 mm × 10 mm × 3 mm, the surface of the samples being prepared by grinding on abrasive paper. Experimental tests consisted of three determinations in different areas on the surface of each sample, using a 9.81 N pressure force and a 12 second measurement time.

Elastic modulus was measured using a CETR UMT-2 Tribometer, by Bruker, Campbell CA, USA. For micro-scratch analysis, a constant load method was used with a load of 5N on a distance of 4 mm, for a single determination. For indentation tests, samples of Ti-based alloys obtained were cut with a dimension of 17 mm × 5 mm × 5 mm, and the surface of the samples was prepared by grinding with high-grain silicon carbide abrasive paper and polished with alumina suspension until obtaining metallic gloss and removal of surface roughness. The investigated samples were fixed to a flat surface of the test device with screws and clamps. Testing was carried out under dry conditions. A Rockwell diamond-type penetrator was used, having an incisor cone with an angle of 120°, applying a force of 5 N.

To determine the elastic modulus as accurately as possible, three determinations were performed for each alloy. From the mean values, using UMT 2 software, provided with the CETR UMT-2 Tribometer, the indentation curves (depth vs. force) of TiMoSi alloys were plotted using the ViEWER program.

In order to obtain the electrochemical parameters for characterizing the corrosion resistance of elaborated Ti-based alloys, the electrochemical tests of linear polarization were used. This method can directly and quantitatively determine the corrosion rate. To determine the corrosion potential of Ti-based alloys developed, samples of 5 mm × 5 mm × 5 mm were used. Prior to being subjected to the corrosion potential determination, the samples were cut and ground to remove impurities and the Ti oxide film that formed on the surface of the alloy, then they were embedded in Teflon.

The electrical component of an electrochemical cell consists of a voltmeter, measuring the potential of the product. A reference electrode was a saturated calomel electrode in a potassium chloride solution whose potential is reproducible and has a value of 242 mV at a temperature of 25 °C. Additionally, a platinum electrode was used as an auxiliary electrode. Data processing was done with the Volta Master 4 -Electrochemical Software, by Radiometer Analytical SAS, Villeurbanne, France, then imported and processed using experimental data processing software.

3. Results and Discussions

Figure 2 shows the structure of Ti-based alloys with titanium alloy grain specifics. Images obtained by optical microscopy for titanium alloys have a dendrite structure for TiMoZrTa alloy systems. TiMoSi alloy systems highlight the formation of β-type equiaxed grains, having different dimensions. The acicular and coarse structures are specific to β-alloys. With TiMoSi, a β structure of cube with a centered volume is visible.

The variation of the α, α + β, and β phases consists in the differences in the chemical composition of the constituents. The high percentage of β (Mo, Ta, and Si) stabilizing elements led to the formation of a β-type structure, highlighted very well in the Ti-based alloys obtained. Zirconium contribution in concentrations below 10% contributes to the refining of the microstructure, thus allowing the formation of a homogeneous and uniformly distributed structure. Therefore, elements with percentages of tantalum (5%–15%), combined with a molybdenum concentration of 15%, contribute to the formation of phase β.

Optical microscopy (Figure 2) reveals a uniform biphasic structure, consisting of a high proportion of solid β solutions, in which the dendrite lamellar structures of the orthorhombic martensite α″ are present. Orthorhombic martensite occurs frequently in titanium alloys containing β-stabilizers of the transition metals category, including molybdenum and tantalum. In the present case, the presence of the α″ phase is due to the decomposition of the β phase during the cooling. Regarding the $Ti_{15}Mo_{0.5}Si$, the small dark spots are due to pitting corrosion after the chemical etching of the polished surface and do not represent any inclusion or other phases.

The hardness measurements highlight the resistance to the penetration action of an external body and provide information on the behavior of the studied materials. In this way, we can analyze the Ti-based alloys developed for the purpose of fitting them into the specific medical application. The hardness measurements made on the titanium alloys, developed by the Vickers method, are a general method for determining the hardness of metallic materials, commonly used in biomaterial testing.

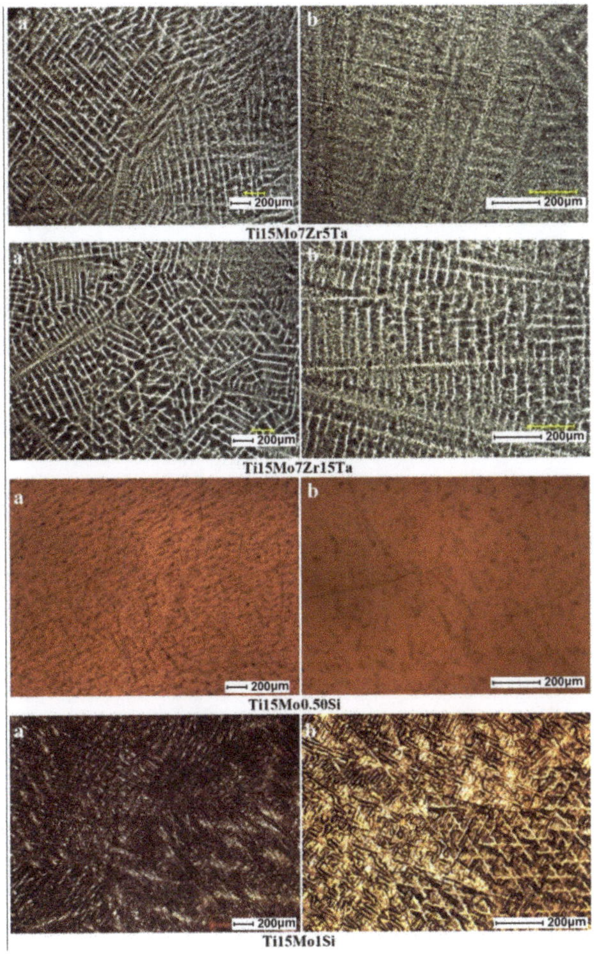

Figure 2. Optical microstructure images of titanium alloys $Ti_{15}Mo_7Zr_5Ta$, $Ti_{15}Mo_7Zr_{15}Ta$, $Ti_{15}Mo_{0.5}Si$, and $Ti_{15}Mo_1Si$ at various magnifications: (**a**) 50×, (**b**) 100×.

Some of the mechanical properties of Ti-based alloys developed are presented in Table 2 (hardness and Young's modulus), compared with two other samples: Commercial pure titanium and Ti_6Al_4V. The values of the elasticity modulus of the titanium alloys developed, resulting from the indentation test, are highlighted in same table.

Table 2. Some mechanical properties of titanium samples.

Alloy	Hardness (HV)	Young Modulus (GPa)
$Ti_{15}Mo_7Zr_5Ta$	379.21	69.02
$Ti_{15}Mo_7Zr_{15}Ta$	390.88	51.93
$Ti_{15}Mo_{0.5}Si$	233.37	19.82
$Ti_{15}Mo_1Si$	188.52	42.84
C.P. Titanium	182	105
Ti6Al4V	541	119

According to the literature [15–19], classical biomaterial alloys like stainless steel and CoCrMo alloys have hardness values between 155 HV and 601 HV. The titanium alloys obtained comprise values ranging from 188.52 to 390.88. Titanium alloys have values approximate to the Ti_6Al_4V alloy (349 HV), which is mostly used in implantology [15–19].

The elasticity module is a very important criterion underlying the choice of metallic materials used in orthopedics and should be as close as possible to human bone (17–30 GPa) to avoid the stress-shielding effect [1–3].

The Ti-based alloys obtained have values ranging from 19.82–69.02 GPa for the modulus of elasticity, measured by indentation tests. The lowest value is the $Ti_{15}Mo_{0.5}Si$ alloy (19.82 GPa), and the highest value are the $Ti_{15}Mo_7Zr_5Ta$ alloys (69.02 GPa). The low elasticity modulus of the investigated alloys is due to the presence of β-stabilizing elements such as Mo, Ta, and Si. According to Table 2, it can be seen that adding elements like Zr and Ta can increase values of modulus of elasticity with approximately 40 GPa. On the other hand, increasing the percentage of Ta from 5% to 15% contributed to a decrease in the modulus of elasticity for TMZT alloys.

As we can see from the data in Table 2, the highest Young's modulus (69 GPa) is visible in the $Ti_{15}Mo_7Zr_5Ta$ systems, due to the presence of α″ phase (martensite). $Ti_{15}Mo_{0.5}Si$ presents the lowest Young's modulus (20 GPa), where on the microstructure, the β boundaries are clearly visible (Figure 2b). These are confirmed also by the literature, that the β phase presents better mechanical properties. Taking into consideration other metallic biomaterials—CoCrMo alloys (210–253 GPa) and stainless steels (190–210 GPa)—the investigated Ti-based alloys exhibited much lower values and closer values to that of the human bone (17–30 GPa). The alloys developed showed a significant improvement for mechanical properties even for titanium alloys: Ti_6Al_4V (100–114 GPa) and C.P. Ti (102–104 GPa).

From the point of view of elasticity, these alloys fulfilled the most important requirement criteria of metallic biomaterials, the values of the modulus of elasticity being close to that of the human bone.

Figure 3 shows the response of alloys during indentation tests in the form of force–depth dependencies.

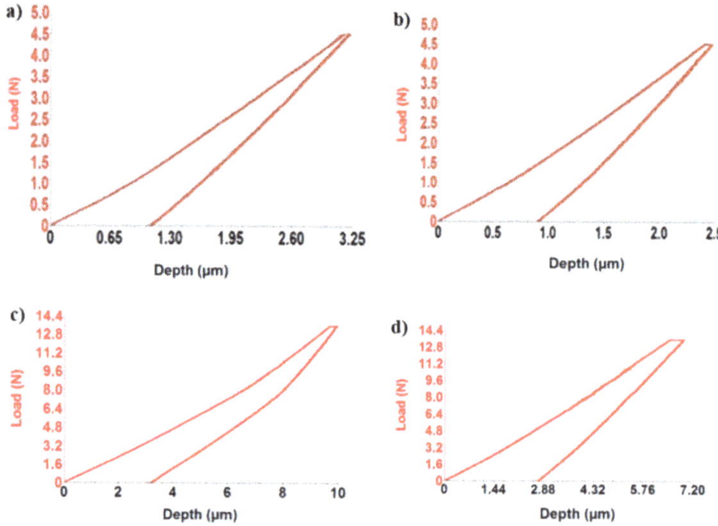

Figure 3. The strength-depth curve of the indentation test for Ti-based alloys developed: (**a**) $Ti_{15}Mo_7Zr_5Ta$, (**b**) $Ti_{15}Mo_7Zr_{15}Ta$, (**c**) $Ti_{15}Mo_{0.5}Si$, and (**d**) $Ti_{15}Mo_1Si$.

The diffractograms obtained (Figure 4) show the β-type structures identified by optical microscopy, taking into account that titanium is an allotropic element, presenting it in different forms: Up to a temperature of 882 °C, having a α-Ti compact hexagonal structure and above 882 °C, β–Ti, having a centered cube structure. In the composition of the investigated alloys, there is a major phase β with a centered cube structure and a secondary phase α″ with an orthorhombic structure (martensite), present only for $Ti_{15}Mo_7Zr_5Ta$ systems.

Phase β, having a centered cubic structure, is fixed by alloying the titanium with transition elements (Mo, Ta, and Si) crystallized in the orthorhombic system and is formed when the content of stabilizing elements phase β decomposed during cooling [20–23].

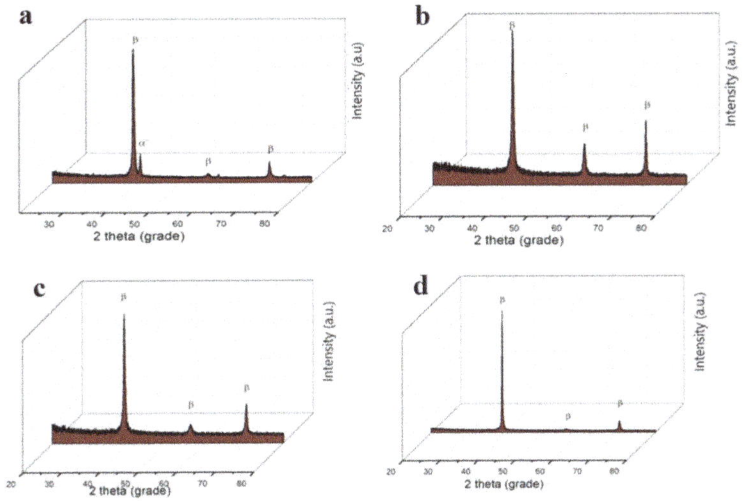

Figure 4. The XRD diffraction patterns of titanium samples: (**a**) $Ti_{15}Mo_7Zr_5Ta$, (**b**) $Ti_{15}Mo_7Zr_{15}Ta$, (**c**) $Ti_{15}Mo_{0.5}Si$, and (**d**) $Ti_{15}Mo_1Si$.

The corrosion of biomaterials is determined by the aggressive corrosive nature of the elements present in body fluids. In order to determine the corrosion potential of titanium alloys developed, it was necessary to investigate them through potentially dynamic and potentiostatic tests in simulated biological environments.

The simplest method of measuring the corrosion rate of a metal involves putting it in contact with the test medium and measuring the amount of material lost by the sample depending on the exposure time.

The polarization resistance method was used to assess the corrosion rate for the Ti-based alloys developed. This method serves to determine the corrosion current and the corrosion potential of the metal or alloy, from the linear polarization curve obtained from relatively small voltages. The corrosion current determined by this method is therefore the current that occurs at the metal/corrosive environment interface when the metal is immersed in the solution and represents the instantaneous corrosion current.

The specific titanium developed samples were introduced into the electrochemical cell, and the selected corrosion medium was Ringer's solution with the composition: NaCl: 8.6 g/L, KCl: 0.3 g/L, $CaCl_2$: 0.48 g/L, in order to investigate their use as potential biomaterials. Measurements were made at 20 °C in naturally aerated solutions. The anodic polarization curves were recorded at a potential sweep rate of 0.5 mV/s over a potential range of ± 200 mV versus the open circuit potential. Corrosion potential (E_{corr} = E (I = 0)), Tafel slopes (βa and βc), polarization resistance (R_p), instantaneous corrosion density (corrosion), and corrosion rate were evaluated using the VoltaMaster 4 software.

Figure 5 shows the linear polarization curves in semilogarithmic coordinates for the samples studied in the Ringer solution, and in Table 3, we present the instantaneous corrosion parameters in the same physiological environment.

Corrosion potential, E_{corr}, measured relative to the potential of the saturated calomel electrode, is the potential at which the oxidation–reduction reactions on the alloy surface are at equilibrium; the rate of the oxidation reaction is equal to the rate of the reduction reaction, and the total current intensity is zero. Increasing the potential to more positive values increases the rate of the oxidation reaction, while the potential shift to negative values is reduced by the oxidation process, and the metal passes. It can be seen that the presence of silicon appears to produce an increase in the corrosion rate.

Table 3. The main parameters of the Ringer solution corrosion process.

Alloy	E_{corr} [mV]	β_a [mV]	β_c [mV]	R_p [kΩ/cm^2]	J_{cor} [μA/cm^2]	V_{cor} [μm/year]
$Ti_{15}Mo_7Zr_5Ta$	−461.20	117.90	−39.50	22.00	1.08	12.67
$Ti_{15}Mo_7Zr_{15}Ta$	−365.40	88.30	−115.00	38.78	0.47	5.49
$Ti_{15}Mo_{0.5}Si$	−266.00	200	−192	14.91	2.13	20.59
$Ti_{15}Mo_1Si$	−309.00	278	−152	8.99	3.96	38.30

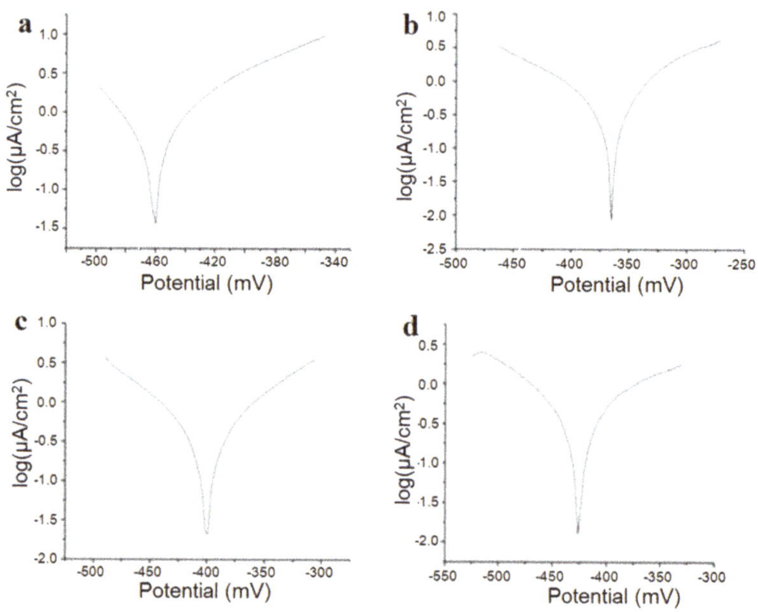

Figure 5. Diagrams of the investigated alloys by corrosion testing in Ringer's solution of the samples: (**a**) $Ti_{15}Mo_7Zr_5Ta$, (**b**) $Ti_{15}Mo_7Zr_{15}Ta$, (**c**) $Ti_{15}Mo_{0.5}Si$, and (**d**) $Ti_{15}Mo_1Si$.

Polarization resistors have high values, which is reflected in very low corrosion rates. The "corrosive" product of these alloys is mainly titanium oxide, TiO_2, which is insoluble and adherent to the surface of the alloy. The oxide layer on the surface protects the alloy against the aggressive action of electrolytic media. Considering this, it can be concluded that in the physiological environment, titanium-based alloys are not corroded but actually suffer a passivation process. Under these conditions, the variable corrosion rate is actually passivation velocity.

Particular attention should be given to the Ti15Mo1Si alloy; in these cases, the polarization resistances were small and the corrosion rates were higher than that for the other alloys (almost double).

The value of molybdenum and tantalum concentrations in the studied alloys are important for corrosion studies. The higher the percentage of tantalum, the better the characteristics of corrosion resistance. Molybdenum has a great influence on the corrosion parameters; thus, for a percentage of 15%, the corrosion resistance is very good [24–27].

4. Conclusions

At the global level, there is continuous concern for the research and development of alloys for medical and biomedical applications. Thus, it is desired to improve both the classic technologies of obtaining implants and the technologies of synthesis of the biomaterials from which they are executed, having as the final aim the promotion of a new generation of multifunctional implants with long-term performance.

Biomaterials are synthetic materials used to replace a part of a living system, or to function in close connection with a living tissue, which is why they must have properties as close to that of the human bone. Young's modulus of the Ti-based alloys obtained values ranging from 19.82 to 69.02 GPa, which is recommended for alloys used in implantology.

For most of the samples, the process that takes place on the surface is the oxidation of the titanium with the formation of an oxide adherent to the surface, which in fact produces the passive alloy. The corrosion rates (in fact passivation rates) are of the same order of magnitude, relatively small (20–30 m/year), for all alloys with uniform composition.

This paper presents the effect of tantalum addition that is beneficial in decreasing the modulus of elasticity as well as the corrosion rate. The addition of silicon has shown that it can present good properties only in a percentage of less than 0.5.

In conclusion, we can state that the characteristics of titanium alloys depend on the alloying elements, influencing the mechanical properties, and the biocompatibility. Elements like molybdenum, zirconium, tantalum, and silicon alloyed with titanium show good properties; in the future, they can be introduced in different medical applications such as dental and orthopedic implants.

Author Contributions: A.V.S. did the formal analysis, writing—reviewing and editing and final validation of results; M.S.B. did the investigation and wrote the original draft; M.N. did the alloys casting and contributed with resources and methodology; A.S. validated upon methodology the results; P.V. supervised the research and did the conceptualization.

Funding: This work was supported by a grant of the Romanian Ministry of Research and Innovation, CCCDI–UEFISCDI, project number PN-III-P1-1.2-PCCDI-2017-0239/ Contract no. 60/2018, within PNCDI III.

Conflicts of Interest: The authors declare no conflict of interest.

References

1. Niinomi, M. Mechanical properties of biomedical titanium alloys. *Mater. Sci. Eng. A* **1998**, *243*, 231–236. [CrossRef]
2. Baltatu, M.S.; Tugui, C.A.; Perju, M.C.; Benchea, M.; Spataru, M.C.; Sandu, A.V.; Vizureanu, P. Biocompatible Titanium Alloys used in Medical Applications. *Rev. Chim.* **2019**, *70*, 1302–1306.
3. Antoniac, I. (Ed.) *Biomateriale Metalice Utilizate la Executia Componentelor Endoprotezelor Totale de Sold (Metallic Biomaterials Used in the Execution of the Components of the Total Hip Stents)*; Printech: Bucharest, Romania, 2007.
4. Zhang, L.C.; Chen, L.Y. A Review on Biomedical Titanium Alloys: Recent Progress and Prospect. *Adv. Eng. Mater.* **2019**, *21*, 1801215. [CrossRef]
5. Kaur, M.; Singh, K. Review on titanium and titanium based alloys as biomaterials for orthopaedic applications. *Mater. Sci. Eng. C* **2019**, *102*, 844–862. [CrossRef] [PubMed]
6. Hong, S.H.; Hwang, Y.J.; Park, S.W.; Park, C.H.; Yeom, J.T.; Park, J.M.; Kim, K.B. Low-cost beta titanium cast alloys with good tensile properties developed with addition of commercial material. *J. Alloy. Compd.* **2019**, *793*, 271–276. [CrossRef]
7. Janson, O.; Gururaj, S.; Pujari-Palmer, S.; Ott, M.K.; Strømme, M.; Engqvist, H.; Welch, K. Titanium surface modification to enhance antibacterial and bioactive properties while retaining biocompatibility. *Mater. Sci. Eng. C* **2019**, *96*, 272–279. [CrossRef] [PubMed]
8. Ludwigson, D.C. Requirements for metallic surgical implants and prosthetic devices. *Met. Eng. Q.* **1965**, *5*, 1–6.
9. Bombac, D.M.; Brojan, M.; Fajfar, P.; Kosel, F.; Turk, R. Review of materials in medical applications. *Mater. Geoenviron.* **2007**, *54*, 471–499.

10. Chen, Q.; Thouas, G.A. Metallic implant biomaterials. *Mater. Sci. Eng. R* **2015**, *87*, 1–57. [CrossRef]
11. Song, Y.H.; Kim, M.K.; Park, E.J.; Song, H.J.; Anusavice, K.J.; Park, Y.J. Cytotoxicity of alloying elements and experimental titanium alloys by WST-1 and agar overlay tests. *Dent. Mater.* **2014**, *30*, 977–983. [CrossRef]
12. Spriano, S.; Yamaguchi, S.; Bainoa, F.; Ferraris, S. A critical review of multifunctional titanium surfaces: New frontiers for improving osseointegration and host response, avoiding bacteria contamination. *Acta Biomater.* **2018**, *79*, 1–22. [CrossRef] [PubMed]
13. Baltatu, I.; Vizureanu, P.; Ciolacu, F.; Achiţei, D.C.; Baltatu, M.S.; Vlad, D. In Vitro study for new Ti-Mo-Zr-Ta alloys for medical use. In *IOP Conference Series: Materials Science and Engineering*; IOP Publishing: Bristol, UK, 2019; Volume 572, p. 012030.
14. Baier, R.E. Surface behavior of biomaterials: The theta surface for biocompatibility. *J. Mater. Sci. Mater. Med.* **2006**, *17*, 1057–1062. [CrossRef] [PubMed]
15. Geetha, M.; Singh, A.K.; Asokamani, R.; Gogia, A.K. Ti based biomaterials, the ultimate choice for orthopaedic implants—A review. *Mater. Sci.* **2009**, *54*, 397–425. [CrossRef]
16. Elias, C.N.; Lima, J.H.C.; Valiev, R.; Meyers, M.A. Biomedical applications of titanium and its alloys. *Biol. Mater. Sci.* **2008**, *60*, 46–49. [CrossRef]
17. Cumpata, C.N. (Ed.) *Implaturiacoperitechimic cu Hidroxiapatitabiologică (Implants Chemically Covered with Biologic Hydroxyapatite)*; Printech: Bucuresti, Romania, 2012.
18. Chelariu, R.; Bujoreanu, L.G.; Roman, C. (Eds.) *Materialemetalicebiocompatibile cu Baza Titan (Metallic Biocompatible Materials Based on Ti)*; Politehnium: Iasi, Romania, 2006.
19. Koizumi, H.; Takeuchi, Y.; Imai, H.; Kawai, T.; Yoneyama, T. Application of titanium and titanium alloys to fixed dental prostheses. *J. Prosthodont. Res.* **2019**, *63*, 266–270. [CrossRef] [PubMed]
20. Long, M.; Rack, H.J. Titanium alloys in total joint replacement—A materials science perspective. *Biomaterials* **1998**, *19*, 1621–1639. [CrossRef]
21. Wang, K. The use of titanium for medical applications in the USA. *Mater. Sci. Eng. A* **1996**, *213*, 134–137. [CrossRef]
22. Correa, D.R.N.; Vicente, F.B.; Donato, T.A.G.; Arana-Chavez, V.E.; Buzalaf, M.A.R.; Grandini, C.R. The effect of the solute on the structure, selected mechanical properties, and biocompatibility of Ti–Zr system alloys for dental applications. *Mater. Sci. Eng. C* **2014**, *34*, 354–359. [CrossRef] [PubMed]
23. Campoccia, D.; Montanaro, L.; Arciola, C.R. A review of the biomaterials technologies for infection-resistant surfaces. *Biomaterials* **2013**, *34*, 8533–8554. [CrossRef] [PubMed]
24. Baltatu, M.S.; Vizureanu, P.; Mareci, D.; Burtan, L.C.; Chirută, C.; Trincă, L.C. Effect of Ta on the electrochemical behavior of new TiMoZrTa alloys in artificial physiological solution simulating in vitro inflammatory conditions. *Mater. Corros. Werkst. Korros.* **2016**, *67*, 1314–1320. [CrossRef]
25. Baltatu, M.S.; Vizureanu, P.; Cimpoesu, R.; Abdullah, M.M.A.B.; Sandu, A.V. The Corrosion Behavior of TiMoZrTa Alloys Used for Medical Applications. *Rev. Chim.* **2016**, *67*, 2100–2102.
26. Kumar, S.; Narayanan, T.S.N.S. Electrochemical characterization of Ti alloy in Ringer's solution for implant application. *J. Alloy. Compd.* **2009**, *479*, 699–703. [CrossRef]
27. Leyens, C.; Peters, M. (Eds.) *Titanium and Titanium Alloys: Fundamentals and Applications*; Wiley-VCH: Weinheim, Germany, 2003; pp. 423–431.

materials

MDPI

Article

Effect of Thermomechanical Treatments on the Phases, Microstructure, Microhardness and Young's Modulus of Ti-25Ta-Zr Alloys

Pedro Akira Bazaglia Kuroda [1,2], Fernanda de Freitas Quadros [1,2], Raul Oliveira de Araújo [2,3], Conrado Ramos Moreira Afonso [4] and Carlos Roberto Grandini [1,2,*]

1 Laboratório de Anelasticidade e Biomateriais, Faculdade de Ciências (FC), Campus de Bauru, Universidade Estadual Paulista (UNESP), Bauru, 17.033-360 SP, Brazil; pedro.kuroda@unesp.br (P.A.B.K.); ff.quadros@unesp.br (F.d.F.Q.)

2 IBTN/BR – Institute of Biomaterials, Tribocorrosion and Nanomedicine - Brazilian Branch, Bauru, 17.033-360 SP, Brazil; raul.araujo@ifsp.edu.br

3 Ciência e Tecnologia de São Paulo, Instituto Federal de Educação (IFSP), Barretos, 14781-502 SP, Brazil

4 Department of Materials Engineering (DEMa), Federal University of Sao Carlos (UFSCar), São Carlos, 13565-905 SP, Brazil; conrado@ufscar.br

* Correspondence: carlos.r.grandini@unesp.br

Received: 10 September 2019; Accepted: 23 September 2019; Published: 30 September 2019

Abstract: Titanium and its alloys currently are used as implants, possessing excellent mechanical properties (more suited than stainless steel and Co-Cr alloys), good corrosion resistance and good biocompatibility. The titanium alloy used for most biomedical applications is Ti-6Al-4V, however, studies showed that vanadium and aluminum cause allergic reactions in human tissues and neurological disorders. New titanium alloys without the presence of these elements are being studied. The objective of this study was to analyze the influence of thermomechanical treatments, such as hot-rolling, annealing and solution treatment in the structure, microstructure and mechanical properties of the Ti-25Ta-Zr ternary alloy system. The structural and microstructural analyses were performed using X-ray diffraction, as well as optical, scanning and transmission electron microscopy. The mechanical properties were analyzed using microhardness and Young's modulus measurements. The results showed that the structure of the materials and the mechanical properties are influenced by the different thermal treatments: rapid cooling treatments (hot-rolling and solubilization) induced the formation of α'' and β phases, while the treatments with slow cooling (annealing) induced the formation of martensite phases. Alloys in the hot-rolled and solubilized conditions have better mechanical properties results, such as low elastic modulus, due to retention of the β phase in these alloys.

Keywords: titanium alloys; biomaterial; heat treatment

1. Introduction

The first alloy developed specifically for use in humans is known as "vanadium steel (1900)", a biomaterial produced for repairing bone fractures and for fixation screws used in orthopedics [1]. Currently, elements such as chromium, tantalum, zirconium, cobalt, titanium, niobium, tungsten, iron and nickel are also used as biomaterials, although the human body tolerates these elements only in small quantities [2]. Materials developed for use as biomaterials must possess certain favorable physical and chemical characteristics and must not induce inflammation, toxic reactions or allergenic symptoms in the host. They must be biocompatible, biofunctional, bioactive, bioinert and sterilizable [3].

Unfortunately, metal implants have been known to fail after prolonged use due to their high elasticity modulus compared to bone, and to their poor resistance to wear and corrosion in body fluids [4,5]. The bonding force between atoms determines Young's modulus. This bond strength is related to the crystalline structure of the material and to the distances between atoms [6,7]. It is possible to change the bond strength between atoms by thermomechanical treatment [8]. Plastic deformations are another treatment that can alter the structure of titanium, thus changing the atomic bond strength and Young's modulus [9]. The ω phase of titanium is known to have the highest modulus of elasticity, followed by the α and the α″ phases; the β phase has the lowest modulus, due to its low atomic bond strength [10]. New alloys are being developed and investigated whose microstructure and mechanical properties can be altered with the addition of various elements, usually β-stabilizers [11–16].

Ti-Ta-Zr alloys are being developed for biomedical applications. Zhou et al. [17] indicate that the most promising of the Ti-Ta binary system alloys for use as a biomaterial is Ti-25Ta. Some authors assert that zirconium is a neutral element in titanium alloys, that is, its addition does not modify the structure of the material [18]. Recently, new research has shown that zirconium helps stabilize the β phase in α + β-type alloys, acting as a β-stabilizer [19–24]. Therefore, alloys of the Ti-25Ta-Zr system may be more promising as biomaterials, hence the addition of zirconium facilitates the β phase stabilization, in reason of the fact that β-type alloys tend to have a lower modulus of elasticity than alpha-type alloys. Besides aiding in the stabilization of the β phase, zirconium contributes to cost savings in the manufacture and commercialization of these alloys [25], because tantalum has a high market value and a high melting point, making production in high ladder more difficult.

The literature does not have great information about the structure, microstructure, phases and mechanical properties of Ti-Ta-Zr alloys. Yan at al [26] developed Ti-15Ta-xZr system alloys (x = 1.5, 5.5, 10.5, and 15.5 wt.%) manufactured by selective laser melting. Structural and microstructural analyses show that the alloys have the α and β phases and excellent elastic modulus values, highlighting the Ti-15Ta-10.5Zr alloy with a value of approximately 43 GPa. Biesiekierski et al. [27] developed new Ti-Ta-Zr system alloys with high tantalum and zirconium contents (Ti-45Zr-10Ta, Ti40Zr-14Ta, Ti-35Zr-18Ta, and Ti-30Zr -22Ta). The results showed that the alloys produced are of β-type and have elastic modulus like commercially pure titanium and Ti-6Al-4V (109-130 GPa). All alloys have good biocompatibility.

Ti-25Ta-10Zr alloy is the only alloy of type α′+ α″ + β reported in the literature. Hardness values are above commercially pure titanium, being hardening by the solid solution. Cytotoxicity assays showed that the alloy does not have a cytotoxic character; on the contrary, the alloy showed perfect cell division, maintaining morphology, indicating good integration between material and cell [28].

The objective of this work was to analyze the influence of thermomechanical treatments, such as hot-rolling and annealing heat treatments, on the structure, microstructure, microhardness and elastic modulus of ternary alloys of the Ti-25Ta-Zr system, with zirconium content varying from 0 to 40% in weight.

2. Materials and Methods

The alloys were melted in an arc-melting furnace [22,28–32]. After melting, the ingots were submitted to a mechanical hot-rolling process at 1273 K, followed by air-cooling.

Because the microstructure of the metal undergoes changes during rolling, an annealing heat treatment was used to reduce the internal stress caused by the mechanical rolling process. The purpose of the annealing treatment was to eliminate residual internal stress and most of the imperfections present in the material and to promote the growth of the grains [33].

Two types of heat treatments were performed. After thermomechanical processing, the samples were submitted to recrystallization heat treatments at 1273 K, remaining at this temperature for 21.6 Ks, followed by slow cooling (SC) (−5 K/min), and another similar treatment was followed by water quenching (RC) (~273 K).

For structural analysis, X-ray diffraction (XRD) analysis (Rigaku model D/Max-2100PC) was used to obtain the diffractograms adopting the powder method, with Cu-Kα radiation, 20 mA current, potential 40 kV, for 3.2 s, range from 20° to 100°, and step of 0.02°, in the fixed time mode.

Before microstructural analysis, samples were submitted to a standard metallographic preparation. Images were obtained using an optical microscope (Olympus model BX51M), a scanning electron microscope (SEM) (Carl Zeiss model EVO-015) and a transmission electron microscope (TEM) (FEI Tecnai G^2 F20). TEM analysis was performed using a FEI Tecnai G2 F20 200 kV microscope with energy-dispersive X-ray spectroscopy (EDS).

Hardness measurements were performed in a microhardness tester (Shimadzu, model HMV-2). In each sample, 20 indentations were made with a 25-g load for 60 s [34]. For dynamic elastic modulus, E (GPa) measurements, the impulse excitation technique (ATCP, Sonelastic®) was used [35].

3. Results

3.1. Phase Identification

The X-ray diffractograms and optical and SEM micrographs for Ti-25Ta-Zr system alloys after hot-rolling are shown in Figure 1; after annealing and with slow cooling in Figure 2; and in Figure 3, the results for samples subjected to annealing with fast cooling, are presented. As shown, in system alloys Ti-25Ta-Zr the crystalline structure was sensitive to thermomechanical treatments.

3.2. Mechanical Characterization

Figure 4 shows the hardness values for Ti-25Ta-Zr system alloys, performed in the samples after each processing condition included in this study. An anomaly for the hardness of Ti-25Ta-30Zr alloy can be observed after annealing with rapid cooling. This anomaly can be associated with the retention of phase [36,37].

To verify the presence of ω phase, TEM measurements were made in the Ti-25Ta-30Zr alloy after annealing with rapid cooling, and the results, presented in Figure 5, show the bright (part (a)) and dark (part (b)) field images, the diffraction pattern (part (c)) and a high-resolution image of the selected area (part (d)). The presence of ω phase can be clearly observed [33].

Figure 6 shows the values of Young's modulus for Ti-25Ta-Zr system alloys, under all conditions investigated in this study. It can be observed that the values of elastic modulus decrease with the zirconium amount for the hot-rolled and annealed with rapid cooling samples. For the annealed with slow cooling, an increase of elastic modulus with the zirconium amount was observed.

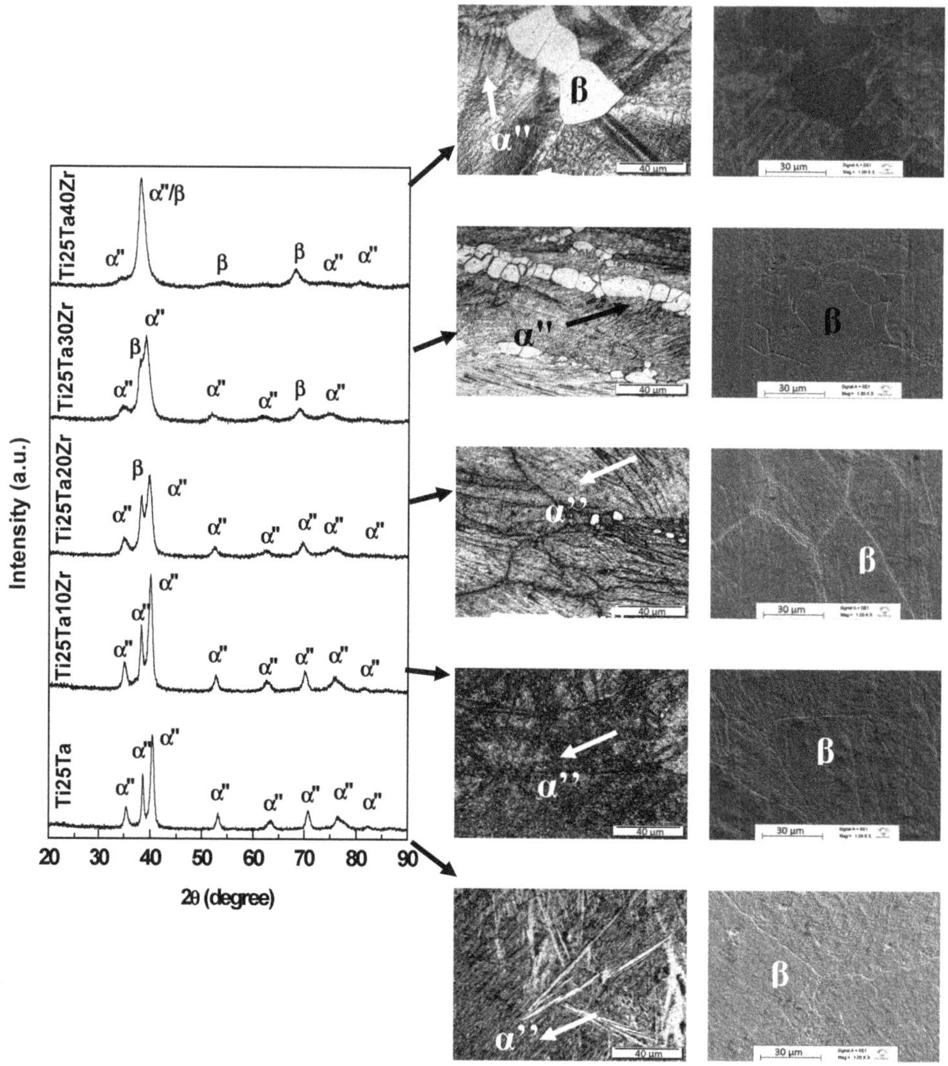

Figure 1. X-ray diffraction patterns and optical and SEM micrographs for Ti-25Ta-xZr system alloys after hot-rolling.

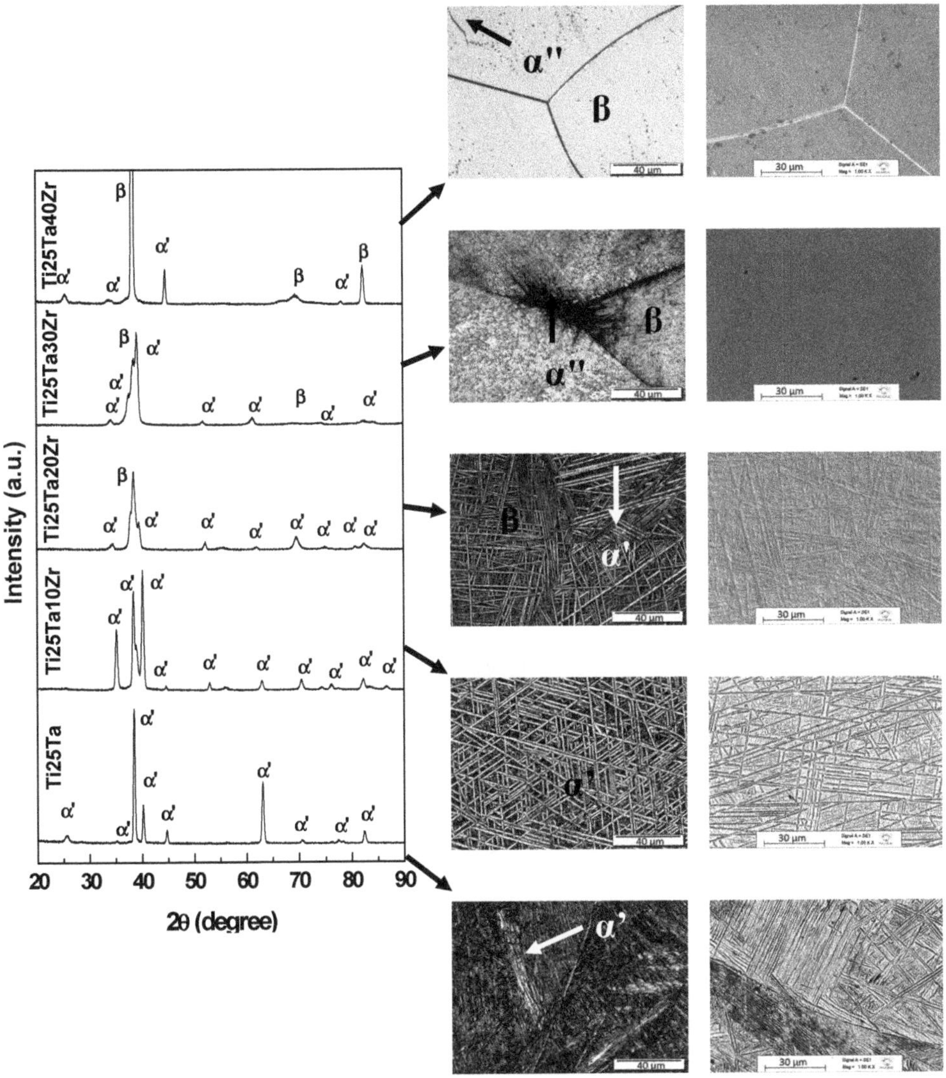

Figure 2. X-ray diffraction patterns and optical and SEM micrographs for Ti-25Ta-xZr system alloys after annealing heat treatment with slow cooling.

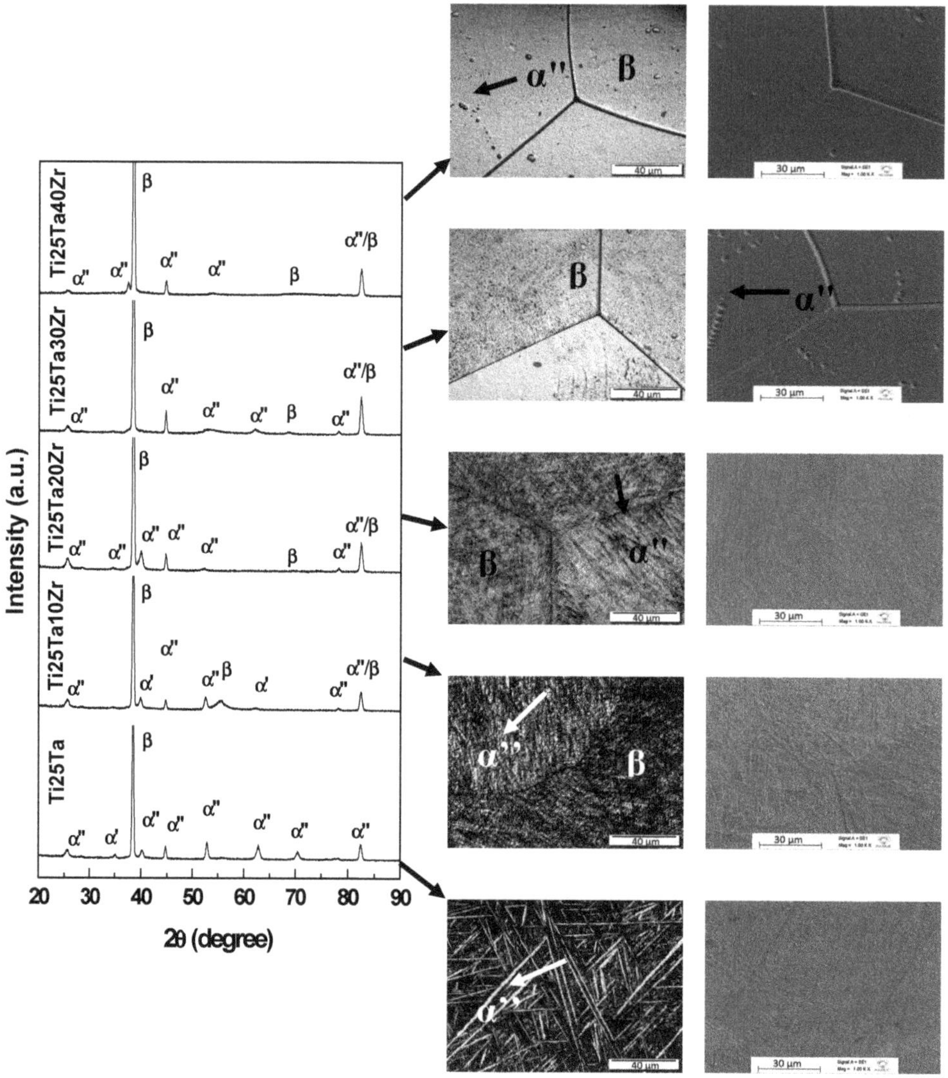

Figure 3. X-ray diffraction patterns and optical and SEM micrographs for Ti-25Ta-xZr system alloys after annealing heat treatment with rapid cooling.

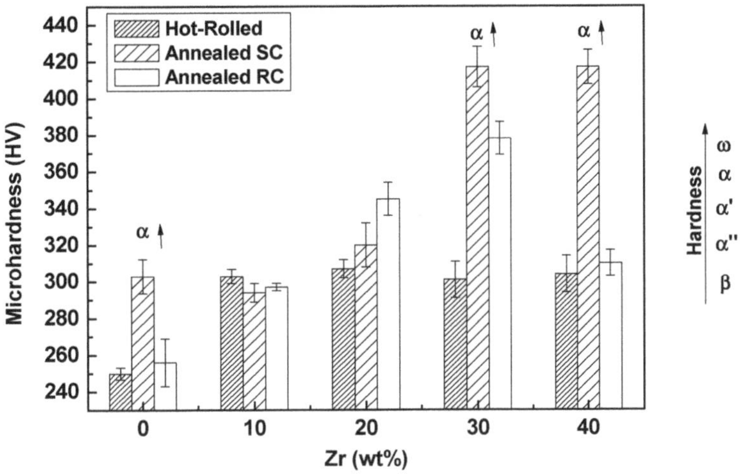

Figure 4. Vickers microhardness values of Ti-25Ta-Zr system alloys after hot-rolling and heat treatments, annealing with slow (SC) and rapid cooling (RC).

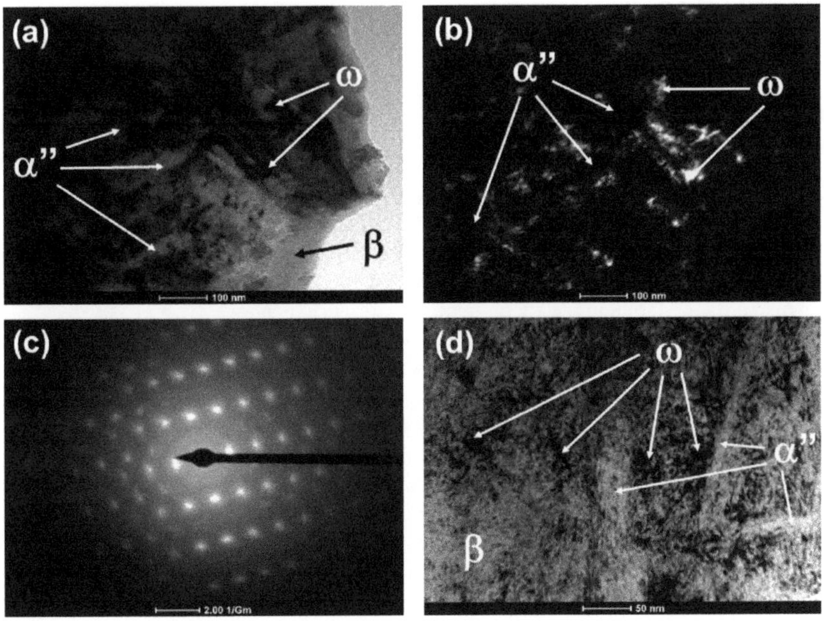

Figure 5. TEM images for Ti-25Ta-30Zr alloy after annealing with rapid cooling: bright-field (**a**), dark-field (**b**), diffraction of the selected area (**c**) and high-resolution image (**d**).

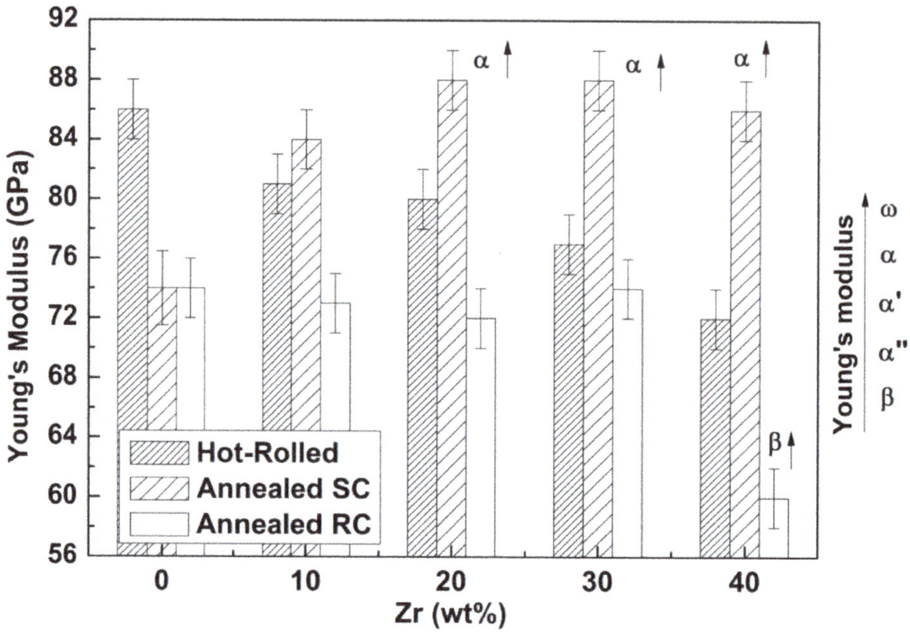

Figure 6. Young's modulus for Ti-25Ta-Zr system alloys, in all studied conditions.

4. Discussion

It can be observed, in Figure 1, that Ti-25Ta and Ti-25aTa-10Zr alloys show only peaks of α″ phase in the hot-rolled condition. For the Ti-25Ta-20Zr alloy, the diffractograms show the coexistence of peaks associated with α″ and β phases, but those of β phase have lower intensity. Ti-25Ta-30Zr and Ti-25Ta-40Zr alloys have peaks that are associated with α″ and β phases, too. It is evident that the zirconium acted as a β-stabilizing element, hence its presence increases the β phase precipitation.

After the annealing heat treatment with slow cooling, the orthorhombic α″ phase changed to the hexagonal close-packed structure, α phase, and the β phase remained in the structure of the material. It can be clearly observed that heat treatment modified the phases of the alloys. Since the annealing was carried out at a temperature above β-transus and cooled slowly, it was provided the necessary thermodynamic conditions for the stabilization of the phases at high temperatures, if they change to stable phases at room temperature, in the case α″→α [38].

The alloys submitted to the annealing treatment with rapid cooling presented the metastable orthorhombic α″ phase. In addition, alloys with a zirconium content above 20 wt% showed a more intense β phase peak at θ ~ 68°, indicating higher fraction and greater stabilization of bcc structure. It is stated in the literature that rapid cooling of titanium alloys from the β field can induce the formation of the metastable α″ martensite since the fast cooling hinders the atomic rearrangement (retaining the β phase) and produces distortions in the crystalline structure [33]. In addition to detection of α″ and β phases, in alloys with 0, 10, 20 and 30 weight percent zirconium, an ω phase peak can be visualized at θ ~ 78°. In the Ti-25Ta-40Zr alloy it was not possible to observe ω phase peaks, showing that zirconium acted as an omega phase suppressor.

Combining the results of XRD, a typical martensitic structure can be seen in Figure 1 to Figure 3, α″ phase, and grains deformed due to plastic deformation in the Ti-25Ta, Ti-25Ta-10Zr and Ti-25Ta-20Zr alloys, in the condition after hot-rolling. After the annealing, there was recrystallization of the material leading to homogenization of the structure and grain size. In these alloys, after recrystallization with slow cooling, an α-type lamellar structure can be seen, with needles emerging from the grain boundaries and thinner needles parallel to the thick needles. In the annealed condition with rapid cooling, there was an increase in grain size, but the martensites phases are more refined, that is, the intra-grain needles are finer with this type of treatment. The images indicate that the cooling rate influences the microstructure in the material—slow cooling induces the formation of the α phase, and rapid cooling promotes the formation of the orthorhombic α″ phase.

In Ti-25Ta-30Zr alloy after hot-rolling, the α″ and β phases coexist; the micrographs exhibit an equiaxial structure of very fine needles distributed in some β-type distorted grains. After annealing with slow cooling, there was an increase in grain size and the α phase needles were within the grain boundaries. In annealing with water cooling, grains of the β-matrix can be observed, with small α″ phase precipitates.

In the alloy with 40 wt% of zirconium, it is possible to observe small grains of the material with morphology characteristic of the β phase, in the after hot-rolling condition. The annealing treatments promoted the growth of these grains. Typical structures of the martensites are difficult to see in micrographs, although the x-ray diffractograms showed peaks of the α 'and α″ phases.

A major problem in the manufacture of titanium prostheses is the mechanical conformation. A biomedical material must be easily shaped to facilitate its handling to produce screws, plates, and other shapes. The microhardness of a material indicates whether it conforms easily (low hardness value) or is a hard, brittle material. Further, biocompatible alloys must have a hardness appropriate to their specific application, because if they have very high hardness in relation to the implanted tissue, it can result in rapid tissue wear. Thus, many studies evaluate the microhardness in titanium alloys with respect to heat treatments, processing and the microstructure [39–43].

Figure 4 shows the hardness values for each process performed in the samples of Ti-25Ta-Zr alloys included in this study. According to the literature, among the phases formed in titanium alloys, the ω phase presents the greatest hardness value, followed by the α, α′, α″ and β [44–46]. Other studies report that β-type alloys may have a greater hardness level than mastensitic phases α′ and α″, because they have a stronger solid solution due to phase stability (ω > α′ > α″ > β > α) [47].

As can be observed, after the hot-rolling process the alloys have an equal or lower hardness value, compared to the values of annealed alloys with slow cooling. From the X-ray diffractograms, it is possible to visualize the effect of hot-rolling, heat treatment and zirconium substitution in the structure of the material. After the hot-rolling, there is a formation of the α″ phase for the alloys with zirconium up to 20%, and in those above 30% of zirconium, the β phase coexists with the α″ phase. The annealing alleviated the internal stress from the mechanical process and modified the α″ to the α phase in the alloys undergoing slow cooling. As mentioned above, α type titanium alloys have greater hardness compared to α″ alloys. Consequently, the alloys in the hot-rolled condition tend to have a lower hardness value compared to the annealed alloys with a slow cooling rate [47].

Ti-25Ta-30Zr and Ti-25Ta-40Zr alloys annealed with slow cooling have high hardness values (~425 HV). This increase can also be explained by the presence of the α and β phases, which may be an obstacle to the dislocations motion.

From hardness measurements, it can be seen that for the lower (SC) cooling rate (5 K/s) condition, there is the formation of β + α phases for all the compositions. For water quenched (RC) condition after heat treatment, it might form precipitates of athermal ω phase due to higher hardness values (from 10 to 30%Zr) reached, but in a lower fraction than SC samples. Rapid cooling condition (RC) resulted in a greater stabilization of β phase for higher zirconium content (40%Zr), suppressing also the ω phase and decreasing the hardness value.

It was observed that the hardness of the hot-rolled alloys and the alloys subjected to the annealing treatment with fast cooling have similar values. The alloys in these two conditions have the same crystalline structure (α″ and β), and therefore their mechanical characteristics are similar. Only in the Ti-25Ta-30Zr alloy (with α″ and β phases), there is an "anomaly" with reference to the hardness value. This high hardness value can be an indication of the formation and highest precipitation of the ω phase in this alloy, as the fast cooling can retain the ω phase, which has the characteristic of hardening and weakening the material.

Due to the high hardness of the Ti-25Ta-30Zr alloy subjected to rapid cooling, a more detailed analysis of the microstructure of the material was performed to see the ω phase as the enhancer. Figure 5 shows the bright and dark field images of TEM, a bright field (BF) image and the selected area diffraction pattern (SAD) showing fine martensite α″ phase needles with a width of approximately 50 nm dispersed in a β phase (bcc) matrix. Nanoscale precipitates of athermal ω phase from 5 to 20 nm are visualized in the TEM micrographs; their nucleation and precipitation are located close to the α″ phase structures. Such athermal ω phase formation occurs in α + β type alloys and when such alloys are subjected to rapid cooling from high temperatures. The omega phase recipients are located intra-grain in the titanium structure, in circular format, visualized in nanometer scale, according to similar results of Niinomi (2016) and Homma (2018) [48,49]. Due to the precipitation of ω phase, Ti-25Ta-30Zr alloy submitted to rapid cooling showed a high value of hardness, as indicated in Figure 4 [50,51].

Figure 6 shows the values of the dynamic elastic modulus of the samples of all Ti-25Ta-Zr alloys used in this study, under all investigated conditions. In this figure, it can be observed that Young's modulus of the alloys studied after hot-rolled condition was approximately 86 GPa for the non-zirconium alloy and reached a minimum value of 72 GPa for the alloy with the highest zirconium content (40 wt%). After hot-rolling, there is an increase in the percentage of the β phase, which decreases the elastic modulus, since the β phase has the smallest modulus value among the phases in titanium alloys. The hot-rolling process generates a high cooling rate when the sample is passed through the mill roller, retaining the β phase.

Annealing with slow cooling, by contrast, increased the value of Young's modulus with the addition of zirconium. The treatment decreased the percentage of the β phase and modified the α″ to α phase due to the recrystallization of the sample. The alloys with the α phase have higher elastic modulus values than α″ alloys. Annealing with rapid cooling decreased the modulus of these alloys, as this type of treatment at high temperatures and rapid cooling induces the formation of the β phase, which has the lowest modulus value in titanium alloys. The Ti-25Ta-40Zr alloy has the lowest modulus of the studied alloys of E = 60 GPa.

It can be noted by the hot-rolled condition, that increasing zirconium fraction leads to the decreasing of the elastic modulus from 86 down to 72 (GPa), due to a combination of stress-induced martensite formation and β phase stabilization with suppression of ω formation. Evaluating the elastic modulus E (GPa) for a lower cooling rate leads to a tendency like the hardness measurements, where the appearance of the α phase in high zirconium alloys leads to an increase in mechanical property values, increasing the modulus from 74 to 88 GPa. Compared with the rapid cooling condition, the modulus only just decreased significantly from 74 down to around 60 GPa for higher zirconium content (40%), due to greater stabilization of β phase.

The Ti-25Ta-xZr alloys produced in the annealed condition with fast cooling have a better modulus compared to Ti-cp and other commercial biomedical metals, such as Ti-6Al-4V, 316L and Co-Cr [12]. Among the produced alloys in this study, the Ti-25Ta-40Zr has the lowest elastic modulus value (60 ± 2) GPa, which represents twice the modulus of elasticity of human cortical bone [52].

Although the Ti25Ta-40Zr alloy is promising in the biomedical field, new mechanical property testing and corrosion analysis are still required, as the surface of metals in contact with body fluids may corrode, reducing implant longevity, and to perform biological biocompatibility to verify if the material has cytotoxic character in the human organism.

5. Conclusions

The results detailed above lead to the following conclusions.

- X-ray diffraction and microscopy reveal that the crystalline structure of the alloys was sensitive to the addition of zirconium and sensitive to thermomechanical treatments.
- The zirconium acted as a β-stabilizing element combined with tantalum in Ti-25Ta-xZr alloys.

- Alloys subjected to a slow-cooling annealing heat treatment have the α phase in the crystalline structure.
- The alloys subjected to a fast-cooling annealing heat treatment have the α″ phase, can retain the β phase with high zirconium content and rapid cooling precipitates the omega phase, but high levels of zirconium suppress this phase.
- Hardness increased due to solid solution hardening and due to ω phase and α-phase precipitation in alloys subjected to slow-cooling annealing heat treatment.
- The Young's modulus decreases with high levels of zirconium due to the stabilization of the β phase.
- The thermomechanical process and heat treatment with rapid cooling retain the β phase, improving mechanical biocompatibility with the decreasing of the elastic modulus values of the materials down to E = 60 GPa for Ti-25Ta-40Zr alloy.

Author Contributions: Investigation, P.A.B.K. and F.d.F.Q.; data analysis, P.A.B.K., F.d.F.Q., R.O.d.A., C.R.M.A. and C.R.G.; writing—original draft preparation, P.A.B.K., F.d.F.Q. and R.O.d.A.; writing—review and editing, C.R.M.A. and C.R.G.; supervision, C.R.G.

Funding: This research was supported by the funding agencies FAPESP (grants #2012/22.742-6 and #2015/09.480-0) and CNPq (grant #307.279/2013-8).

Acknowledgments: The authors would like to thank the Faculdade de Ciências de Bauru, UNESP, for the XRD and SEM measurements.

Conflicts of Interest: The authors declare no conflicts of interest.

References

1. Meyers, M.A.; Chen, P.Y.; Lin, A.Y.M.; Seki, Y. Biological materials: Structure and mechanical properties. *Prog. Mater. Sci.* **2008**, *53*, 1–206. [CrossRef]
2. Noumbissi, S.; Scarano, A.; Gupta, S. A Literature Review Study on Atomic Ions Dissolution of Titanium and Its Alloys in Implant Dentistry. *Materials* **2019**, *12*, 368. [CrossRef] [PubMed]
3. Ratner, B.D.; Bryant, S.J. Biomaterials: Where We Have Been and Where We Are Going. *Annu. Rev. Biomed. Eng.* **2004**, *6*, 41–75. [CrossRef] [PubMed]
4. Niinomi, M. Biologically and Mechanically Biocompatible Titanium Alloys. *Mater. Trans.* **2008**, *49*, 2170–2178. [CrossRef]
5. Geetha, M.; Singh, A.K.; Asokamani, R.; Gogia, A.K. Ti based biomaterials, the ultimate choice for orthopaedic implants—A review. *Prog. Mater. Sci.* **2009**, *54*, 397–425. [CrossRef]
6. Zhou, Y.L.; Niinomi, M.; Akahori, T. Effects of Ta content on Young's modulus and tensile properties of binary Ti–Ta alloys for biomedical applications. *Mater. Sci. Eng. A* **2004**, *371*, 283–290. [CrossRef]
7. Lee, Y.T.; Welsch, G. Young's modulus and damping of Ti 6Al 4V alloy as a function of heat treatment and oxygen concentration. *Mater. Sci. Eng. A* **1990**, *128*, 77–89. [CrossRef]
8. Froes, F.H. Titanium Alloys: Thermal Treatment and Thermomechanical Processing. In *Encyclopedia of Materials: Science and Technology*, 2nd ed.; Buschow, K.H.J., Flemings, M.C., Kramer, J.E., Veyssiere, P., Cahn, W.R., Ilschner, B., Mahajan, S., Eds.; Elsevier: Oxford, UK, 2001; pp. 9369–9373.
9. Weiss, I.; Semiatin, S.L. Thermomechanical processing of beta titanium alloys—An overview. *Mater. Sci. Eng. A* **1998**, *243*, 46–65. [CrossRef]
10. Lütjering, G.; Williams, J.C.; Gysler, A. Microstructure and Mechanical Properties of Titanium Alloys. In *Microstructure and Properties of Materials*; World Scientific: Washington, DC, USA, 1998; pp. 1–77.
11. Oldani, C.; Dominguez, A. Titanium as a Biomaterial for Implants. In *Recent Advances in Arthroplasty*; Fokter, S.K., Ed.; IntechOpen: Rijeka, Croatia, 2012; pp. 149–162.
12. Kaur, M.; Singh, K. Review on titanium and titanium based alloys as biomaterials for orthopaedic applications. *Mater. Sci. Eng. C* **2019**, *102*, 844–862. [CrossRef]
13. Kolli, R.; Devaraj, A. A Review of Metastable Beta Titanium Alloys. *Metals* **2018**, *8*, 506. [CrossRef]
14. Kirmanidou, Y.; Sidira, M.; Drosou, M.E.; Bennani, V.; Bakopoulou, A.; Tsouknidas, A.; Michalakis, K. New Ti-Alloys and Surface Modifications to Improve the Mechanical Properties and the Biological Response to Orthopedic and Dental Implants: A Review. *BioMed Res. Int.* **2016**, *2016*, 21. [CrossRef] [PubMed]

15. Li, Y.; Yang, C.; Zhao, H.; Qu, S.; Li, X.; Li, Y. New developments of Ti-based alloys for biomedical applications. *Materials* **2014**, *7*, 1709–1800. [CrossRef] [PubMed]
16. Wang, W.; Poh, C.K. *Titanium Alloys in Orthopaedics, in Titanium Alloys—Advances in Properties Control*; InTech: Rijeka, Croatia, 2013; pp. 1–20.
17. Zhou, Y.-L.; Niinomi, M. *Ti–25Ta* Alloy with the Best Mechanical Compatibility in Ti–Ta Alloys for Biomedical Applications. *Mater. Sci. Eng. C* **2009**, *29*, 1061–1065. [CrossRef]
18. Wang, K. The use of titanium for medical applications in the USA. *Mater. Sci. Eng. A* **1996**, *213*, 134–137. [CrossRef]
19. Correa, D.R.N.; Kuroda, P.A.B.; Grandini, C.R. Structure, Microstructure, and Selected Mechanical Properties of Ti-Zr-Mo Alloys for Biomedical Applications. *Adv. Mater. Res.* **2014**, *922*, 75–80. [CrossRef]
20. Correa, D.R.N.; Vicente, F.B.; Araújo, R.O.; Lourenço, M.L.; Kuroda, P.A.B.; Buzalaf, M.A.R.; Grandini, C.R. Effect of the substitutional elements on the microstructure of the Ti-15Mo-Zr and Ti-15Zr-Mo systems alloys. *J. Mater. Res. Technol.* **2015**, *4*, 180–185. [CrossRef]
21. Araújo, R.O.d.; Buzalaf, M.A.R.; Grandini, C.R. Preparation and Characterization of Ti-10Mo-xZr Alloys for Biomedical Applications. *Mater. Sci. Forum* **2016**, *869*, 940–945. [CrossRef]
22. Kuroda, P.A.B.; Buzalaf, M.A.R.; Grandini, C.R. Effect of molybdenum on structure, microstructure and mechanical properties of biomedical Ti-20Zr-Mo alloys. *Mater. Sci. Eng. C* **2016**, *67*, 511–515. [CrossRef] [PubMed]
23. Correa, D.R.N.; Kuroda, P.A.B.; Lourenço, M.L.; Fernandes, C.J.C.; Buzalaf, M.A.R.; Zambuzzi, W.F.; Grandini, C.R. Development of Ti-15Zr-Mo alloys for applying as implantable biomedical devices. *J. Alloys Compd.* **2018**, *749*, 163–171. [CrossRef]
24. Correa, D.R.N.; Kuroda, P.A.B.; Lourenço, M.L.; Buzalaf, M.A.R.; Mendoza, M.E.; Archanjo, B.S.; Grandini, C.R. Microstructure and selected mechanical properties of aged Ti-15Zr-based alloys for biomedical applications. *Mater. Sci. Eng. C* **2018**, *91*, 762–771. [CrossRef]
25. Cui, C.; Hu, B.; Zhao, L.; Liu, S. Titanium alloy production technology, market prospects and industry development. *Mater. Des.* **2011**, *32*, 1684–1691. [CrossRef]
26. Yan, L.; Yuan, Y.; Ouyang, L.; Li, H.; Mirzasadeghi, A.; Li, L. Improved mechanical properties of the new Ti-15Ta-xZr alloys fabricated by selective laser melting for biomedical application. *J. Alloys Compd.* **2016**, *688*, 156–162. [CrossRef]
27. Biesiekierski, A.; Ping, D.; Li, Y.; Lin, J.; Munir, K.S.; Yamabe-Mitarai, Y.; Wen, C. Extraordinary high strength Ti-Zr-Ta alloys through nanoscaled, dual-cubic spinodal reinforcement. *Acta Biomater.* **2017**, *53*, 549–558. [CrossRef] [PubMed]
28. Quadros, F.D.F.; Kuroda, P.A.B.; Sousa, K.D.S.J.; Donato, T.A.G.; Grandini, C.R. Preparation, structural and microstructural characterization of Ti-25Ta-10Zr alloy for biomedical applications. *J.Mater. Res. Technol.* **2019**, *8*, 4108–4114. [CrossRef]
29. Kuroda, P.A.B.; Buzalaf, M.A.R.; Grandini, C.R. Preparation, microstructural characterization, and selected mechanical properties of Ti-20Zr-2.5Mo and Ti-20Zr-7.5Mo used as biomaterial. *Mater. Sci. Forum* **2016**, *869*, 946–951. [CrossRef]
30. Martins, J.R.S., Jr.; Araújo, R.O.; Donato, T.D.; Arana-Chavez, V.; Buzalaf, M.R.; Grandini, C.R. Influence of Oxygen Content and Microstructure on the Mechanical Properties and Biocompatibility of Ti–15 wt%Mo Alloy Used for Biomedical Applications. *Materials* **2014**, *7*, 232–243. [CrossRef] [PubMed]
31. Kuroda, P.A.B.; Lourenço, M.L.; Correa, D.R.N.; Grandini, C.R. Thermomechanical treatments influence on the phase composition, microstructure, and selected mechanical properties of Ti–20Zr–Mo alloys system for biomedical applications. *J. Alloys Compd.* **2019**, *812*, 152108. [CrossRef]
32. Luz, A.R.; Santos, L.S.; Lepienski, C.M.; Kuroda, P.B.; Kuromoto, N.K. Characterization of the morphology, structure and wettability of phase dependent lamellar and nanotube oxides on anodized Ti-10Nb alloy. *Appl. Surf. Sci.* **2018**, *448*, 30–40. [CrossRef]
33. Collings, E.W. *The Physical Metallurgy of Titanium Alloys*; ASM International: Cleveland, OH, USA, 1989.
34. ASTM. *E92-82—Standard Test Method for Vickers Hardness of Metallic Materials*; ASTM International: Conshohocken, PA, USA, 2003.
35. ASTM. *E1876–01—Standard Test Method for Dynamic Young's Modulus, Shear Modulus, and Poisson's Ratio by Impulse Excitation of Vibration*; ASTM International: Conshohocken, PA USA, 2002.

36. Gray, G.T.; Morris, C.E.; Lawson, A.C. *Omega Phase Formation in Titanium and Titanium Alloys*; Los Alamos National Laboratory: Los Alamos, NM, USA, 1992; pp. 1–9.
37. Tane, M.; Nakano, T.; Kuramoto, S.; Niinomi, M.; Takesue, N.; Nakajima, H. ω Transformation in cold-worked Ti–Nb–Ta–Zr–O alloys with low body-centered cubic phase stability and its correlation with their elastic properties. *Acta Mater.* **2013**, *61*, 139–150. [CrossRef]
38. Davis, R.; Flower, H.M.; West, D.R.F. The decomposition of Ti-Mo alloy martensites by nucleation and growth and spinodal mechanisms. *Acta Metall.* **1979**, *27*, 1041–1052. [CrossRef]
39. Callister, W.D., Jr. *Fundamentos da Ciência e Engenharia de Materiais*, 2nd ed.; LTC: Rio de Janeiro, Brazil, 2006.
40. Ho, W.F.; Ju, C.P.; Chern Lin, J.H. Structure and properties of cast binary Ti–Mo alloys. *Biomaterials* **1999**, *20*, 2115–2122. [CrossRef]
41. Delvat, E.; Gordin, D.M.; Gloriant, T.; Duval, J.L.; Nagel, M.D. Microstructure, mechanical properties and cytocompatibility of stable beta Ti-Mo-Ta sintered alloys. *J. Mech. Behav. Biomed. Mater.* **2008**, *1*, 345–351. [CrossRef] [PubMed]
42. Kao, Y.L.; Tu, G.C.; Huang, C.A.; Liu, T.T. A study on the hardness variation of α- and β-pure titanium with different grain sizes. *Mater. Sci. Eng. A* **2005**, *398*, 93–98. [CrossRef]
43. Sukedai, E.; Yukihiro, T.; Miyaji, D.; Matsumoto, H.; Nishizawa, H.; Hashimoto, H. Aging behavior of Ti–Mo alloys heavily compressed in ultra-high strain rate mode. *Mater. Sci. Eng. A* **2004**, *387* (Suppl. C), 249–253. [CrossRef]
44. Leyens, C.; Peters, M. *Titanium and Titanium Alloys: Fundamentals and Applications*; Wiley-VCH: New York, NY, USA, 2005.
45. Zhou, Y.-L.; Niinomi, M. Microstructures and mechanical properties of Ti–50 mass% Ta alloy for biomedical applications. *J. Alloys Compd.* **2008**, *466*, 535–542. [CrossRef]
46. Gabriel, S.B.; de Almeida, L.H.; Nunes, C.A.; Dille, J.; Soares, G.A. Maximisation of the ratio of microhardness to the Young's modulus of Ti–12Mo–13Nb alloy through microstructure changes. *Mater. Sci. Eng. C* **2013**, *33*, 3319–3324. [CrossRef]
47. Lee, C.M.; Ju, C.P.; Lin, J.H.C. Structure–property relationship of cast Ti–Nb alloys. *J. Oral Rehabil.* **2002**, *29*, 314–322. [CrossRef] [PubMed]
48. Niinomi, M.; Nakai, M.; Hendrickson, M.; Nandwana, P.; Alam, T.; Choudhuri, D.; Banerjee, R. Influence of oxygen on omega phase stability in the Ti-29Nb-13Ta-4.6Zr alloy. *Scr. Mater.* **2016**, *123*, 144–148. [CrossRef]
49. Homma, T.; Arafah, A.; Haley, D.; Nakai, M.; Niinomi, M.; Moody, M.P. Effect of alloying elements on microstructural evolution in oxygen content controlled Ti-29Nb-13Ta-4.6Zr (wt%) alloys for biomedical applications during aging. *Mater. Sci. Eng. A* **2018**, *709*, 312–321. [CrossRef]
50. Aleixo, G.T.; Lopes, E.S.; Contieri, R.; Cremasco, A.; Afonso, C.R.M.; Caram, R. Effects of Cooling Rate and Sn Addition on the Microstructure of Ti-Nb-Sn Alloys. *Solid State Phenom.* **2011**, *172*, 190–195. [CrossRef]
51. Afonso, C.R.M.; Chaves, J.M.; Florêncio, O. Effect of Rapid Solidification on Microstructure and Elastic Modulus of β Ti-xNb-3Fe Alloys for Implant Applications. *Adv. Eng. Mater.* **2017**, *19*, 1600370. [CrossRef]
52. Wu, D.; Isaksson, P.; Ferguson, S.J.; Persson, C. Young's modulus of trabecular bone at the tissue level: A review. *Acta Biomater.* **2018**, *78*, 1–12. [CrossRef] [PubMed]

materials

MDPI

Article

Investigation of Copper Alloying in a TNTZ-Cu_x Alloy

Lee Fowler [1], Arno Janse Van Vuuren [2], William Goosen [2], Håkan Engqvist [1], Caroline Öhman-Mägi [1,*] and Susanne Norgren [1,3]

1 Division of Applied Material Science, Department of Engineering Sciences, The Ångström Laboratory, Uppsala University, Box 534, 75121 Uppsala, Sweden; lee.fowler@angstrom.uu.se (L.F.); hakan.engqvist@angstrom.uu.se (H.E.); susanne.norgren@angstrom.uu.se (S.N.)
2 Centre for High Resolution Transmission Electron Microscopy, Department of Physics, Nelson Mandela University, 6031 Port Elizabeth, South Africa; arno.jansevanvuuren@mandela.ac.za (A.J.V.V.); william.goosen@mandela.ac.za (W.G.)
3 Sandvik, Lerkrogsvägen 13, 12680 Stockholm, Sweden
* Correspondence: caroline.ohman@angstrom.uu.se

Received: 16 October 2019; Accepted: 6 November 2019; Published: 8 November 2019

check for updates

Abstract: Alloying copper into pure titanium has recently allowed the development of antibacterial alloys. The alloying of biocompatible elements (Nb, Ta and Zr) into pure titanium has also achieved higher strengths for a new alloy of Ti-1.6 wt.% Nb-10 wt.% Ta-1.7 wt.% Zr (TNTZ), where strength was closer to Ti-6Al-4V and higher than grade 4 titanium. In the present study, as a first step towards development of a novel antibacterial material with higher strength, the existing TNTZ was alloyed with copper to investigate the resultant microstructural changes and properties. The initial design and modelling of the alloy system was performed using the calculation of phase diagrams (CALPHAD) methods, to predict the phase transformations in the alloy. Following predictions, the alloys were produced using arc melting with appropriate heat treatments. The alloys were characterized using energy dispersive X-ray spectroscopy in scanning transmission electron microscopy (STEM-EDS) with transmission Kikuchi diffraction (TKD). The manufactured alloys had a three-phased crystal structure that was found in the alloys with 3 wt.% Cu and higher, in line with the modelled alloy predictions. The phases included the α-Ti (HCP-Ti) with some Ta present in the crystal, Ti_2Cu, and a bright phase with Ti, Cu and Ta in the crystal. The Ti_2Cu crystals tended to precipitate in the grain boundaries of the α-Ti phase and bright phase. The hardness of the alloys increased with increased Cu addition, as did the presence of the Ti_2Cu phase. Further studies to optimize the alloy could result in a suitable material for dental implants.

Keywords: titanium alloy; microstructures; biomaterial; TNTZ

---

1. Introduction

Commercially pure titanium (CP-Ti) and the Ti-6Al-4V (Ti-64) alloy are standard materials for medical implants, but problems with both have come to light with in vivo use. The former has a lower yield strength, which for grades 1–4, ranges from 170–480 MPa, respectively [1]. The latter contains vanadium which could be toxic [2] and aluminium which has been linked to Alzheimer's disease [3]. These problems have inspired development of β-Ti [4–7] and α + β-Ti [8] alloys to replace these. Of the novel alloys to date, the Ti-Nb-Ta-Zr (TNTZ) system has shown promising properties for biocompatibility and strength comparable to Ti-64 [9–11]. Depending on alloying compositions, this alloy can be manipulated to have Young's moduli lower than Ti-64, which is an appropriate step towards achieving elasticity similar to cortical bone in future [12]. All these Ti-alloys are however vulnerable to biofilm formation and patients could require antibiotic treatments in the event of infection [13,14]. In lieu of the growing problem with antibiotic resistance [15], antibacterial alloys could be useful for the biomaterials field.

Excessive addition of Cu in Ti-Cu_x binary alloys could however lead to toxicity [13] or material embrittlement, which is a disadvantage for load bearing biomaterials [16]. Therefore careful microstructural design is required so that the mechanical properties can be optimized for the intended application. While these findings are descriptive of a binary alloy of Ti-Cu_x, and Cu alloying has been performed on Ti-13Nb-13Zr-10Cu [17], it is envisioned that a similar antibacterial ability may be engineered into other quaternary alloys, i.e. TNTZ [11].

The alloying of Cu to TNTZ could lead to a novel alloy with several advantages, but microstructural, mechanical and biological properties still require careful study for optimization. Binary systems such as Ti-Cu show clear microstructural dependence on crystal relationships, chemical phases present, chemical-migration and -ordering [18]. Donthula et al. [18] and Contieri et al. [19] in particular have described the actively driven eutectoid transformation of β-Ti to α-Ti and Ti_2Cu, which elucidates why β-Ti is not found in rapidly quenched alloys of this variety [20]. In contrast, studies on TNTZ without Cu present, show β-Ti and α-Ti microstructure with metastable β-phases present [21]. These two alloy systems are micro-structurally dissimilar, which further motivates the investigation into Cu addition in the TNTZ alloy systems. For these reasons the present study aims to determine the effects of Cu addition to an existing alloy of TNTZ [11], and characterize the material.

2. Materials and Methods

2.1. Computational Modelling of Alloys

The impact of Cu to the Ti-Nb-Ta-Zr system was modelled using computational thermodynamic modelling based on the CALPHAD approach [22], using the Thermo-Calc software (Thermo-Calc software AB, Solna, Sweden) and the SSOL5 database, available from www.thermocalc.se.

Since the TNTZ alloy [11] has low additions of Zr and Nb, it is hypothesized that these will remain in the solid solution of the α and β phases, respectively. The Cu, being a, β-eutectoid stabilizer, is also expected to create a eutectoid microstructure of lathes, but with increasing Cu additions, a Cu-rich phase is predicted to precipitate. It is likely that this phase will precipitate preferentially at the grain boundaries, which could lead to embrittlement. In the literature there are two contradictory predictions of the Ti-Cu phase diagram where the first Cu-rich phase is either Ti_2Cu [23] or Ti_3Cu [24]. The thermodynamic modelling done in this work is based on the first description [15], which excludes Ti_3Cu which is a metastable phase, based on the observations by Zhang et al. [25]. Therefore, the binary Ti-Cu system was taken from the 1996 description by Kumar et al. [23].

The results should be regarded as an initial prediction of the phases and transition temperatures. This is due to the fact that the Ta-Cu binary, the Ta-Nb-Cu ternary and the Ti-Ta-Cu ternary systems have not been thermodynamically assessed and thus are lacking in the SSOL5 database. Nevertheless, the predictions given by the calculations are useful as a starting point for alloy development and to guide the experimental work. The Ti-1.7 wt.% Nb-10.1 wt.% Ta-1.6 wt.% Zr (TNTZ) has been modelled previously [11] and gave only α and β phases. The equilibrium phases as a function of temperature were modelled for this alloy with increasing Cu additions (0 wt.% Cu, 1 wt.% Cu, 3 wt.% Cu, 5 wt.% Cu and 10 wt.% Cu). In Figure 1a,b the phase fractions in the alloys with 1 wt.% Cu and 5 wt.% Cu addition are shown. Given the prerequisites mentioned, the Ti_2Cu forms at 656 °C in the 1 wt.% Cu alloy (Figure 1a) in thermodynamic equilibrium; However, since the mole fraction is very low it is not likely to nucleate due to kinetic reasons. Nevertheless, when the phase fraction of Ti_2Cu increases with Cu addition, already at 3 wt.% Cu and here at 5 wt.% Cu (Figure 1b), the phase fraction is considerable. The modelling resulted in the phase transition temperatures given in Table 1, where transus in this case, is the temperature above which the phase is no longer stable.

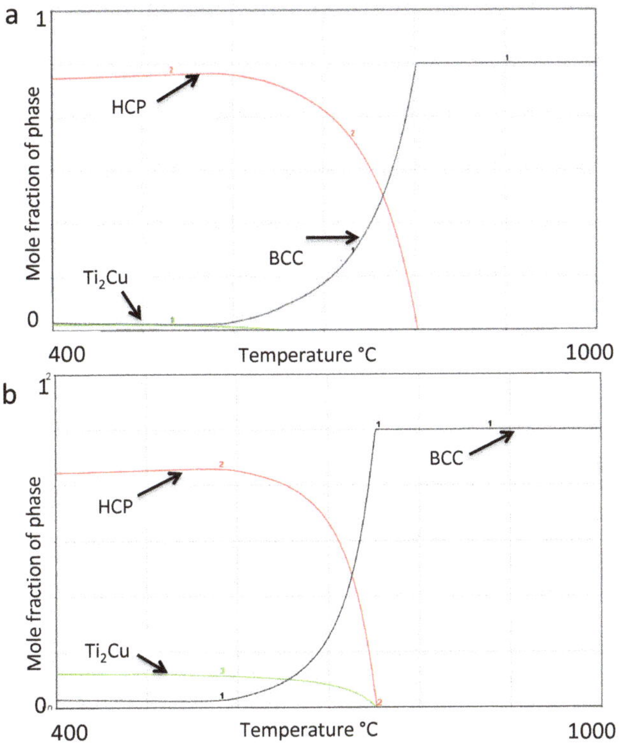

Figure 1. Mole fraction of phases as a function of temperature for (**a**) the Ti-Nb-Ta-Zr-1 wt.% Cu alloy and (**b**) the Ti-Nb-Ta-Zr-5 wt.% Cu alloy. Composition of the alloys can be found in Table 3.

Table 1. Calculated β- and Ti_2Cu-transus temperatures for the investigated alloys.

Cu Addition [wt.%]	Calc. β-transus [Celsius]	Calc. Ti_2Cu-transus [Celsius]
0	814	N/A
1	802	655
3	777	728
5	753	752
10	746	859

The wt.% Cu added to each alloy introduces a change in the Gibb's free energy for each of the predicted phases [26], and depending on the resultant driving force (ΔG_v), the phase development will proceed as specified according to phase reactions (Table 2). The cooling rate will furthermore determine the microstructure, where metastable martensitic phases (α' and ω) could form and change the resultant material properties during a rapid quenching.

Table 2. Reaction equations for stable phases and microstructural development.

$\Delta G_{Ti_2Cu} < \Delta G_\alpha$	$\Delta G_{Ti_2Cu} = \Delta G_\alpha$	$\Delta G_{Ti_2Cu} > \Delta G_\alpha$
$\beta \rightarrow \alpha$ $\beta + \alpha \rightarrow \alpha + Ti_2Cu$	$\beta \rightarrow Ti_2Cu + \alpha$	$\beta \rightarrow Ti_2Cu$ $\beta + Ti_2Cu \rightarrow Ti_2Cu + \alpha$

2.2. Production of Alloys

Alloys of Ti-Nb-Ta-Zr-Cu_x were produced in the range from 0 to 10 wt.% Cu (Table 3). Pre-alloyed Ti-Nb-Ta-Zr (Sandvik AB, Stockholm, Sweden) and 99.9999% pure copper rods (365327-21.5G, Sigma Aldrich, MO, USA) were used to produce the investigated alloys. Alloys were re-melted 5 times in an arc furnace, then melted into rods in the same furnace (Series 5 Bell Jar, Centorr Vacuum industries, Nashua, NH, USA). Partial homogenisation was achieved by the turning-over of the melted alloys between each of the five melting events. Complete homogenisation was achieved by heat treatments of the alloys at 988 °C, which is above all the calculated β-transus temperatures, for 48 h, then 747 °C for 18 h followed by a rapid quench. The second temperature was chosen based on the solution treatment for the Ti_2Cu. The annealing was done in vacuumed ampoules, at a pressure of 1.333 mbar, to reduce the oxygen content in the alloys. All alloys were quenched in salt brine water. Thereafter all samples were embedded in Bakelite resin (PolyFast, Stuers, Ballerup, Denmark) and cut into slices using an aluminium oxide disk (50A13, Struers) before further analysis. Metallographic preparation included grinding according to the three-step preparation developed by Vander Voort [27], which was appropriately adapted (Table 4).

Table 3. Nominal weight percentages of elements in the TNTZ-Cu_x alloys.

Designation	Cu	Ti	Nb	Zr	Ta
0 Cu	0	86.6	1.7	1.6	10.1
1 Cu	1	85.74	1.68	1.58	10.0
3 Cu	3	84	1.65	1.55	9.8
5 Cu	5	82.26	1.62	1.52	9.6
10 Cu	10	77.94	1.53	1.44	9.09

Table 4. The 3-step metallographic preparation of TNTZ-Cu_x. All products sourced from Struers, except H_2O_2, which was sourced from BASF SE (Ludwigshafen, Germany).

Steps	1–Grind	2–Rough Polish	3–Final Polish
Surface	SiC–120 P	MD–Dur cloth	MD–Floc cloth
Abrasive	-	6 µm diamond suspension	OP-S Si-Colloids and H_2O_2 (5:1) solution
Lubricant	water	DP lubricant red	-
Speed (rpm)	250 contra	200 contra	150 contra
Duration (min)	until planar	30 min	30 min

2.3. Calorimetric Measurements of Phase Transformations

The β-transus and phase transformation temperatures were measured by differential scanning calorimetry (DSC) using a Netzsch STA 409 CD (NETZSCH-Gerätebau GmbH, Selb, Germany). Aluminium oxide crucibles were used. The 1, 3, 5 and 10 wt.% Cu samples were chosen for these measurements, which have gone through the above-mentioned heat treatments with rapid cooling. The rate of the temperature change was 10 °C/min. The phase transformation temperatures were determined using the onset method.

2.4. X-ray Diffraction

X-ray diffraction patterns were recorded in the Bragg-Brentano geometry using a Bruker TWIN-TWIN diffractometer (D8 Advance, AXS GmbH, Karlsruhe, Germany) with Ni-filtered Cu Kα radiation (Kα1 = 1.540598 Å). Samples were polished to 6 µm using a diamond suspension (DiaDuo-2, Struers). Crystalline phases were studied in EVA software version 4.3 (Bruker, Billerica, MA, USA). The identified phases from the ICDD database PDF–4+ 2019 [28] included PDF# 04-003-1382 (Ti_2Cu), PDF# 00-044-1294 (HCP-Ti) and PDF# 03-065-9616 (HCP Ti-Ta).

2.5. Microstructural Studies

The microstructure of the samples was studied in scanning electron microscopy (SEM), focused ion beam (FIB) and scanning transmission electron microscopy (STEM) using a Zeiss Merlin (Oberkochen, Germany), an FEI Helios Nano-Lab (Brno, Czech Republic), and a JEOL 2100 TEM/STEM (Tokyo, Japan), respectively.

The Zeiss SEM and JEOL TEM/STEM were equipped with INCA AZtec Energy Dispersive X-ray Spectroscopy systems (EDS, Oxford Instruments, High Wycombe, UK) while the SEM and FIB-SEM each were equipped with back scatter, in-lens and Everhart-Thornley detectors. The JEOL TEM/STEM was additionally equipped with an annular dark field and bright field detector (JEOL, Tokyo, Japan).

For crystallographic investigations, Transmission Kikuchi Diffraction (TKD) was done with a custom made sample holder, in a JEOL 7001F SEM instrument (JEOL, Tokyo, Japan) equipped with a Schottky FEG and electron backscatter diffraction system (EBSD, Oxford Instruments) coupled to and INCA Aztec system (EDS, Oxford Instruments).

2.6. Hardness Studies

The hardness of the alloys was measured using an EMCO Test Duravision Vickers Hardness tester (Prufmaschinen GmbH, Kuchl, Austria). The machine was calibrated with a standard Vickers sample, before testing the samples. The samples were polished to grit of P400 with silicon carbide grinding paper (Struers). The applied mass for the hardness tester was set to 9.8 centinewton for all the alloys.

3. Results

3.1. Phase Calculations

The phase transformation temperatures, as determined by differential scanning calorimetry, were 829 °C, 751 °C, 746 °C and 744 °C for the 1, 3, 5 and 10 wt.% Cu alloys, respectively. The measurements were in good agreement with the calculated values for the β-transus temperatures at 746 °C (10 wt.% Cu alloy) and 753 °C (5 wt.% Cu alloy). The discrepancy between the calculated and measured values increased at lower Cu additions.

The thermodynamic prediction of phases for the alloys is given in Figure 1a,b as a function of temperature, and predicted HCP-Ti (α) and BCC-Ti (β), and additionally Ti_2Cu for Cu additions of 1% and higher. At the heat treatment temperature of 747 °C, the calculated mol% of the phases is given in Table 5, where the 10 wt.% Cu alloy was predicted to have no α-Ti phase present, while alloys below 5 wt.% Cu were predicted to have no Ti_2Cu phase present.

Table 5. Calculated mol% of phases of α, β and Ti_2Cu in the investigated alloys at 747 °C.

	Calculated Phases [mol%]		
Cu Addition [wt.%]	**α**	**β**	**Ti_2Cu**
0	76	24	0
1	37	63	0
3	35	65	0
5	14	85	2
10	0	85	15

Given the prerequisites mentioned earlier, the Ti_2Cu forms at 656 °C in the 1 wt.% Cu alloy (Figure 1a), which is below the annealing temperature. In addition, the predicted phase fraction at lower temperature is very small, thus the phase is not likely to nucleate on quenching due to kinetic reasons.

3.2. XRD and Microstructure

With X-ray diffraction studies on the 0 wt.% Cu and 1 wt.% Cu alloys, only the α-phase was determined to be present (Figure 2). However, SEM imaging for the same alloys (Figure 3b,c) indicated that the materials had two crystal phases present, as predicted (i.e. within the α + β region). The bright phase is most likely remaining β-phase since Ta, Nb and Cu (where Cu was given by the calculations) are β-stabilisers and the heat treatment temperatures were at 747 °C, thus within the α-β region.

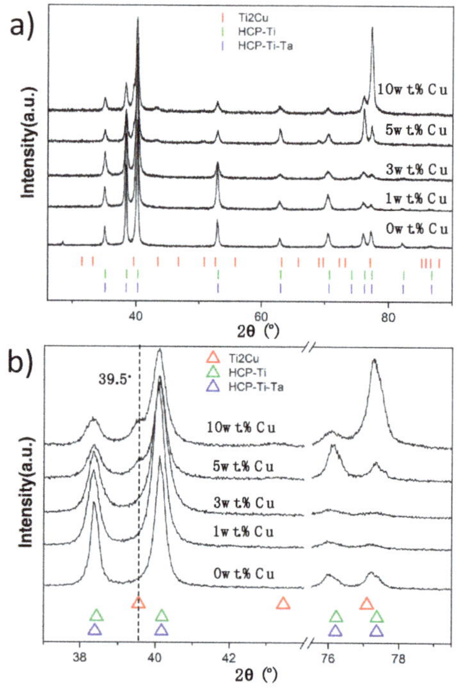

Figure 2. X-ray diffraction on TNTZ-Cu$_x$ alloys: (**a**) Diffraction for all alloys including references from Ti_2Cu (04-003-1382), HCP- (Ti-Ta) (03-065-9616) and HCP-Ti (00-044-1294). (**b**) X-ray diffraction pattern showing the 2θ angular ranges for the alloys from 38°–41° and 76°–78°. Note the Ti_2Cu peak at 39.5°.

In the 3 wt.% Cu alloy a third phase was predicted and observed in SEM imaging (Figure 3e). Diffraction peaks were not observed for this phase but it was assumed to be the predicted intermetallic Ti_2Cu. However, a very low amount was observed at large grain boundaries (GBs), which was in line with the prediction. β-Ti was not detected in the X-ray diffractogram (Figure 2) for 3 wt.% Cu alloy either, but is present in SEM images (Figure 3e). When the Cu addition was increased, the microstructure became coarser grained with thicker and disrupted bright (β) phase lathes. This coarsening is observed when comparing the 3 wt.% Cu to the 5 wt.% Cu alloy (Figure 3e,d, respectively). Likewise the phase fraction of the bright Ta, Cu-rich phase increased from 3 to 5 wt.% Cu.

The 5 wt.% Cu alloy clearly had three phases, while only small precipitates were observed in the 3 wt.% Cu. The X-ray diffraction patterns displayed an increase in Ti_2Cu phase with increase in Cu concentration from 5 wt.% Cu to 10 wt.% Cu (Figure 2b).

By comparing the 10 wt.% Cu and 5 wt.% Cu samples, it is clear that the microstructure is much coarser in the 10 wt.% Cu and that a higher volume fraction of the Ti_2Cu phase is found in (Figure 3a and inset). The phase diagram for 10 wt.% Cu also predicts that the Ti_2Cu phase exists, above the β-transus predicted to be at 747 °C and this is supported by the micrographs, which show large "globular" structures of the predicted Ti_2Cu phase present (Figure 3a inset). These precipitates were found exclusively at the GBs between α-Ti and β-Ti, for all alloys (Figure 3).

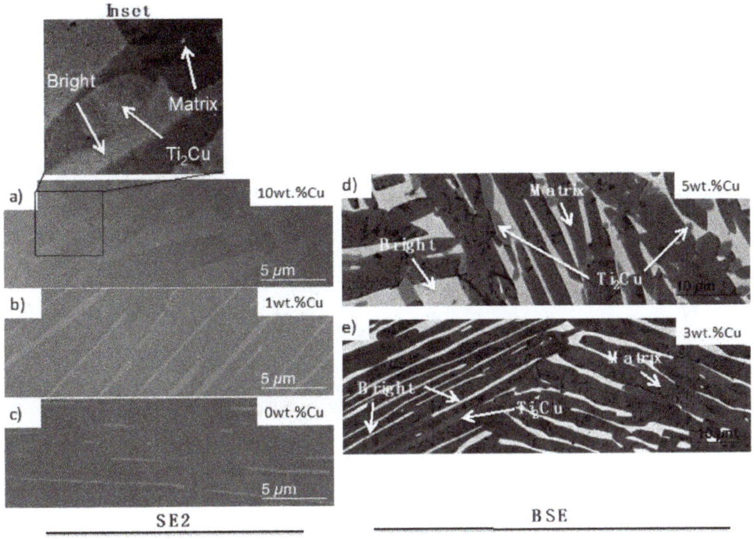

Figure 3. SEM micrographs showing (**a**) 10 wt.% Cu alloy with Inset showing three crystal phases, (**b**) 1 wt.% Cu alloy, (**c**) 0 wt.% Cu alloy (**d**) 5 wt.% Cu alloy with 3 crystal phases shown (**e**) 3 wt.% Cu alloy with three crystal phases shown.

3.3. Chemical and Crystal Phase Analysis

A change in phase development with the precipitation of the Cu-rich phase (Ti_2Cu) was observed in the 3 wt.% Cu and 5 wt.% Cu samples and thus they were the focus of further study.

The 3 wt.% Cu alloy had a microstructure similar to the lower Cu content alloys with thin lathes, but by using backscattered electron imaging, smaller precipitates were discovered at the GBs of the larger α-Ti grains (Figure 3e). These areas were studied further by preparation of focused ion beam (FIB) lamella, STEM-EDS and transmission Kikuchi diffraction (TKD). Regions of Cu-rich precipitates were observed, with adjacent crystals containing Ti and Ta (Figure 4a). The grains with the brightest contrast, which probably was β-Ti considering the heat treatment temperature, were also slightly coarser grained in the 3 wt.% Cu alloy compared to those with lower Cu content. The phases were assigned as a matrix phase of α-Ti, Ti_2Cu and a bright phase, where the bright phase could not be assigned to a known crystal phase using TKD (Figure 5). The 5 wt.% Cu alloy had a microstructure of irregular lathes compared to those with lower Cu content (Figure 3). The lathes that formed were not straight-line structures as in the 3 wt.% Cu, but instead lathes disrupted by Cu-rich globules, formed along the length of the bright β-phase.

Using TKD, the Cu-rich phase and the matrix phase were designated as Ti_2Cu and α (HCP-Ti), respectively (Figure 6). Assignment of a known crystal to the bright phase was challenging using TKD for this alloy as well (Figure 6). The formation of the Cu-rich phase- in 3 wt.% Cu and 5 wt.% Cu occurred selectively at the GBs between the α-Ti and the bright phase. Spectroscopic comparison of the 3 wt.% Cu and 5 wt.% Cu showed that the former alloy contained a crystal with more Cu in a "globular" shaped crystal, surrounded by a crystal with more Ta and Ti (Figure 4a). The 5 wt.% Cu contained a thin crystal enriched with Cu and Ta and surrounded by crystals of Ti with lower concentrations of Cu with Ta (Figure 4b). Using TKD coupled to EDX mapping on a different lamella, the 3 wt.% Cu alloy showed a bright phase crystal with Cu, Ta and Ti (Figure 5f–h).

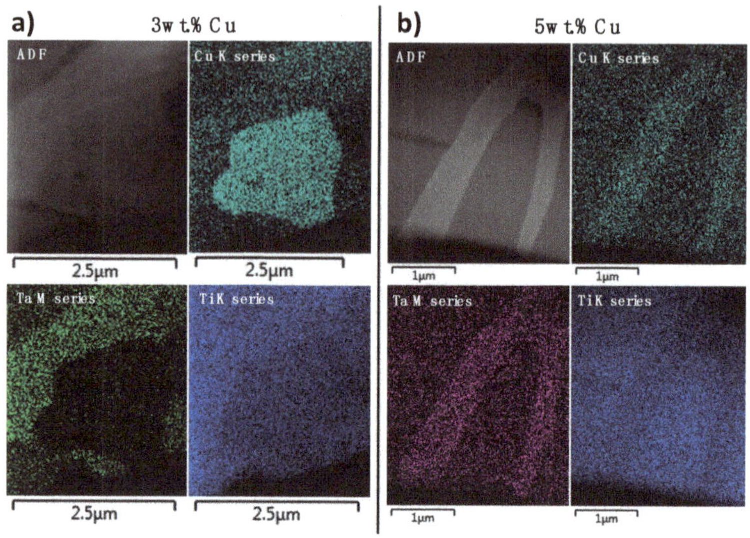

Figure 4. STEM-EDS maps on the crystal boundary showing 3 crystal phases for (**a**) 3 wt.% Cu alloy and (**b**) 5 wt.% Cu alloy, with associated Annular dark field detector image, Cu K series map, Ta M series map and Ti K series map.

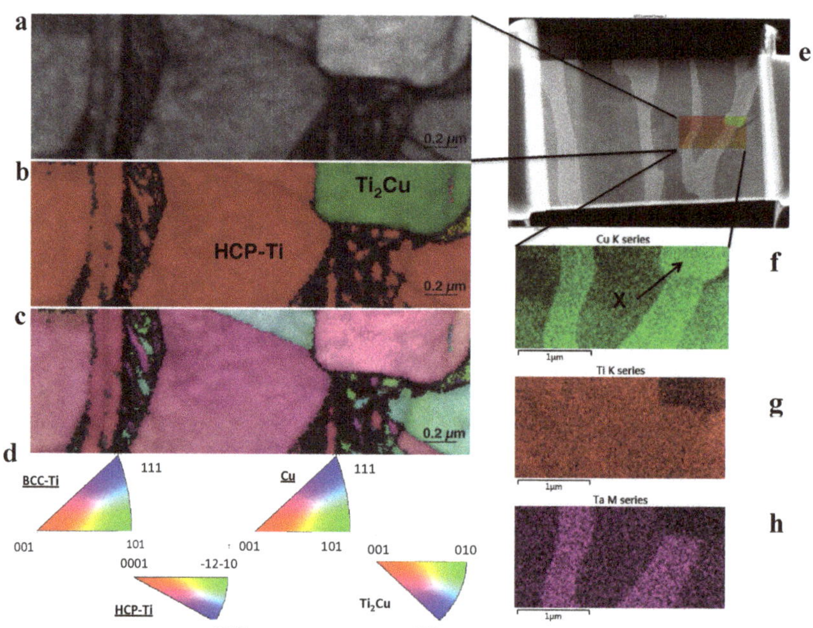

Figure 5. 3 wt.% Cu alloy studied using TKD and EDS. From the TKD study (**a**) displays band contrast, (**b**) a phase map and (**c**) IPF Z map of the same area with (**d**) associated pole figures. (**e**) An electron image of the area investigated with both techniques, (**f–h**) EDS maps of: (**f**) Cu K series (where X indicates Ti_2Cu), (**g**) Ti K series and (**h**) Ta M series.

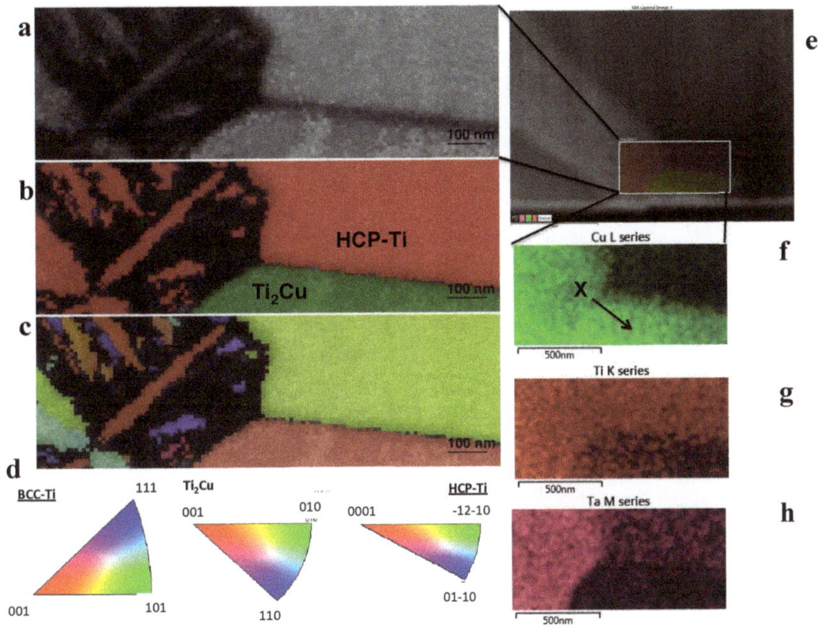

Figure 6. 5 wt.% Cu alloy studied using TKD and EDS. From the TKD study (**a**) displays band contrast, (**b**) a phase map and (**c**) IPF Z map of the same area with (**d**) associated pole figures. (**e**) An electron image of the area investigated with both techniques, (**f–h**) EDS maps of: (**f**) Cu L series (where X indicates Ti_2Cu), (**g**) Ti K series and (**h**) Ta M series.

3.4. Hardness

The hardness (Figure 7) of the 0 wt.% Cu alloy (135 ± 3 Hv) was significantly lower than the 1 wt.% Cu alloy (198 ± 9 Hv, $p = 0.0005$), which in turn was significantly lower than the 3 wt.% Cu, 5 wt.% Cu and 10 wt.% Cu alloys ($p < 0.035$). No statistically significant difference in hardness was found among the 3 wt.% Cu (238 ± 15 Hv), 5 wt.% Cu (232 ± 7 Hv) and 10 wt.% Cu (238 ± 19 Hv) alloys ($p > 0.975$).

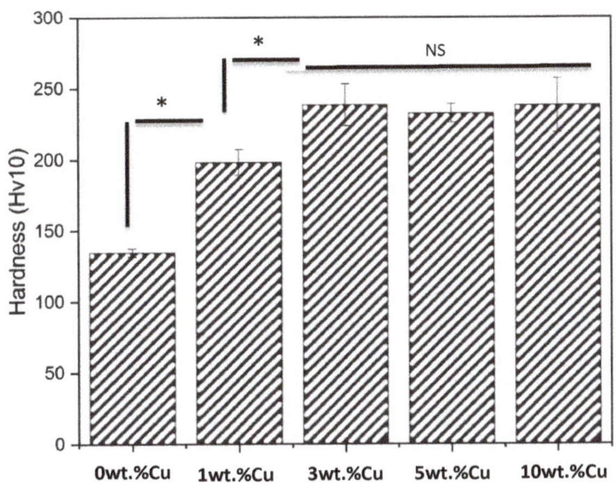

Figure 7. Vickers hardness of the TNTZ-Cu_x alloys ("*" indicates statistical significance of $p < 0.05$, as per Tukey Anova test, while "NS" indicates no statistically significant difference).

4. Discussion

The present study investigated the addition of copper to a TNTZ alloy and its effect on the microstructure, with the scope of developing a biomedical alloy with potential antibacterial ability in future. Predictions of the stable phases in the alloy system - that are to be regarded as an initial approach to the development of TNTZ-Cu_x alloys-revealed 3-phases in equilibrium (Figure 1).

Comparison of the predicted β-transus temperatures to the experimentally observed values, revealed discrepancies in the data at 829 °C, 751 °C, 746 °C and 744 °C for the 1, 3, 5 and 10 wt.% Cu alloys, respectively. The measurements were in good agreement with the calculated values of 746 °C and 753 °C, for the 10 wt.% Cu and 5 wt.% Cu alloys. However, the discrepancy between the calculated and measured values increased as the Cu content decreased. The reason for the discrepancies could be the absence of the Ti-Ta-Cu system in the database. An additional cause for variance in the discrepancies could be due to the reduction in the effective Cu content, since Cu is bonded in the intermetallic (Ti_2Cu) phase, which was identified by diffraction for the 5 and 10 wt.% Cu alloys. A further reason could be that the β-stabilizers of Ta and Nb are soluble in the intermetallic phases, in addition to Cu. It is also uncertain whether the metastable Ti_3Cu [24] is present for the lower Cu compositions, thus further research is required.

The changing copper concentrations in the TNTZ materials also changed the phase development by affecting the "driving force" (change in Gibb's free energy), leading to one of three phase development scenarios (Table 2). For the 5 wt.% Cu alloy, the β- and Ti_2Cu-transus temperatures are within 1 °C of each other and could lead to precipitation of phases according to $\beta \rightarrow Ti_2Cu + \alpha$. For Cu concentrations below 5 wt.% Cu, the phase development will likely proceed according to according to the reaction $\Delta G_{Ti2Cu} > \Delta G_{\alpha}$. When the Cu concentration is below the 5 wt.% Cu, the phase development will likely proceed according to the reaction $\Delta G_{Ti2Cu} < \Delta G_{\alpha}$. Cu addition also affects the volume fraction of the phases that develop and thermodynamic modelling predicted that no α-phase would nucleate for the 10 wt.% Cu alloy quenched from 747 °C (Table 5). Predictions were also made that no Ti_2Cu would nucleate in the quenched 3 wt.% Cu alloy (Table 5). A possible reason for this difference is that the quenching rates from 747 °C, might have been too slow, and thus the α-phase nucleated in the 10 wt.% Cu alloy. The same process could have caused the Ti_2Cu to nucleate in the 3 wt.% Cu alloy. Alternatively, deviations in the furnace temperature prior to quenching might also be responsible for the nucleation of the phases. Further investigations in modelling and rapid quenching could elucidate the cause for the differences between predictions and experiments.

The 0 wt.% Cu TNTZ was heat treated at 747 °C, and predicted to have a microstructure consisting of α (76.4%) and β (23.6%), with a hardness of 135 ± 3 Hv. In a previous study [11] the alloy was found to be a α (50%) and β (50%) alloy with hardness of 340 HVN. The differences found were probably due to various forging treatments of the alloy in the previous study [11]. The addition of 1 wt.% Cu did not cause a third phase to precipitate, presumably due to the fact that the Ti_2Cu phase only forms at temperatures lower than 747 °C (Figure 1a). Therefore the 0 wt.% Cu and 1 wt.% Cu alloys are confirmed as two-phased materials via diffraction (Figure 2) and microscopy studies (Figure 3).

The 3 and 5 wt.% Cu alloys both had a 3-phased (Figure 3d,e) crystal structure, even though calculation of phases at 747 °C predicts a 2-phase structure (α and β) for the 3 wt.% Cu (Table 5). This indicates that precipitation could have taken place below 747 °C prior to the quenching into salt water. Alternatively, since 3 wt.% Cu is lower than the 5 wt.% Cu, the precipitation of α might have occurred according to the reaction $\beta + \alpha \rightarrow \alpha + Ti_2Cu$, which is in line with kinetics for active eutectoid transformations described in studies on Ti-Cu [18,19]. The Ti_2Cu was not observed in the XRD pattern of the 3 wt.% Cu but this could be due to the volume fraction being below 2 wt.% of Ti_2Cu, which is the detection limit for the X-ray diffraction technique [20]. The 5 wt.% Cu however showed the Ti_2Cu in the diffractogram (Figure 2) and these were present as "globular" and irregular crystals in the microstructure. The bright phase was observed as "globular" shaped in the 5 and 10 wt.% Cu alloys.

The standard crystal structures assigned using TKD were HCP-Ti [29] and Ti_2Cu [30], since these crystals gave the most appropriate match to the experimental data. The indexing of 3- and 5 wt.% Cu in TKD gave similar chemical phases of the Ti-Ta (bright β phase), the α (HCP-Ti) phase as well as (Ti_2Cu) phase. The β phase was nano-crystalline and therefore contained many internal GBs, resulting in difficulty for the computational phase assignment routines in TKD to assign these GB pixels to specific phases [31], resulting in an inability to index the crystals. The α (HCP-Ti) and Ti_2Cu however were indexed and matched well to the standard crystals. EDX-maps of the crystals in SEM (Figures 5 and 6) and STEM (Figure 4) provide a view to the "globular" structure of the Cu-rich phases in the 5 wt.% Cu, while lathes of Cu-rich crystals were discerned in the 3 wt.% Cu (Figure 3e). This difference in the structure for the Cu-rich phase could be due to particle coarsening, at higher Cu concentrations, via diffusion in the matrix of the alloy [32]. The maps also indicated that the 5 wt.% Cu had thin crystals containing Ta and Cu with Ti in adjacent phases (Figure 4b). The thin crystal could be the bright phase enriched in Ta surrounded by the Ti-rich α-Ti matrix. The 3 wt.% Cu had crystals enriched in Cu (Figure 4a) with Ta surrounding the crystals. These Cu-rich crystals could be the Ti_2Cu phase that was surrounded by the Ta-rich bright phase found in the study.

The 10 wt.% Cu alloy had a 3-phased microstructure (Figure 3) with α, Ti_2Cu and β (bright) phase, which contradicts thermodynamic modelling predictions, where the α phase was calculated as absent (Table 5). The prediction of only β and Ti_2Cu after a rapid quench is not experimentally observed in microstructural transformations of active β-eutectoid alloys, of which Ti-Cu is one such example [18,33]. This microstructure contains "globular" precipitates of a Cu-rich phase (Ti_2Cu) that seems to form between the α and β phase (Figure 3a inset), similar to 3- and 5 wt.% Cu. The bright (β) phase seems to have changed to a "globular" shaped structure with the increase in wt.% Cu, similar in shape to the matrix (α) phases. While the 10 wt.% Cu was not studied in TKD, it is probable that the β (bright) phase would index the same as in the 3- and 5 wt.% Cu. All transformation reactions in these alloys containing the bright phase seem to preclude β phase formation, and it is possible that the reason could be martensitic phase transformations, on account of the rapid cooling [34].

The Cu addition to the TNTZ material drives the alloy to rapid transformations via the eutectoid reactions. For the 5 wt.% Cu and 3 wt.% Cu, the bright phase could not be indexed as α, and might be a martensitic phase. This is observed in Ti-Ta (1–10 at.% Ta) and the phase identified was orthorhombic α'' [34] according to the reaction $\beta \rightarrow \alpha'' \rightarrow \alpha$. This reaction is thought to occur exclusively at temperatures below 927 °C (1200 K) and involves mechanical shearing and shuffling of atoms to achieve the transformation [35]. The exact shearing planes are however presently being debated and could involve twinning shear in one or both of the planes of {332}$11\bar{3}$ [36] and 111{112} [37].

The hardness increased significantly with the addition of Cu (Figure 7). This could be a consequence of the Cu atoms dissolving into the α and β crystal phases as a solid solution mixture. The hardness of the 3 wt.% Cu, 5 wt.% Cu and 10 wt.% Cu was not significantly different but was significantly higher than 0 wt.% Cu and 1 wt.% Cu, which could indicate that the saturation limit of solid solution alloying had been reached.

5. Conclusions

In the present study, the effect of Cu addition to an existing TNTZ alloy was investigated. The alloys in the range of 3 to 10 wt.% Cu all had Ti_2Cu present, while in alloys with less than 3 wt.% Cu, Ti_2Cu was not observed. From 5 to 10 wt.% Cu, the alloys showed the presence of Ti_2Cu in increasingly "globular" structures with increase in Cu concentration. An associated effect of increasing to Cu content from 0 to 3 wt.% Cu was that the hardness increased, but no additional increase was achieved from 3 to 10 wt.% Cu. The hardness could be a result of solid solution strengthening, but might also be affected by martensitic transformations. While the material has reasonable hardness, the potential antibacterial ability of the material requires assessment in future. Therefore further studies are envisioned for this alloy system to optimize the mechanical, antibacterial and corrosion properties for the purpose of producing a suitable antibacterial implant material.

Author Contributions: Conceptualization, C.-Ö.M. and S.N.; Formal analysis, L.F., A.J.V.V. and W.G.; Funding acquisition, H.E., C.-Ö.M. and S.N.; Investigation, L.F., A.J.V.V. and W.G.; Methodology, L.F., C.-Ö.M. and S.N.; Project administration, H.E. and C.-Ö.M.; Resources, H.E. and C.-Ö.M.; Supervision, H.E., C.-Ö.M. and S.N.; Validation, L.F., H.E., C.-Ö.M. and S.N.; Writing—original draft, L.F. and S.N.; Writing—review & editing, L.F., W.G., H.E., C.-Ö.M. and S.N. The supervision, planning, conceptualization, funding, project administration, resources, writing and reviewing was the responsibility of H.E, C.-Ö.M. and S.N. The initial and final writing of the manuscript, production of alloys, analysis and characterization was the responsibility of L.F. and A.J.V.V assisted with TEM lamellas, as well as STEM-EDS studies of the samples. The TKD studies and analysis of the samples was performed by W.G. All authors assisted in the writing and reviewing of the manuscript.

Funding: This research was funded by the Swedish Foundation for International Cooperation in Research and Higher Education grant number GA SA2017-7127 (STINT), the South African National Research Foundation grant number STNT170905261815 (NRF), the INSPIRE Scholarship of the ERASMUS Mundus program of the EU (no grant number available) as well as the Axel Hultgren Fund (no grant number available).

Acknowledgments: A special thanks is made to the Centre for High Resolution Transmission electron Microscopy at Nelson Mandela University and Johan Westraadt for fruitful discussions.

Conflicts of Interest: The authors declare no conflict of interest.

References

1. Mouritz, A. Titanium alloys for aerospace structures and engines. In *Introduction to Aerospace Materials*; Woodhead Publishing: Cambridge, UK, 2012; pp. 202–223. [CrossRef]
2. Rehder, D. Vanadium. Its role for humans. *Met. Ions Life Sci.* **2013**, *13*, 139–169. [CrossRef] [PubMed]
3. Klotz, K.; Weistenhöfer, W.; Neff, F.; Hartwig, A.; van Thriel, C.; Drexler, H. The Health Effects of Aluminum Exposure. *Dtsch. Arztebl. Int.* **2017**, *114*, 653–659. [CrossRef] [PubMed]
4. Ji, X.; Gutierrez-Urrutia, I.; Emura, S.; Liu, T.; Hara, T.; Min, X.; Ping, D. Tsuchiya, Twinning behavior of orthorhombic-α" martensite in a Ti-7.5Mo alloy. *Sci. Technol. Adv. Mater.* **2019**, *20*, 401–411. [CrossRef]
5. Sandu, A.V.; Baltatu, M.S.; Nabialek, M.; Savin, A.; Vizureanu, P. Characterization and Mechanical Proprieties of New TiMo Alloys Used for Medical Applications. *Materials* **2019**, *12*, 2973. [CrossRef] [PubMed]
6. Ho, W.F.; Ju, C.P.; Lin, J.H.C. Structure and properties of cast binary Ti–Mo alloys. *Biomaterials* **1999**, *20*, 2115–2122. [CrossRef]
7. Bai, W.; Xu, G.; Tan, M.; Yang, Z.; Zeng, L.; Wu, D.; Liu, L.; Zhang, L. Diffusivities and Atomic Mobilities in bcc Ti-Mo-Zr Alloys. *Materials* **2018**, *11*, 1909. [CrossRef] [PubMed]
8. Ferrandini, P.L.; Cardoso, F.F.; Souza, S.A.; Afonso, C.R.; Caram, R. Aging response of the Ti–35Nb–7Zr–5Ta and Ti–35Nb–7Ta alloys. *J. Alloy. Compd.* **2007**, *433*, 207–210. [CrossRef]
9. Afonso, C.R.M.; Ferrandini, P.L.; Ramirez, A.J.; Caram, R. High resolution transmission electron microscopy study of the hardening mechanism through phase separation in a β-Ti–35Nb–7Zr–5Ta alloy for implant applications. *Acta Biomater.* **2010**, *6*, 1625–1629. [CrossRef]
10. Elmay, W.; Laheurte, P.; Eberhardt, A.; Bolle, B.; Gloriant, T.; Patoor, E.; Prima, F.; Laille, D.; Castany, P.; Wary, M. Stability and elastic properties of Ti-alloys for biomedical application designed with electronic parameters. *EPJ Web Conf.* **2010**, *6*, 29002. [CrossRef]
11. Stenlund, P.; Omar, O.; Brohede, U.; Norgren, S.; Norlindh, B.; Johansson, A.; Lausmaa, J.; Thomsen, P.; Palmquist, A. Bone response to a novel Ti–Ta–Nb–Zr alloy. *Acta Biomater.* **2015**, *20*, 165–175. [CrossRef]
12. Niinomi, M.; Liu, Y.; Nakai, M.; Liu, H.; Li, H. Biomedical titanium alloys with Young's moduli close to that of cortical bone. *Regen. Biomater.* **2016**, *3*, 173–185. [CrossRef] [PubMed]
13. Heidenau, F.; Mittelmeier, W.; Detsch, R.; Haenle, M.; Stenzel, F.; Ziegler, G.; Gollwitzer, H. A novel antibacterial titania coating: Metal ion toxicity and in vitro surface colonization. *J. Mater. Sci. Mater. Med.* **2005**, *16*, 883–888. [CrossRef] [PubMed]
14. Arciola, C.R.; Campoccia, D.; Speziale, P.; Montanaro, L.; Costerton, J.W. Biofilm formation in Staphylococcus implant infections. A review of molecular mechanisms and implications for biofilm-resistant materials. *Biomaterials* **2012**, *33*, 5967–5982. [CrossRef] [PubMed]
15. Klevens, R.M.; Morrison, M.A.; Nadle, J.; Petit, S.; Gershman, K.; Ray, S.; Harrison, L.H.; Lynfield, R.; Dumyati, G.; Townes, J.M.; et al. Active Bacterial Core surveillance (ABCs) MRSA Investigators, Invasive methicillin-resistant Staphylococcus aureus infections in the United States. *JAMA* **2007**, *298*, 1763–1771. [CrossRef] [PubMed]

16. Liu, J.; Li, F.; Liu, C.; Wang, H.; Ren, B.; Yang, K.; Zhang, E. Effect of Cu content on the antibacterial activity of titanium–copper sintered alloys. *Mater. Sci. Eng. C* **2014**, *35*, 392–400. [CrossRef] [PubMed]
17. Ke, Z.; Yi, C.; Zhang, L.; He, Z.; Tan, J.; Jiang, Y. Characterization of a new Ti-13Nb-13Zr-10Cu alloy with enhanced antibacterial activity for biomedical applications. *Mater. Lett.* **2019**, *253*, 335–338. [CrossRef]
18. Donthula, H.; Vishwanadh, B.; Alam, T.; Borkar, T.; Contieri, R.J.; Caram, R.; Banerjee, R.; Tewari, R.; Dey, G.K.; Banerjee, S. Morphological evolution of transformation products and eutectoid transformation(s) in a hyper-eutectoid Ti-12 at% Cu alloy. *Acta Mater.* **2019**, *168*, 63–75. [CrossRef]
19. Contieri, R.J.; Lopes, E.S.N.; Caram, R.; Devaraj, A.; Nag, S.; Banerjee, R. Effects of cooling rate on the microstructure and solute partitioning in near eutectoid Ti–Cu alloys. *Philos. Mag.* **2014**, *94*, 2350–2371. [CrossRef]
20. Fowler, L.; Janson, O.; Engqvist, H.; Norgren, S.; Öhman-Mägi, C. Antibacterial investigation of titanium-copper alloys using luminescent Staphylococcus epidermidis in a direct contact test. *Mater. Sci. Eng. C* **2019**, *97*, 707–714. [CrossRef]
21. Banerjee, R.; Nag, S.; Stechschulte, J.; Fraser, H.L. Strengthening mechanisms in Ti–Nb–Zr–Ta and Ti–Mo–Zr–Fe orthopaedic alloys. *Biomaterials* **2004**, *25*, 3413–3419. [CrossRef]
22. Lukas, H.L.; Fries, S.G.; Sundman, B. Computational Thermodynamics—The Calphad Method. 2007. Available online: https://app.knovel.com/hotlink/toc/id:kpCTTCM005/computational-thermodynamics/computational-thermodynamics (accessed on 28 July 2019).
23. Kumar, K.C.H.; Ansara, I.; Wollants, P.; Delaey, L. Thermodynamic optimisation of the Cu-Ti system. *Zeitschrift Für Metallkunde* **1996**, *87*, 666–672.
24. Canale, P.; Servant, C. Thermodynamic Assessment of the Cu–Ti System Taking into Account the New Stable Phase CuTi3. *MEKU* **2002**, *93*, 273–276. [CrossRef]
25. Zhang, E.; Ren, J.; Li, S.; Yang, L.; Qin, G. Optimization of mechanical properties, biocorrosion properties and antibacterial properties of as-cast Ti–Cu alloys. *Biomed. Mater.* **2016**, *11*, 065001. [CrossRef] [PubMed]
26. Porter, D.A.; Easterling, K.E. *Thermodynamics and Phase Diagrams. Phase Transformations in Metals and Alloys*; Van Nostrand Reinhold Co. Ltd.: Berkshire, UK, 1981; pp. 1–57.
27. Voort, G.V. Materials Characterization & Testing: Microstructure of Titanium and its Alloys. *Ind. Heat.* **2006**, *73*, 77–80, 82–84.
28. PDF-4+ 2019, ICDD. (n.d.). Available online: http://www.icdd.com/index.php/pdf-4/ (accessed on 10 May 2019).
29. Spreadborough, J.; Christian, J.W. The Measurement of the Lattice Expansions and Debye Temperatures of Titanium and Silver by X-ray Methods. *Proc. Phys. Soc.* **1959**, *74*, 609–615. [CrossRef]
30. Mueller, M.H.; Knott, H.W. The crystal structures of Ti_2Cu, Ti_2Ni, Ti_4Ni_2O, and Ti_4Cu_2O. *Trans. Metall. Soc. AIME* **1963**, *227*, 674–678.
31. Bingert, J.; Mason, T.; Kaschner, G.; Maudlin, P.; Gray, G. Anisotropic plasticity modeling incorporating EBSD. In *Electron Backscatter Diffraction in Materials Sciences*; Plenun Publishing: New York, NY, USA, 2000; pp. 213–229.
32. Porter, D.A.; Easterling, K.E. Particle coarsening. In *Phase Transformations in Metals and Alloys*; Van Nostrand Reinhold Co. Ltd.: Berkshire, UK, 1981; pp. 314–315.
33. Devaraj, A.; Nag, S.; Muddle, B.C.; Banerjee, R. Competing Martensitic, Bainitic, and Pearlitic Transformations in a Hypoeutectoid Ti-5Cu Alloy. *Metall. Mat. Trans. A* **2011**, *42*, 1139–1143. [CrossRef]
34. Dobromyslov, A.V.; Dolgikh, G.V.; Dutkevich, Y.; Trenogina, T.L. Phase and structural transformations in Ti-Ta alloys. *Phys. Met. Metallogr.* **2009**, *107*, 502–510. [CrossRef]
35. Hennig, R.G.; Lenosky, T.J.; Trinkle, D.R.; Rudin, S.P.; Wilkins, J.W. Classical potential describes martensitic phase transformations between the α, β, and ω titanium phases. *Phys. Rev. B* **2008**, *78*, 054121. [CrossRef]
36. Tobe, H.; Kim, H.Y.; Inamura, T.; Hosoda, H.; Miyazaki, S. Origin of {332} twinning in metastable β-Ti alloys. *Acta Mater.* **2014**, *64*, 345–355. [CrossRef]
37. Xing, H.; Sun, J. Mechanical twinning and omega transition by ⟨111⟩ {112} shear in a metastable β titanium alloy. *Appl. Phys. Lett.* **2008**, *93*, 031908. [CrossRef]

materials MDPI

Article

Development of Antibacterial Ti-Cu$_x$ Alloys for Dental Applications: Effects of Ageing for Alloys with Up to 10 wt% Cu

Lee Fowler [1], Nomsombuluko Masia [2,3], Lesley A. Cornish [3], Lesley H. Chown [3], Håkan Engqvist [1], Susanne Norgren [1] and Caroline Öhman-Mägi [1,*]

1 Division of Applied Materials Science, Department of Engineering Sciences, The Ångström Laboratory, Uppsala University, Box 534, 751 21 Uppsala, Sweden; lee.fowler@angstrom.uu.se (L.F.); hakan.engqvist@angstrom.uu.se (H.E.); susanne.norgren@angstrom.uu.se (S.N.)
2 Advanced Materials Division, Mintek, Randburg 2124, South Africa; nomsombulukoh@mintek.co.za
3 School of Chemical and Metallurgical Engineering, and DST-NRF Centre of Excellence in Strong Materials, hosted by the University of the Witswatersrand, Johannesburg 2000, South Africa; lesley.cornish@wits.co.za (L.A.C.); lesley.chown@wits.co.za (L.H.C.)
* Correspondence: caroline.ohman@angstrom.uu.se

Received: 29 October 2019; Accepted: 28 November 2019; Published: 3 December 2019

Abstract: Peri-implantitis, a disease caused by bacteria, affects dental implants in patients. It is widely treated with antibiotics, however, with growing antibiotic resistance new strategies are required. Titanium-copper alloys are prospective antibacterial biomaterials, with the potential to be a remedy against peri-implantitis and antibiotic resistance. The aim of this study was to investigate Ti-Cu$_x$ alloys, exploring how Cu content (up to 10 wt%) and ageing affect the material properties. Electron microscopy, X-ray diffraction, hardness testing, bacteriological culture, and electrochemical testing were employed to characterize the materials. It was found that alloys with above 3 wt% Cu had two phases and ageing increased the volume fraction of Ti_2Cu. An un-aged alloy of 5 wt% Cu showed what could be Ti_3Cu, in addition to the α-Ti phase. The hardness gradually increased with increased Cu additions, while ageing only affected the alloy with 10 wt% Cu (due to changes in microstructure). Ageing resulted in faster passivation of the alloys. After two hours the aged 10 wt% Cu alloy was the only material with an antibacterial effect, while after six hours, bacteria killing occurred in all alloys with above 5 wt% Cu. In conclusion, it was possible to tune the material and antibacterial properties of Ti-Cu$_x$ alloys by changing the Cu concentration and ageing, which makes further optimization towards an antibacterial material promising.

Keywords: titanium alloys; copper; Ti_2Cu; Ti_3Cu; biomaterial; antibacterial

1. Introduction

Peri-implantitis is a disease caused by bacterial population of an implant surface, which results in local inflammation of soft tissue and ultimately implant loosening and failure [1]. The loss of bone tissue is often associated with the inflammation surrounding the implants, leading to revision surgeries often being troublesome to successfully accomplish [1]. The prevalence of peri-implantitis in the general population is estimated to be between 19% [2] and 28% [3]. Despite the discrepancy in prevalence data, it is generally agreed that the prevalence is under-estimated at present [1,3]. Nevertheless, the importance of the problem is clear and increasing, considering that the dental implant industry is estimated to have reached USD 5 billion in 2018 [1]. The bacteria responsible for the disease include: *Peptostreptococcus micros, Porphyromonas ginigivalis and Treponema denticola*, among others [4,5]. The fact that these bacteria have virulent characteristic behavior [6], coupled with the increasing problems with antibiotic resistance [7], corroborate that an effective antibacterial solution is needed.

The effectiveness of bacterial colonization lies in the creation of biofilms [8], which have the ability to survive antibiotic treatments due to a multifaceted strategy of slower growth in the biofilm, poor penetration of antibiotics, and genetic variations in the bacteria [9,10]. The gram-positive *Staphylococcus epidermidis* is a biofilm forming bacteria, and is one of the most common bacteria in implant infections [11], making the bacteria a suitable strain for testing the antibacterial ability of biomaterials.

To date, much research has been focused on various approaches to reduce the bacterial burden correlated to implants, which include: surface treatments [12], the use of photochemical reactions [13], alloying of silver to titanium [14], and recently, the alloying of copper to titanium [15–17]. The antibacterial effect of Cu additions was found to be superior compared to silver [18]. The Ag appears to be effectual only at elevated temperatures [19], whereas Cu was found—using in vitro tests—to have antibacterial ability at ambient conditions [20]. For this reason, Cu is being investigated as an antibacterial element with increased frequency in thin films [21,22], ionic form [23,24], and in alloys [17,25,26]. In particular, Zhang et al. [27] have investigated the addition of Cu in titanium and found that there are discrepancies in the literature regarding required copper content and the mechanisms causing bacterial reduction. While success in antibacterial applications has been achieved with copper, the molecular process of bacterial death is not clear. However, it is hypothesized to be caused by Cu ions creating reactive oxygen species (ROS) via the Fenton reaction ($Cu^+ + H_2O_2 \rightarrow Cu^{2+} + OH^- + \cdot OH$) [24], where ROS plays a role in the prohibition of bacteria 16SrRNA replication, resulting in bacteria death [28].

While it is important to understand the biological interactions in these systems, the antibacterial effect is only one aspect of successfully functioning antibacterial biomaterials. The mechanical properties are also important and should be considered, since implants experience mechanical loading in vivo. When excessive Cu is alloyed to Ti, intermetallic compounds (Ti_2Cu [29] and Ti_3Cu [30]) form, which can lead to brittleness [31]. While heat treatments of Ti alloys does allow some degree of ductility to be achieved in initially brittle materials, the resultant effect on microstructure and antibacterial properties must be assessed and understood. This inter-relationship between the various choices of compositions and heat treatments, and their impacts on mechanical properties and antibacterial effects, provide an opportunity for optimization of the final alloy. Therefore, the aim of the present study was to investigate these aspects (microstructural, mechanical and antibacterial) for different Ti-Cu alloys, to contribute to the understanding of these materials, and facilitate their future use as antibacterial biomaterials. The investigation only focused on alloys in the range 0 to 10 wt% Cu, since Cu ions released in this range of alloys did not show notable toxicity effects in previous works [15,23].

2. Materials and Methods

2.1. Production and Preparation of Alloys

Alloys of titanium and copper were produced in a range from 0 to 10 wt% Cu (Table 1). Titanium grade 4 (Sandvik AB, Stockholm, Sweden), "CPTi", and 99.9999% pure copper rods (365327-21.5G, Sigma Aldrich, St. Louis, MO, USA) were used to produce the alloy mixtures. Alloys were re-melted five times in an arc furnace, then re-melted and cast into rods in the same furnace (Series 5 Bell Jar, Centorr Vacuum industries, Nashua, New Hampshire, USA). Partial homogenization was achieved by turning over the melted alloys between each of the five melting events. Complete homogenization was achieved by heat treatments of the rods at 900 °C for 18 h, then 798 °C for 24 h and then finally, for half of the produced alloys, ageing at 400 °C for six hours. The annealing was done in vacuumed ampoules to reduce the oxygen contamination in the alloys, at a pressure of 1.333 mbar. All alloys were quenched into salt brine water by breaking the ampoules, to allow faster cooling. The alloy samples were cut into slices for analysis using a silicon carbide disk (150s, Struers, Ballerup, Denmark), to minimize cutting damage. Following sectioning, the samples were embedded in Bakelite resin (PolyFast, Stuers, Ballerup, Denmark) for metallographic preparation.

Table 1. Ti-Cu$_x$ samples used in the study.

Nominal wt% Cu	CPTi	1 wt% Cu	3 wt% Cu	5 wt% Cu	10 wt% Cu
Alloys heat treated at 798 °C (un-aged)	CPTi-798	1Cu798	3Cu798	5Cu798	10Cu798
Alloys aged at 400 °C	CPTi-400	1Cu400	3Cu400	5Cu400	10Cu400

The metallographic preparation of the samples included grinding and polishing for subsequent microstructural studies. The three-step metallographic procedure developed by Vander Voort [32] was adapted for the present study, further details are given in Table 2.

Table 2. Three-step metallographic preparation of Ti-Cu (All products sourced from Struers, except H2O2 sourced from BASF SE, Ludwigshafen, Germany).

Steps	1–Grind	2–Rough Polish	3–Final Polish
Surface	SiC–320P	MD–Dur cloth	MD–Floc cloth
Abrasive	-	6 µm diamond suspension	OP-S Si-Colloids and H_2O_2 (5:1) solution
Lubricant	Water	DP Lubricant Red	-
Speed (rpm)	200 contra	150 contra	150 contra
Duration (min)	Until planar	15 min	15 min

2.2. Microstructural Studies

The microstructures of the materials were studied by scanning electron microscopy (SEM)—using a LEO1550 SEM (ZEISS group, Oberkochen, Germany) equipped with an INCA AZtec Energy Dispersive X-ray Spectroscopy (EDX) system (Oxford Instruments, High Wycombe, UK). Images were taken with an Everhart-Thornley (HE-SE2 detector, ZEISS group, Oberkochen, Germany) and backscattered electron detector (BSD). To ensure that the microstructure was discernible in the SEM, the samples were etched in Kroll's Etchant: 2 mL HF, 4 mL HNO_3 and 100 mL distilled water, with etching times of 30 s using cotton swabbing [33].

The energy dispersive x-ray spectroscopy (EDX) analysis of the 5Cu798 sample was performed on a Merlin SEM (ZEISS group, Oberkochen, Germany) equipped with an Ultim-Max 100 mm^2 Silicon Drift Detector (Oxford Instruments, High Wycombe, UK) at a voltage of 15 kV and 8.5 mm working distance.

2.3. Hardness Studies

The hardness of the Ti-Cu$_x$ alloys was measured using a Vickers Hardness tester (Duravision EMCO, Prufmaschinen GmbH, Kuchl, Austria) with a 9.8 centinewton load. The machine was calibrated before testing with a standard Vickers sample, and an average was taken from three indentations.

2.4. Phase Identification

All samples studied by X-ray diffraction were polished as specified by Vander Voort [32] (Table 2). X-ray diffraction (XRD) was performed on a D8 Advance TWIN-TWIN diffractometer (Bruker, AXS GmbH, Karlsruhe, Germany) with Cu Kα radiation (Kα_1 = 1.540598 Å with a Ni filter) and using the Bragg-Bretano experimental set-up. The detector system was a LynxEye XE PSD detector (Bruker, AXS GmbH, Karlsruhe, Germany). The EVA software suite (2015, Bruker) was used for phase analysis, while diffractogram plotting was done in Origin (2018b, OriginLab Corp., Northhampton, MA, USA). All crystallographic data were retrieved from the ICDD database PDF–4 + 2019 [34] and included PDF# 00-044-1294 (HCP-Ti), PDF# 00-015-0717 (Ti_2Cu) [35] and PDF# 00-055-0296 (Ti_3Cu).

2.5. Bacterial Luminescence by Direct Contact Test

The bacteria direct contact test has already been described [17], so only a summary is provided here. Staphylococcus epidermidis (XEN43) is a genetically modified bacterial strain that is bioluminescent due to the luxABCDE gene being bio-engineered into the bacterial genome [36,37]. Overnight bacterial inoculum of XEN43 was prepared a priori then seeded on the surface of sterile 5 mm diameter Ti-Cu_x alloys (both aged and un-aged) and allowed to attach. Tryptic soy broth (TSB, Fluke-Sigma Aldrich, Stockholm, Sweden) was then carefully added to the test wells. Periodic measurements of luminescence were recorded for the samples on a Hidex plate CHAMELEON V (425-106, Multilabel counter, Turku, Finland), and the mean was used to determine the antibacterial rate (R) after two- and six-hours of exposure, using Equation 2 [38]:

$$\mathrm{R} = \frac{N_{control} - N_{sample}}{N_{control}} \times 100\% \tag{1}$$

where $N_{control}$ = mean luminescence from the CPTi sample, and N_{sample} = mean luminescence for the Ti-Cu_x alloys [38].

2.6. Corrosion Testing

The corrosion testing was performed on all the samples at 37 ± 1 °C in a phosphate buffered saline (PBS) solution (containing 8 g/L sodium chloride, 0.2 g/L potassium chloride, 1.44 g/L sodium hydrogen phosphate, and 0.24 g/L potassium di-hydrogen phosphate) maintained at a pH of 7.4.

The sample preparation for the corrosion studies included mounting the samples (CPTi and Ti–Cu_x) of approximately 1 cm^2 in area, in non-conductive epoxy resin with a copper wire soldered to the samples for electrical connection. The samples were ground to 120 grit surface finishes using a SiC paper, then rinsed with de-mineralized water and degreased in acetone.

The electro-chemical testing was performed using a computer-driven Potentiostat (AutoTafel, ACM Instruments, Cark, Cumbria, England). Two graphite rods acted as counter-electrodes and a Haber-Luggin capillary made the junction with a saturated calomel reference electrode (SCE). All potential values were with respect to the SCE. Nitrogen was bubbled continuously during testing in order to remove all the oxygen and maintain an anaerobic condition.

After each sample was immersed in the solution, the open circuit potential against time curve was recorded for up to 4 h to determine the open-circuit potential (OCP). When the potential had reached a sufficiently stable value at four hours, a cyclic polarization scan was recorded from −250 mV to 1500 mV versus the corrosion potential at a scanning speed of 10 mV/min. The scan direction was not reversed.

2.7. Statistical Analysis

The statistical analysis for the bacterial luminescence measurements was done in Origin software (2018b, OriginLab Corp., Northhampton, MA, USA) at two and six hours, using a One-way ANOVA with a Tukey HSD Post-Hoc test, and with Levene's homogeneity of variance test. The same statistical test was also performed for the hardness with the exception that a Brown-Forsythe test for homogeneity of variance was performed. All tests had a statistical significance setting of $p = 0.05$.

3. Results

3.1. Microstructural Studies

Microstructures for the alloys were compared to determine any differences due to ageing. The microstructures of the alloys (both aged and un-aged) with less than 3 wt% Cu had a single-phase structure of α-Ti (the hcp solid solution of titanium) (Figure 1). However, the 1Cu798 showed peaks for the 2θ diffraction angle of the Ti_3Cu phase, at 20.9° and 23.4° (Figure 2). Those with equal to and greater than 3 wt% Cu were all two-phased: Ti_2Cu and α-Ti (Figure 1). X-ray diffraction studies confirmed these findings (Figure 2), where HCP-Ti and Ti_2Cu peaks were identified, but peaks for the Ti_3Cu crystals were also observed at the 2θ diffraction angles of 20.9° and 23.4° for the alloys: 1Cu798, 3Cu798, 5Cu798, and 3Cu400.

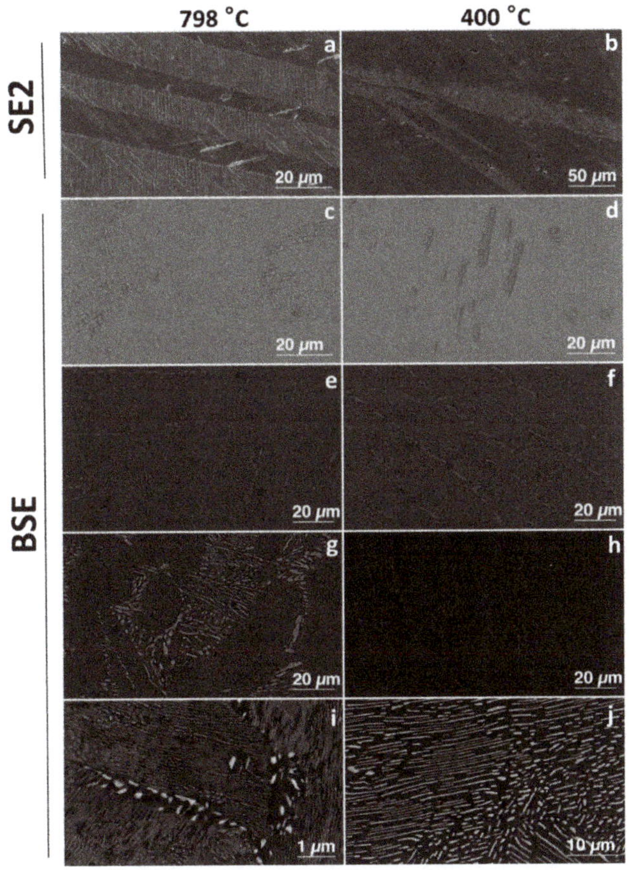

Figure 1. Secondary electron images of: (**a**) CPTi-798, (**b**) CPTi-400, and backscattered electron images of: (**c**) 1Cu798, (**d**) 1Cu400, (**e**) 3Cu798, (**f**) 3Cu400, (**g**) 5Cu798, (**h**) 5Cu400, (**i**) 10Cu798, and (**j**) 10Cu400.

Comparison of the 10 wt% Cu before and after ageing displayed that a larger volume fraction of Ti_2Cu precipitated due to this heat treatment. While an increase in Ti_2Cu with ageing was also found for the 5 and 3 wt% Cu, the volume fraction of precipitated Ti_2Cu was lower. In addition, these alloys had lamellar microstructures of Ti_2Cu and α-Ti (Figure 1).

3.2. EDX Study of Precipitates

Precipitates in the 5Cu798 sample were studied to determine the presence of variations in the Cu content for the individual phases in the alloy. Since it is known that the Ti_2Cu and Ti_3Cu crystals have an atomic % Cu concentration of 33% and 25%, respectively [29,30], point analysis of individual crystals was performed to ascertain which phase was present. For sample area 1 (Figure 3a), the atomic concentrations for precipitates 1 and 2 were 33.2 ± 1.1 and 34.1 ± 0.3, respectively (Table 3). For sample area 2 (Figure 3b), the atomic concentrations for precipitates 3 and 4 were 31.5 ± 2.6 and 24.6 ± 0.5, respectively (Table 3).

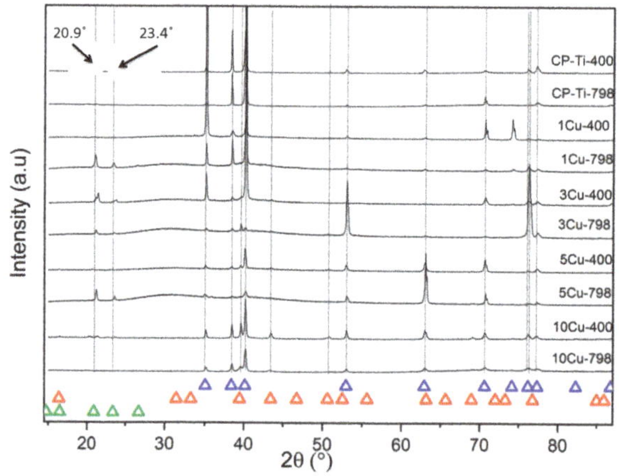

Figure 2. X-ray diffraction patterns of Ti-Cu_x alloys heat-treated at 798 °C, some aged at 400 °C: blue triangles indicate HCP-Ti, red triangles indicate Ti_2Cu and green triangles indicate Ti_3Cu. Note the peaks for Ti_3Cu at 20.9° and 23.4°.

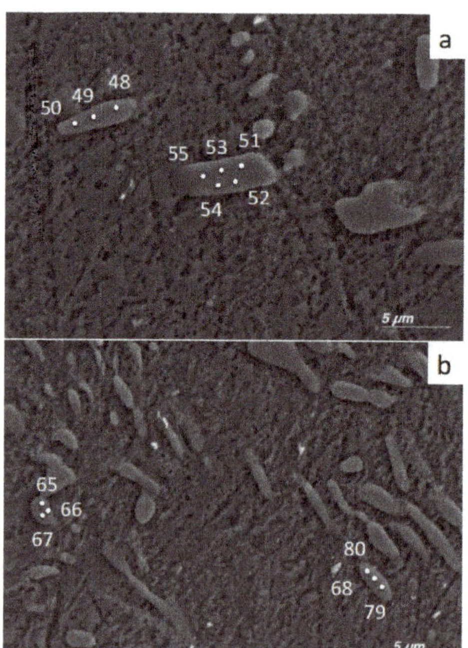

Figure 3. Positions of EDX spectral analyses of sample 5Cu798: (**a**) Sample area 1 with precipitate 1 (48, 49, 50) and precipitate 2 (51, 52, 53, 54, 55). (**b**) Sample area 2 with precipitate 3 (65, 66, 67) and precipitate 4 (68, 79, 80).

Table 3. EDX point spectral analyses of sample areas 1 and 2 (point spectral analysis correspond to points in Figure 3).

Precipitate (EDX Point Analyzed)	Atomic % Copper (Balance % Titanium)	Possible Phases
Precipitate 1 (48, 49, 50)	33.2 ± 1.1	Ti_2Cu
Precipitate 2 (51, 52, 53, 54, 55)	34.1 ± 0.3	Ti_2Cu
Precipitate 3 (65, 66, 67)	31.5 ± 2.6	Ti_2Cu
Precipitate 4 (68, 79, 80)	24.6 ± 0.5	Ti_3Cu

3.3. Bacterial Luminescence

At two hours, no significance in luminescence was found among the samples, except between CPTi-798 and the 1Cu400 and 3Cu400 samples, which had significantly higher values (Figure 4). However, the antibacterial rate of 10Cu400 (R = 12%) was higher than the other aged alloys at this time point. All other alloys had a negative R at this time point, indicating that more bacteria were found on these alloys than on the respective control.

For the un-aged alloys, it was found that after six-hours of exposure (Figure 5) the R for 10Cu798 (42%) was higher than that for 5Cu798 (7%). The alloys with lower Cu content still had negative R-values. The 10Cu798 alloy also had significantly lower luminescence than alloys 1Cu798, 3Cu798 and 1Cu400 ($p < 0.022$). At this time point, the 10Cu400 had lower luminescence than the samples CPTi-798, 1Cu798, 1Cu400, 3Cu798, and 3Cu400 ($p < 0.036$). Within the aged alloys, it was found that R for 10Cu400 (45%) was greater than that for the 5Cu400 (15%). As for the un-aged alloys, the 3Cu400 and 1Cu400 had negative R-values.

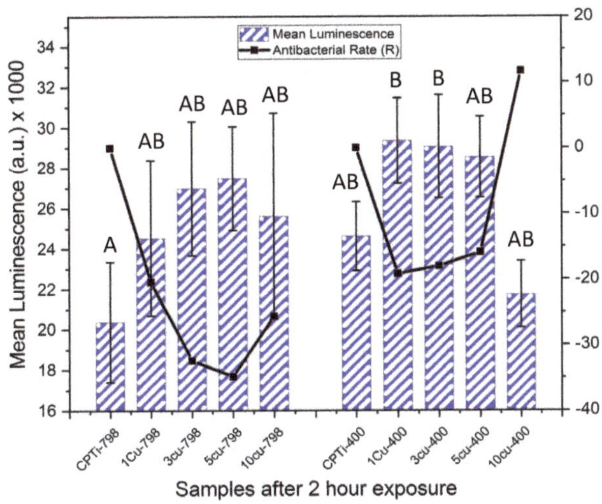

Figure 4. Mean luminescence counts (bars for standard deviation) of XEN 43 bacteria after two hours for all alloys, the negative controls were (CPTi), and the corresponding antibacterial rates are shown in black. Levene's Test: $F(9,20) = 1.83$, $p = 0.123$. One-way Anova: $F = 3.176$ ($p = 0.015$) with Tukey Test. Note: The same letters (e.g, "A" compared to "A") denotes being non-significantly different, while different letters (e.g. "A" compared to "B") denotes being statistically significantly different.

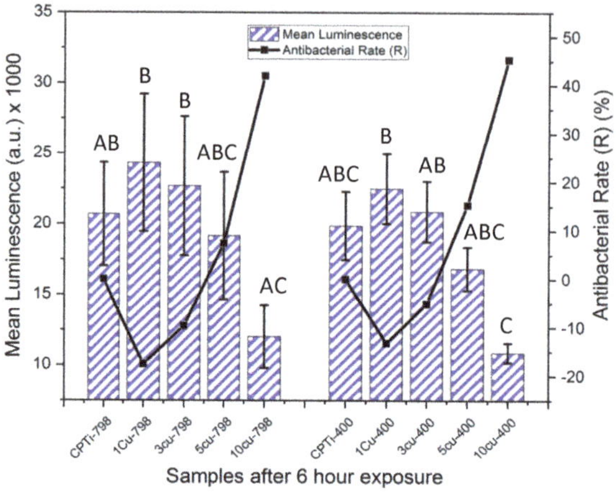

Figure 5. Mean luminescence counts (bars for standard deviation) of XEN 43 bacteria after six hours for all alloys, the negatives controls were CPTi, and the corresponding antibacterial rates are shown in black. Levene's Test: F (9,20) = 2.21, $p = 0.066$. One-way Anova: F = 5.697 ($p < 0.001$) with Tukey Test. Note: The same letters (e.g, "A" compared to "A") denotes being non-significantly different, while different letters (e.g. "A" compared to "B") denotes being statistically significantly different.

3.4. Hardness Tests

The effects of alloying Cu to Ti and ageing on hardness were investigated (Figure 6). The 10Cu798 alloy had the highest hardness (350 ± 12 Hv) of the alloys while the CPTi-400 alloy had the lowest (120 ± 3 H_V). The increased Cu addition to the CPTi correlated with a gradual augmentation in the Vicker's hardness, except for the 10Cu798 alloy with a hardness twice that of the 5Cu798 alloy (171 ± 7 H_V, $p < 0.0001$ in comparison with all other alloys). Ageing did not significantly affect the hardness of the alloys, except for 10Cu798, which was 1.9 times ($p < 0.0001$) harder than 10Cu400 (182 ± 1 H_V), and CPTi798 which had a 14% ($p < 0.021$) increase in H_V after ageing.

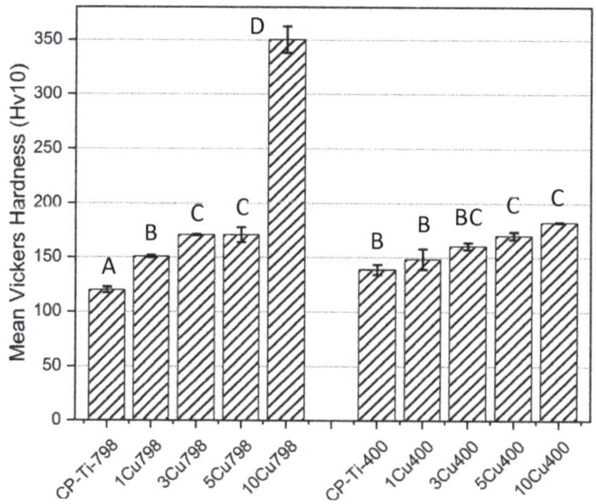

Figure 6. Mean Vickers hardness (bars for standard deviation) of all samples. Brown-Forsythe: F (9,20) = 1.413, $p = 0.2476$. One-way Anova: F = 368.92 ($p < 0.001$) with Tukey test. Note: The same letters (e.g, "A" compared to "A") denotes being non-significantly different, while different letters (e.g. "A" compared to "B") denotes being statistically significantly different.

3.5. Corrosion Tests

Corrosion studies were performed to determine the effect of the copper additions on the corrosion resistance, as well as the effect of ageing. The alloys were plotted in a single graph (Figure 7). The alloys aged at 400 °C showed a distinct corrosion profile after the anodic reaction took place. The aged alloys tended to passivate rapidly after the anodic reaction commenced. The un-aged alloys (quenched from 798 °C) clearly had a different corrosion profile trend, which indicated a gradual change in the passivation after the anodic reaction.

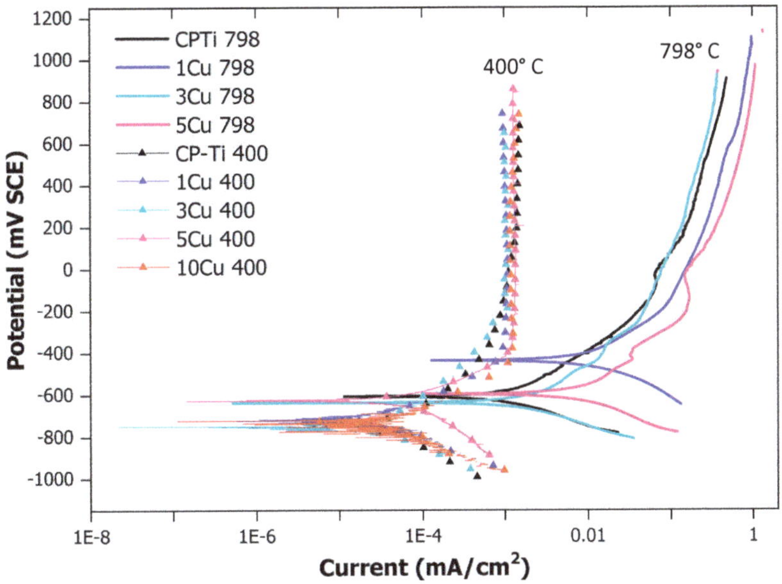

Figure 7. Corrosion plots for the aged (at 400 °C) and un-aged (quenched from 798 °C) alloys. The aged alloys are indicated with triangles to indicate the profile. Note: the 10Cu798 alloy was not included in the corrosion study.

4. Discussion

The present investigation focused on alloying up to 10 wt% of Cu into commercially pure titanium (a well-known implant material), where Cu has shown antibacterial potential in previous studies [17,20,38,39]. Its purpose was to determine the resultant microstructural and antibacterial characteristics of the materials, as a function of Cu addition and ageing heat treatments.

Microstructural and X-ray diffraction observations indicated no β-Ti in any of the alloys, despite a rapid quench (Figure 2). The reasons for this could be the active eutectoid transformation kinetics that drive transformation from $\beta - \text{Ti} \rightarrow \alpha - \text{Ti} + \text{Ti}_2\text{Cu}$ [40]. The two phases of α-Ti (HCP-Ti) and Ti_2Cu were observed as predicted in phase calculations [17]. The CPTi and 1Cu–except the 1Cu798 alloy that could have Ti_3Cu present–alloys were single-phase α-Ti alloys (Figures 1 and 2), while in alloys with higher Cu content (aged and un-aged) both α-Ti and Ti_2Cu were found. The ageing had a major effect on the microstructure and the Ti_2Cu precipitates had a larger volume fraction in the aged alloys. The ageing of the 10Cu798 alloy coarsened the Ti_2Cu from small precipitates to larger lamellar microstructures (Figure 1). A similar change occurred with ageing in the other two-phased alloys (5Cu798 and 3Cu798), but with coarsening of uniform lamellae of Ti_2Cu in the α-Ti phase.

α-Ti and Ti_2Cu were observed by X-ray diffraction, in addition to Ti_3Cu at 2θ angles of 20.9° and 23.4° for the 1Cu798, 3Cu798, 3Cu400, and 5Cu798 alloys. EDX measurements on a 5Cu798 sample confirmed that some of the precipitates had the expected 33 at% Cu for Ti_2Cu [35], while at least one precipitate had compositions closer to 25 at% Cu, which is the expected composition for Ti_3Cu [30]. However, the results might have been affected by analyzing small areas, which resulted in collecting the signal from the surrounding phases. Although this study did not make use of a standard reference material for the analysis, the results gave insight into the variation of composition for the precipitates with Cu in the Ti-Cu_x alloys. It should however be noted that the precipitates with compositions closer to Ti_3Cu were in lower volume fractions—for the 5Cu798 alloy—relative to the precipitates with compositions close to 33 at% Cu. This lower volume fraction could be the reason for the absence of Ti_3Cu by X-ray diffraction in the 10 wt% Cu alloys, since it could have been below the detection limit for the technique. However, this was only a preliminary study on one alloy and further studies are needed to confirm the general presence of Ti_3Cu in Ti-Cu_x alloys.

As well as the microstructure, the hardness was also affected by Cu additions and ageing, and the 10Cu798 was twice as hard as the 10Cu400 alloy. The mechanism causing this could be the coarser lamellar structure in the 10Cu400 [41]. Ageing had a hardening effect on the CPTi-798 alloy as well, because the hardness was significantly higher for the CPTi-400 alloy.

The bacteria test gave an interesting result after two hours of exposure for the alloys, and 10Cu400 had an antibacterial rate R of 11%, while all the other Ti-Cu_x alloys had higher mean luminescence than CPTi. The reason for the higher luminescence could be that the bacteria underwent a stress response, which caused a rapid increase in population [23]. Comparison of the aged alloys at two hours and at six hours gave a similar trend for antibacterial rate (R). The 10Cu400 alloys seemed to be effectual at bacteria reduction already from two hours and thereafter increased, and at six hours the R was 46%. The 5Cu798 and 5Cu400 alloys after six hours had R values of 7% and 15%, respectively, although this was not significantly different from alloys with lower Cu content. The greater R at six hours for both aged (45%) and un-aged (42%) 10 wt% Cu alloys agreed with previous findings where a higher Cu content has been reported to yield a stronger antibacterial material [17,38]. The difference in R for the 10 wt% Cu alloys at two hours could be due to the larger amounts of Ti_2Cu in the aged alloy leading to higher Cu ion release rates as has been observed in studies on Ti-6Al-4V-5Cu alloys [28]. However, further studies on ion release from Ti-Cu_x are needed to support this conclusion. Additionally, direct contact of bacteria with a Cu rich surface—such as the 10Cu400 with higher volume fraction Ti_2Cu—could also have played a role in the antibacterial effect [24]. Thus, Cu ions and direct contact killing contributed to the antibacterial effect, and future studies could investigate how the amount and size of the Ti_2Cu phase influences copper ion release and the contact killing of bacteria.

The corrosion results indicated that alloying with Cu and ageing affected the corrosion rates and passivation rates. Alloying and ageing induced faster passivation for higher Cu contents, and decreased the corrosion rates. These results are in agreement with the findings of Zhang et al. [38] for an aged 4 wt% Cu alloy.

A limitation to the present study is that only alloys in the range of 0 to 10 wt% Cu were investigated. It has been demonstrated that a higher Cu content yields a better antibacterial response [17], however, an excess of copper can be harmful to the human body [42]. Previous studies [15,23] have reported that the Cu ions released from alloys with up to 10 wt% Cu did not have any notable toxic effects and were therefore the focus of the present study. Moreover, only two heat treatments were studied due to a necessary delimitation of the study. There are other heat treatments that could be tested, and future studies should further optimize the alloying process. The investigation on the presence of Ti_3Cu was a preliminary study and the results should not be viewed as conclusive due to only one alloy being characterized, the low number of precipitates characterized and the low resolution of the EDX technique. Another limitation is that the antibacterial ability of the alloys was only tested with one type of bacteria. However, the bacterium used, Staphylococcus epidermidis, is one of the most prevalent bacteria in implant infections and its use additionally allowed comparison with previous studies [23].

5. Conclusions

Both Ti_2Cu and Ti_3Cu were detected in Cu alloys using X-ray diffraction. A preliminary EDX study on the un-aged 5 wt% Cu alloy also indicated the presence of Ti_3Cu, however with the Ti_2Cu in majority. While the study of the Ti_3Cu was not quantitative in this investigation, it is recommended that future works focus on determination of this phase.

The hardness was reduced after ageing of the 10Cu798 alloy, as the microstructure changed from small precipitates to coarser lamellar microstructures. The alloys had good antibacterial rates above 5 wt% Cu (aged and un-aged) after six-hours of exposure, and the two-hour exposure was probably too short a duration for bacterial reduction to occur in most of the alloys. Ageing the alloys only ensured faster antibacterial rates after 2 hours, when the concentration of Cu was greater than 10 wt% Cu. The addition of Cu to Ti and ageing increased the corrosion resistance of the alloys, which could protect the biomaterial in vivo. However, since the ageing also lowered the hardness of the higher Cu content alloys, care should be taken to avoid over-ageing, and hence softening.

In conclusion, the un-aged 10 wt% Cu alloy was considered a suitable candidate material, which provided a good antibacterial effect, with superior hardness and corrosion protection. The ideal composition and heat treatment of these materials will however depend on the specific application envisioned, and will require further optimization.

Author Contributions: Conceptualization, H.E., S.N. and C.Ö.-M; Formal analysis, L.F., N.M., S.N. and C.Ö.-M; Funding acquisition, L.A.C., L.H.C., S.N. and C.Ö.-M; Investigation, L.F. and N.M.; Methodology, S.N. and C.Ö.-M; Project Administration, L.A.C., L.H.C., H.E., S.N. and C.Ö.-M; Resources, L.A.C., H.E., S.N. and C.Ö.-M; Supervision, L.A.C., L.H.C., H.E., S.N. and C.Ö.-M; Validation, L.F. and N.M.; Visualization, L.F. and N.M.; Writing-original draft, L.F.; Writing-review & editing, L.F., N.M., L.A.C., L.H.C., H.E., S.N. and C.Ö.-M.

Funding: This research was funded by the South African-Sweden Bilateral Scientific Research Cooperation program funded by the Swedish Foundation for International Cooperation in Research and Higher Education (STINT, GA SA2017-7127) and the South African National Research Foundation (NRF funding through STNT170905261815).

Acknowledgments: The authors thank Victoria Sternhagen for help with SEM imaging.

Conflicts of Interest: The authors declare no conflict of interest.

References

1. Alani, A.; Kelleher, M.; Bishop, K. Peri-implantitis. Part 1: Scope of the problem. *Br. Dent. J.* **2014**, *217*, 281–287. [CrossRef] [PubMed]
2. Atieh, M.A.; Alsabeeha, N.H.M.; Faggion, C.M.; Duncan, W.J. The frequency of peri-implant diseases: A systematic review and meta-analysis. *J. Periodontol.* **2013**, *84*, 1586–1598. [CrossRef] [PubMed]
3. Fransson, C.; Lekholm, U.; Jemt, T.; Berglundh, T. Prevalence of subjects with progressive bone loss at implants. *Clin. Oral Impl. Res.* **2005**, *16*, 440–446. [CrossRef] [PubMed]
4. Norowski, P.A.; Bumgardner, J.D. Biomaterial and antibiotic strategies for peri-implantitis: A review. *J. Biomed. Mater. Res.* **2009**, *88B*, 530–543. [CrossRef]
5. Hultin, M.; Gustafson, A.; Hallström, H.; Johansson, L.-Å.; Ekfeldt, A.; Klinge, B. Microbiological findings and host response in patients with peri-implantitis. *Clin. Oral Impl. Res.* **2002**, *13*, 349–358. [CrossRef] [PubMed]
6. Ladomersky, E.; Petris, M.J. Copper tolerance and virulence in bacteria. *Metallomics* **2015**, *7*, 957–964. [CrossRef]
7. Hawkey, P.M. The growing burden of antimicrobial resistance. *J. Antimicrob. Chemother.* **2008**, *62*, i1–i9. [CrossRef]
8. Arciola, C.R.; Campoccia, D.; Speziale, P.; Montanaro, L.; Costerton, J.W. Biofilm formation in Staphylococcus implant infections. A review of molecular mechanisms and implications for biofilm-resistant materials. *Biomaterials* **2012**, *33*, 5967–5982. [CrossRef]
9. Stewart, P.S. Mechanisms of antibiotic resistance in bacterial biofilms. *Int. J. Med. Microbiol.* **2002**, *292*, 107–113. [CrossRef]

10. Vasudevan, S.; Joseph, H.A.; Swamy, S.S.; Solomon, A.P. *Antibiotic Resistance in Biofilms*; ACS Symposium Series; Rathinam, N.K., Sani, R.K., Eds.; American Chemical Society: Washington, DC, USA, 2019; pp. 205–224. [CrossRef]
11. Zhang, Y.-Q.; Ren, S.-X.; Li, H.-L.; Wang, Y.-X.; Fu, G.; Yang, J.; Qin, Z.-Q.; Miao, Y.-G.; Wang, W.-Y.; Chen, R.-S.; et al. Genome-based analysis of virulence genes in a non-biofilm-forming Staphylococcus epidermidis strain (ATCC 12228). *Mol. Microbiol.* **2003**, *49*, 1577–1593. [CrossRef]
12. Spriano, S.; Bosetti, M.; Bronzoni, M.; Vernè, E.; Maina, G.; Bergo, V.; Cannas, M. Surface properties and cell response of low metal ion release Ti-6Al-7Nb alloy after multi-step chemical and thermal treatments. *Biomaterials* **2005**, *26*, 1219–1229. [CrossRef]
13. Unosson, E.; Tsekoura, E.K.; Engqvist, H.; Welch, K. Synergetic inactivation of Staphylococcus epidermidis and Streptococcus mutans in a TiO2/H2O2/UV system. *Biomatterials* **2013**, *3*, e26727. [CrossRef]
14. Unosson, E.; Morgenstern, M.; Engqvist, H.; Welch, K. In vitro antibacterial properties and UV induced response from Staphylococcus epidermidis on Ag/Ti oxide thin films. *J. Mater. Sci. Mater. Med.* **2016**, *27*, 49. [CrossRef]
15. Bai, B.; Zhang, E.; Dong, H.; Liu, J. Biocompatibility of antibacterial Ti–Cu sintered alloy: In vivo bone response. *J. Mater. Sci. Mater. Med.* **2015**, *26*, 265. [CrossRef]
16. Zhang, E.; Li, F.; Wang, H.; Liu, J.; Wang, C.; Li, M.; Yang, K. A new antibacterial titanium–copper sintered alloy: Preparation and antibacterial property. *Mater. Sci. Eng. C.* **2013**, *33*, 4280–4287. [CrossRef]
17. Fowler, L.; Janson, O.; Engqvist, H.; Norgren, S.; Öhman-Mägi, C. Antibacterial investigation of titanium-copper alloys using luminescent Staphylococcus epidermidis in a direct contact test. *Mater. Sci. Eng. C.* **2019**, *97*, 707–714. [CrossRef]
18. Codiţă, I.; Caplan, D.M.; Drăgulescu, E.-C.; Lixandru, B.-E.; Coldea, I.L.; Dragomirescu, C.C.; Surdu-Bob, C.; Bădulescu, M. Antimicrobial activity of copper and silver nanofilms on nosocomial bacterial species. *Roum. Arch. Microbiol. Immunol.* **2010**, *69*, 204–212.
19. Russel, A.D.; Hugo, W.B. 7 Antimicrobial activity and action of silver. *Progr. Med. Chem.* **1994**, *31*, 351–370. [CrossRef]
20. Michels, H.T.; Noyce, J.O.; Keevil, C.W. Effects of temperature and humidity on the efficacy of methicillin-resistant Staphylococcus aureus challenged antimicrobial materials containing silver and copper. *Lett. Appl. Microbiol.* **2009**, *49*, 191–195. [CrossRef]
21. Stranak, V.; Wulff, H.; Rebl, H.; Zietz, C.; Arndt, K.; Bogdanowicz, R.; Nebe, B.; Bader, R.; Podbielski, A.; Hubicka, Z.; et al. Deposition of thin titanium–copper films with antimicrobial effect by advanced magnetron sputtering methods. *Mater. Sci. Eng. C* **2011**, *31*, 1512–1519. [CrossRef]
22. Jäger, E.; Schmidt, J.; Pfuch, A.; Spange, S.; Beier, O.; Jäger, N.; Jantschner, O.; Daniel, R.; Mitterer, C. Antibacterial silicon oxide thin films doped with zinc and copper grown by atmospheric pressure plasma chemical vapor deposition. *Nanomaterials* **2019**, *9*, 255. [CrossRef]
23. Fowler, L.; Engqvist, H.; Öhman-Mägi, C. Effect of Cu ion concentration on bacteria and cells. *Materials* **2019**, *12*, 3798. [CrossRef]
24. Mathews, S.; Hans, M.; Mücklich, F.; Solioz, M. Contact killing of bacteria on copper is suppressed if bacterial-metal contact is prevented and is induced on iron by copper ions. *Appl. Environ. Microbiol.* **2013**, *79*, 2605–2611. [CrossRef]
25. Ke, Z.; Yi, C.; Zhang, L.; He, Z.; Tan, J.; Jiang, Y. Characterization of a new Ti-13Nb-13Zr-10Cu alloy with enhanced antibacterial activity for biomedical applications. *Mater. Lett.* **2019**, *253*, 335–338. [CrossRef]
26. Liu, C.; Zhang, E. Biocorrosion properties of antibacterial Ti–10Cu sintered alloy in several simulated biological solutions. *J. Mater. Sci. Mater. Med.* **2015**, *26*, 142. [CrossRef]
27. Zhang, E.-L.; Fu, S.; Wang, R.-X.; Li, H.-X.; Liu, Y.; Ma, Z.-Q.; Liu, G.-K.; Zhu, C.-S.; Qin, G.-W.; Chen, D.-F. Role of Cu element in biomedical metal alloy design. *Rare Met.* **2019**, *38*, 476–494. [CrossRef]
28. Li, M.; Ma, Z.; Zhu, Y.; Xia, H.; Yao, M.; Chu, X.; Wang, X.; Yang, K.; Yang, M.; Zhang, Y.; et al. Toward a molecular understanding of the antibacterial mechanism of copper-bearing titanium alloys against staphylococcus aureus. *Adv. Healthc. Mater.* **2016**, *5*, 557–566. [CrossRef]
29. Kumar, K.C.H.; Ansara, I.; Wollants, P.; Delaey, L. Thermodynamic optimisation of the Cu-Ti system. *Zeitschrift Für Metallkunde* **1996**, *87*, 666–672.
30. Canale, P.; Servant, C. Thermodynamic assessment of the Cu–Ti system taking into account the new stable phase CuTi3. *MEKU* **2002**, *93*, 273–276. [CrossRef]

31. Appel, F.; Oehring, M. γ-titanium aluminide alloys: Alloy design and properties. In *Titanium and Titanium Alloys*, 1st ed.; Leyens, C., Peters, M., Eds.; Wiley-VCH Verlag GmbH & Co. KGaA: Weinheim, Germany, 2003; pp. 89–152. [CrossRef]
32. Voort, G.V. Materials characterization & testing: Microstructure of titanium and its alloys. *Ind. Heat.* **2006**, *73*, 82–84.
33. Nadai, L.; Katona, B.; Terdik, A.; Bognar, E. Chemical etching of titanium samples. *Periodica Polytechnica Mech. Eng.* **2013**, *57*, 57. [CrossRef]
34. PDF-4+ 2019, ICDD. (n.d.). Available online: http://www.icdd.com/index.php/pdf-4/ (accessed on 10 May 2019).
35. Mueller, M.H.; Knott, H.W. The crystal structures of Ti2 Cu, Ti2 Ni, Ti4 Ni2 O, and Ti4 Cu2 O. *Trans. Metallurg. Soc. AIME* **1963**, *227*, 674–678.
36. Vuong, C.; Kocianova, S.; Yu, J.; Kadurugamuwa, J.L.; Otto, M. Development of real-time in vivo imaging of device-related *staphylococcus epidermidis* infection in mice and influence of animal immune status on susceptibility to infection. *J. Infect. Dis.* **2008**, *198*, 258–261. [CrossRef]
37. Wiedmer, D.; Petersen, F.C.; Lönn-Stensrud, J.; Tiainen, H. Antibacterial effect of hydrogen peroxide-titanium dioxide suspensions in the decontamination of rough titanium surfaces. *Biofouling* **2017**, *33*, 451–459. [CrossRef]
38. Zhang, E.; Ren, J.; Li, S.; Yang, L.; Qin, G. Optimization of mechanical properties, biocorrosion properties and antibacterial properties of as-cast Ti–Cu alloys. *Biomed. Mater.* **2016**, *11*, 065001. [CrossRef]
39. Bai, B.; Zhang, E.; Liu, J.; Zhu, J. The anti-bacterial activity of titanium-copper sintered alloy against Porphyromonas gingivalis in vitro. *Dent. Mater. J.* **2016**, *35*, 659–667. [CrossRef]
40. Donthula, H.; Vishwanadh, B.; Alam, T.; Borkar, T.; Contieri, R.J.; Caram, R.; Banerjee, R.; Tewari, R.; Dey, G.K.; Banerjee, S. Morphological evolution of transformation products and eutectoid transformation(s) in a hyper-eutectoid Ti-12 at% Cu alloy. *Acta Mater.* **2019**, *168*, 63–75. [CrossRef]
41. Soffa, W.A.; Laughlin, D.E. High-strength age hardening copper–titanium alloys: Redivivus. *Progr. Mater. Sci.* **2004**, *49*, 347–366. [CrossRef]
42. Bull, P.C.; Thomas, G.R.; Rommens, J.M.; Forbes, J.R.; Cox, D.W. The Wilson disease gene is a putative copper transporting P–type ATPase similar to the Menkes gene. *Nat. Genet.* **1993**, *5*, 327–337. [CrossRef]

materials

MDPI

Article

Formation and Properties of Biomedical Ti-Ta Foams Prepared from Nanoprecursors by Thermal Dealloying Process

Grzegorz Adamek [1,*], Mikolaj Kozlowski [1], Mieczyslawa U. Jurczyk [2], Przemyslaw Wirstlein [3], Jakub Zurawski [4] and Jaroslaw Jakubowicz [1]

1 Institute of Materials Science and Engineering, Poznan University of Technology, Jana Pawla II 24, 61-138 Poznan, Poland

2 Division Mother's and Child's Health, Poznan University of Medical Sciences, Polna 33, 60-535 Poznan, Poland

3 Department of Gynecology and Obstetrics, Division of Reproduction, Poznan University of Medical Sciences, Polna 33, 60-535 Poznan, Poland

4 Department of Immunobiochemistry, Chair of Biology and Environmental Sciences, Poznan University of Medical Sciences, Rokietnicka 8, 60-806 Poznan, Poland

* Correspondence: grzegorz.adamek@put.poznan.pl; Tel.: +48-61-665-3665

Received: 26 July 2019; Accepted: 19 August 2019; Published: 22 August 2019

Abstract: The paper presents a promising method of preparation of titanium-based foams by the thermal dealloying method. The first step of this study was the Ti-Ta-Mg based nanopowder preparation using the mechanical alloying (MA) process performed at room temperature. The next step was forming the green compacts by cold pressing and then sintering with magnesium dealloying from the titanium-based alloy structure. The mechanism of the porous structure formation was based on the removal of magnesium from the titanium alloy at a temperature higher than the boiling point of magnesium (1090 °C). The influence of the Mg content on the formation of the porous Ti-30Ta foam has been investigated. The sintering stage was performed in vacuum. During the dealloying process, the magnesium atoms diffuse from the middle to the surface of the sample and combine to form vapors and then evaporate leaving pores surrounded by the metallic scaffold. The porosity, the mechanical properties as well as biocompatibility have been investigated. The titanium-based foam of high porosity (up to 76%) and the pore size distribution from nano- to micro-scale have been successfully prepared. For the medical applications, the Ti-Ta metallic foams have shown a positive behavior in the MTT test. The as-shown results clearly exhibit a great potential for thermal dealloying in the preparation of porous structures.

Keywords: titanium-based foams; mechanical alloying; thermal dealloying

1. Introduction

Titanium based materials such as alloys, composites or even pure Ti are the most popular materials for hard tissue implants because of their excellent corrosion resistance, good mechanical properties and superior biocompatibility [1]. Titanium appears in two allotropic forms: Low temperature α and high temperature (above 882 °C) β with hexagonal and cubic crystal structures, respectively. By careful chemical composition and the elimination of toxic elements, it is possible to prepare alloys with excellent biocompatibility and mechanical properties. Alloys with the β structure are the preferred biomaterials because of a much lower Young's modulus and better corrosion resistance compared to the commercially pure titanium or the most popular Ti-6Al-4V alloy.

Alloys with additions of Zr, Nb and Ta show good biocompatibility and these elements belong to the vital group in the tissue reaction (they belong to the non-toxic and non-allergic group of elements) [2–6]. The titanium alloys composed of such elements have been intensely developed for more than a decade [2–9]. There are many papers reporting on binary Ti-Ta [10–16] alloys. Tantalum has excellent corrosion resistance and appears to be a good alloying element to the titanium alloys. Ti-Ta binary alloys have recently attracted interest due to a good combination of high strength and low modulus [10]. For example, Ti-30Ta alloy has an elastic modulus of 77 GPa and the tensile strength of 713MPa [10].

One of the new methods of preparation of porous metals developed by our group is thermal dealloying of Mg. In the previous paper, we described the formation of a metallic foam by removing Mg during the sintering process and the development of a scaffold based on the tantalum-based alloys and composites [16,17]. In the current paper, we focus on the preparation and characteristics of Ti-Ta β alloys with a high content of magnesium. The alloys developed in this work will be used as precursors for the preparation of metallic foams/scaffolds. Such structures are characterized by high porosity of up to 70%. In order to achieve that in the process of thermal dealloying of magnesium, a relatively high Mg content in the precursor alloy is necessary. In this study, the authors prepared ternary alloys based on Ti-30Ta with 30–50 wt% of magnesium.

2. Materials and Methods

2.1. Materials Preparation

The nanostructured ternary alloys were prepared by mechanical alloying (MA) in an Spex SPEX 8000 Mixer Mill filled with argon gas. The ball to powder weight ratio (B/P) was 20:1. The composition of alloys was as follows: Ti-30Ta-(30 ÷ 50)Mg. The initial powders were: Ti – 325 mesh, purity 99.5%, Ta – 100 mesh, purity 99.98%, and Mg – 325 mesh, purity 99.8% (all from AlfaAesar). The process batch yield was defined as the ratio of the weight of the recovered powder and the weight of the starting powder measured using a precision balance (0.001 g repeatability). These measurements were performed in the glove box (LabMaster 130) filled with automatically controlled argon atmosphere ($O_2 < 2$ ppm and $H_2O < 1$ ppm).

The consolidation step was as follows: The powder after MA was portioned, placed into a steel die and uniaxially pressed at a pressure of 1000 MPa. The green compacts were 10 mm in diameter and 5–6 mm in height. The sintering/dealloying step was carried out in the tube furnace (Nabertherm, Lilentahl, Germany) at 1300 °C for 10.8 ks, in 10^{-2} Pa vacuum to remove the magnesium vapors from the material and prevent excessive oxidation. Then, the sinters were slowly chilled to the room temperature together with the furnace.

2.2. Phase and Structural Analysis

The phase constitution and the crystallographic structure of the Ti-based powders were analyzed by the X-ray powder diffraction (XRD) with Cu Kα radiation (Panalytical, Empyrean model, Almelo, Netherlands). The conditions of the XRD measurements were as follows: voltage 45 kV, anode current 40 mA, 2theta range 30°–90°, time per step 60.214 s, step size 0.0245°, $\lambda = 1.54$ Å.

The XRD data were used to calculate the average crystallite size and the lattice strain using the Williamson-Hall plot and the Panalytical HighScore 3.0a (Panalytical, Almelo, Netherlands) software. All computations, based on the XRD data, were made after subtracting the background and the peak fittings. The refinement of the diffraction data was performed using the Rietveld refinement technique with the Maud 2.91 software (by Luca Luttorotti, Univ. of Trento, Italy). The instrumental peak broadening was evaluated using an Si reference sample. The Marquardt Least Squares based on the Levenberg-Marquardt algorithm was used in order to minimize the difference between the experimental and the calculated patterns. The refinement analyses were carried out using space groups and crystallographic information from the Crystallography Open Database (β-Ti: COD 9012924).

2.3. Microstructure and Porosity Analysis

A scanning electron microscope (SEM, Mira 3, Tescan, Brno, Czech Republic) was applied for the investigations of the morphology and the porosity of the sinters.

The microstructure of the obtained materials was studied by the CM 20 Super Twin (Philips, Amsterdam, Netherlands) transmission electron microscope (TEM) microscope providing 0.24 nm resolution at an acceleration voltage of 200 kV. The powders for the TEM observation were ultrasonically distributed in ethanol in the time of 60 s and one drop of such a suspension was transferred onto the EMS support film on the grids (square mesh standard thickness FF100–Cu 100 mesh).

2.4. Measurements of Mechanical Properties

The compressive strength and elastic moduli were measured using a 4483 Instron mechanical testing machine of the measuring range of up to 20 kN with a constant crosshead speed (the average strain rate of 0.001 s^{-1}). The mechanical properties were measured on eight samples of each series. The Vickers hardness (HV) and the indentation modulus (E_{IT}) were also measured at loading using the Picodentor HM500 (Fischer, Sindelfingen, Germany) nanoindentation tester. The indentations were made using a diamond indenter at the force of 300 mN for 20 s.

2.5. MTT Assay

Normal human osteoblast (CC-2538) (NHost) and human periodontal ligament fibroblasts (CC-7049) (HPLF) (Lonza Group Ltd., Basel, Switzerland) were used for the in vitro cytocompatibility tests. The tests were obtained under static conditions. The samples were sterilized by autoclaving at 120 °C for 15 min and separately located in 24-well microplates. The cells were cultured at a concentration of 5000 cells/well in 1 mL of culture medium on each sample at 37 °C in a 5% CO_2 incubator for 1, 3 and 4 days. The proliferation of the cells in the conditioned mediums was expressed as a percentage of the value of relative viability of the cells (RVC) of the reference medium. The pure bulk microcrystalline titanium samples were used for preparation of the reference medium.

The statistical significance was analyzed using Kruskal-Wallis One Way Analysis of Variance on Ranks with multiply repetition option SigmaStat 3.5 (Systat Software Inc., Erkrath, Germany) with U-Mann Whitney test. The significance level was p-value < 0.05.

3. Results and Discussion

Mechanical alloying is a ball-milling technique used in the preparation of non-equilibrium nanomaterials (alloys and composites). In this work, the authors show the possibility of developing the β-Ti alloys containing more than 30 wt% of magnesium, which is a very high value. In all probability, it was possible to achieve using a high B/P ratio (high milling energy).

In this work, the preparation by mechanical alloying was shown of the β-Ti alloys, containing 30, 40 and 50 wt% of magnesium. The composition of alloys was as follows: Ti-30Ta-(30 ÷ 50)Mg wt%. The changes in the phase constitution during MA were observed using XRD and the morphology of the powders using SEM. Figure 1 shows the XRD data for the Ti-30Ta-30Mg alloy after milling. After 5 h of milling, there were well visible peaks of Ti, Ta and still visible peaks of Mg on the XRD spectra. Such a short time allowed us to achieve just a mixture of the initial powders. An increase in the milling time resulted in a decreased intensity of the XRD peaks of the initial powders and the appearance of the new XRD peaks corresponding to β-Ti. After 40 h of MA only the β-Ti peaks were observed. During the MA process, a phase transformation was observed: the peaks corresponding to the beta stabilized element (Ta) disappeared on the XRD spectra after a long milling time and the intensity of the peaks corresponding to Mg and α-Ti decreased. However, after approx. 20 h, the peaks corresponding to the β-Ti structure became visible. The intensity of these peaks increased and between the 25th and 30th hour, the intensities of the α-Ti and β-Ti peaks (between the 2 theta 35th–40th degree) became equal. After 40 h of milling for the alloy containing 30% of Mg, there were only peaks corresponding to β-Ti.

Figure 1. X-ray powder diffraction (XRD) spectra of the Ti-30Ta-30Mg and scanning electron microscope (SEM) micrographs of the powders at different milling times.

When the content of Mg increased, an additional phase was visible (Figure 2). The XRD spectra for the alloy with 40 wt% and 50 wt% of Mg has been shown in Figure 2. Mechanical alloying with such a high content of Mg resulted in the appearance of an extra phase (Ti-Mg with hcp structure).

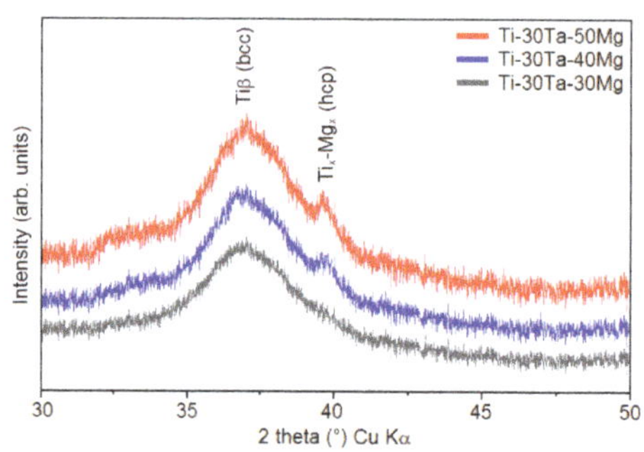

Figure 2. XRD spectra of the Ti-30Ta-(30-50)Mg alloys after 40 h of MA process.

The SEM micrographs of powders at the subsequent MA stages (Figure 1) show that all of the powders are characterized by fractured and porous morphology and are composed of irregular particles/agglomerates typical of mechanically alloyed materials. After a long milling time, we can observe the following trend: with the increasing milling time, the size of the particles decreases from over 200 μm to approx. 50 μm after 5 h and 40 h of milling, respectively, which is natural to observe during MA.

One of the advantages of using Mg as an alloying element to Ti is very good yield characteristics. Zadra showed that using a small amount (0.5%) of Mg results in approx. 65% yield after 2 h of milling [18]. If we compare the results at hand with the authors' previous work [19], it could be concluded that the yield increases with the increasing Mg content. In this work, the authors use a relatively high content of Mg. Such alloys are characterized by outstanding yield results even after a relatively long milling time. After 40 h of the MA process, more than 94% of the powder yield was achieved for all chemical compositions.

The microstructure was investigated by TEM. Using the mechanical alloying process, we can obtain amorphous, nanocrystalline and ultra-fine-grained materials. Figure 3 shows the example TEM micrograph and the grain size distribution of the Ti-30Mo-30Mg alloys. The results for all alloys together with the structural analysis data have been shown in Table 1. The applied algorithm achieves high goodness of fit (GOD).

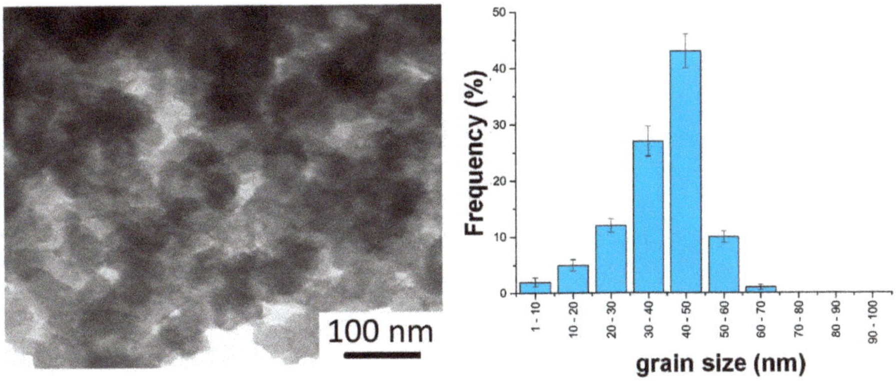

Figure 3. Example of TEM micrograph and the grain size distribution of the Ti-30Ta-30Mg alloy.

Table 1. Crystallographic data of the powders after mechanical alloying.

Alloy	D [nm]	a [Å]	V [Å^3]	Q [%]	$Q_{Ti\text{-}Mg\ hcp}$ [%]	R_{wp} [%]	R_{exp} [%]	GOD
Ti-30Ta + 30Mg	44 ± 8	3.3556	37.78	100	-	4.81	2.85	1.45
Ti-30Ta + 40Mg	53 ± 6	3.3659	38.13	100	-	5.66	3.23	1.70
Ti-30Ta + 50Mg	87 ± 12	3.3778	38.54	95.92	4.08	6.58	3.49	1.98

D—crystallite size, a—lattice parameter, V—lattice volume, Q—amount of phase, R_{wp}—weighted pattern residual indicator, R_{exp}—expected residual indicator, GOD—goodness of fit.

In all the prepared alloys, we can observe the nanocrystalline grains. On the TEM micrographs, homogenous round-shaped nanograins are visible. Alloys containing 50% of Mg are made of crystals of larger diameter (approx. 70–90 nm) than those with 30% of Mg (approx. 30–50 nm). The fracturing could be more difficult with a higher content of Mg as a result of smaller accumulation of the lattice defects and the newly obtained Ti-Mg phase. It is obvious that the chemical composition influences the lattice parameters. An increase in the content of Mg clearly increases the cell volume, which also depends on the type of other alloying elements but in a relatively less significant way. The TEM analyses were compared with the Williamson-Hall method. The main reasons for the incompatibility could be the incompatible profile fit after the refinement of the XRD data. The most reliable technique is TEM. The results are similar for all the used techniques and the size of the crystals is confirmed. The obtained nanostructure may be helpful in the consolidation stage. The large number of grain boundaries (typical of nanomaterials) is a relatively simple way leading to the diffusion and evaporation of Mg during the high temperature dealloying process.

The sintering stage was performed in vacuum. When the temperature rises above the Mg boiling point, the magnesium atoms start to separate from the titanium alloy structure. This moment is the start of the thermal dealloying process during which the porosity increases. When the dealloying process is underway, the magnesium atoms diffuse from the middle to the surface of the sample and combine to form vapors and then evaporate leaving open spaces surrounded by the metallic scaffold. Thermal dealloying of magnesium results in the formation of interconnected pores with a relatively wide range of pore size distribution (from nano- to micro-scale). An example of microscopic images of broken samples along with the pore size distribution have been shown in Figure 4. There are a lot of large (5 to over 90 μm) interconnected pores elongated in shape that are approx. 50% of the total pores. Another group of pores is much smaller (approx. 0.1 to 2 μm) and globular in shape, located mostly in areas between the larger pores. The total porosity is approx. 60%, 72% and 76% for the alloys with 30%, 40%, and 50% of Mg, respectively. There is a trend of porosity increase with the increasing Mg content, however, the increase is not proportional to the changes of the chemical composition.

The pores are interconnected and will determine mechanical properties, which is very attractive in the biomaterial-related aspect.

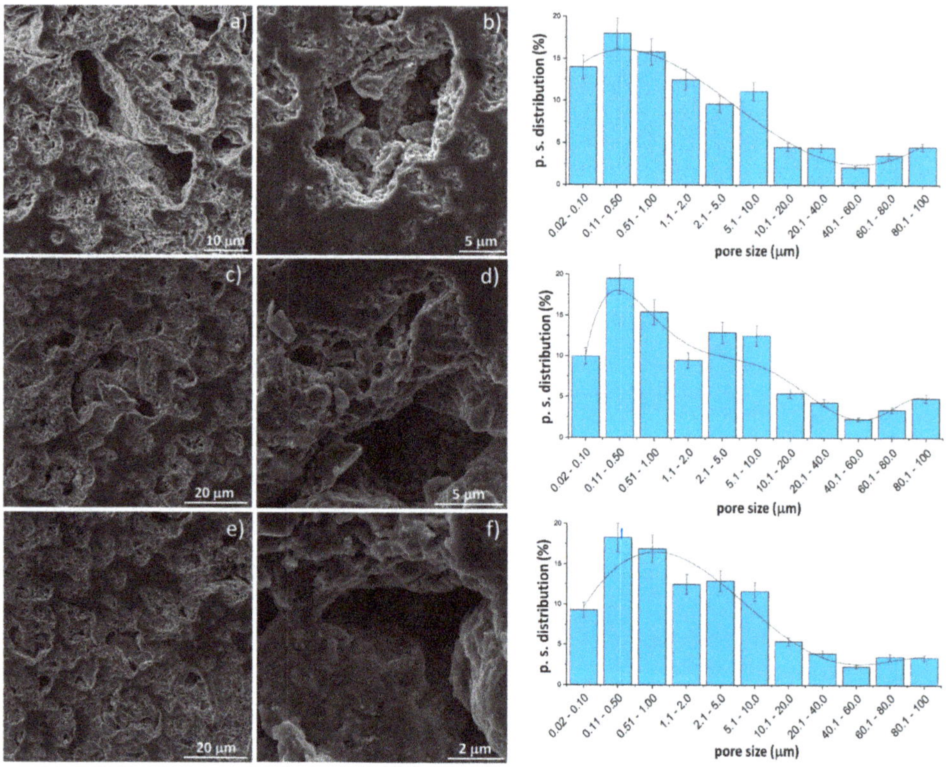

Figure 4. SEM image of: (**a**,**b**) Ti-30Ta (+initial 30 Mg), (**c**,**d**) Ti-30Ta (+initial 40 Mg), (**e**,**f**) Ti-30Ta (+initial 50 Mg) metallic foams.

The mechanical properties were measured in the compression tests and using a nanoindentation tester (Table 2). The authors chose those two techniques to measure the properties at different aspects: the compression test will provide information about the whole sample including the porosity while in the nanoindentation test, the influence of porosity is limited.

Table 2. Mechanical properties of alloys after sintering/dealloying process.

Alloy.	Compression Test		Nanoindentation Test	
-	Compression Strength (MPa)	Elastic Modulus (GPa)	Hardness (HV)	Elastic Modulus (GPa)
Ti-30Ta (+initial 30 Mg)	14.60 ± 0.91	0.65 ± 0.03	307.08 ± 23.9	120.45 ± 8.97
Ti-30Ta (+initial 40 Mg)	10.81 ± 1.03	0.56 ± 0.03	299.60 ± 17.10	119.20 ± 8.76
Ti-30Ta (+initial 50 Mg)	9.96 ± 0.93	0.53 ± 0.03	305.05 ± 19.01	117.73 ± 7.89

When comparing the elastic modulus results obtained with the compression test and nanoindentation test, it could be seen that the values are totally different. The modulus measured by the nanoindentation test is much higher than that obtained by the compression test. This is because the porosity level for nanoindentation is not fully taken into account, as is in the compression measurements. There is a high probability that using the nanointender we can measure the moduli appropriate for the bulk alloys. It was averagely 120 GPa for all samples, which is close to pure Ti and higher than for the Ti-30Ta alloy. However, in the investigations of porous materials, it is more important to include the influence of porosity. The moduli measured by the compression machine oscillate between 0.65 and 0.5 GPa. Those values are insignificantly higher than those measured for the Ta-Ti foams, presented in the authors' previous work [16], which confirms the conclusion of negligibility of the impact of the chemical composition on the porosity in highly porous materials (60%–80%). The hardness of the prepared foams was approx. 300 HV, which was higher than pure titanium and comparable to some of the titanium alloys. The main factor influencing the hardness appears to be the solid solution strengthening.

Cytotoxic activity was analyzed using the MTT assay. The obtained RVC results have been shown in Figure 5. The reference medium conditioned with pure bulk microcrystalline titanium was represented by 100%. After 24 h, the RVC for all the tested materials was lower (above 60%) than that of the reference sample. The possible reason for such a situation could be the richer chemical composition of new materials than reference sample and consequent longer adaptation time of cells to the medium. However, after three and four days, the proliferation of both the osteoblasts and fibroblasts cells in all of the media conditioned with the tested porous alloys rebalanced the rate of the cell proliferation (approx. 98%–110%) in the medium conditioned with reference titanium sample. Such results allow a conclusion that all the prepared Ti based foams are non-toxic.

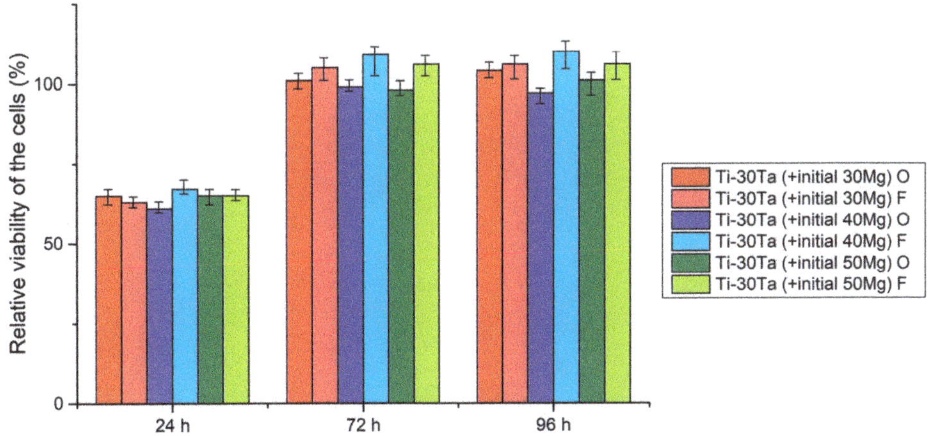

Figure 5. Results of the MTT assay performed at 24 h, 72 h and 96 h. (O)—NHost and (F)—HPLF cells.

4. Conclusions

In the paper, the authors have shown a development and characteristics of the Ti beta alloys contain a high content of magnesium. These new materials were the precursors for the preparation of titanium-based metallic foams using a new thermal dealloying technique. They will find application in the production of hard tissue implants. Based on this study, the following conclusions can be drawn:

- Using a beta stabilizer such as Ta, it is possible to achieve a beta titanium structure in the presence of Mg during mechanical alloying;
- Using magnesium as the alloying element to titanium alloys, cold welding utilizing a shaker-type mill and a high B/P ratio does not appear to be an issue;
- Mechanical alloying is beneficial in the synthesis of the nanocrystalline Ti-Ta-Mg alloys;
- Thermal dealloying of magnesium results in the formation of a porous interconnected structure with wide range of the pore size (from nano- to micro-scale);
- The new foams present attractive mechanical and biological properties and may be good candidates for hard tissue implant application.

Author Contributions: Conceptualization and writing—G.A., J.J.; methodology—G.A.; samples preparation—G.A.; mechanical measurements—G.A. and M.K.; structure and microstructure analysis—G.A.; MTT assay—M.U.J., P.W. and J.Z., supervised the project—G.A.

Funding: The work has been financed by National Science Centre Poland under the project no. DEC-2014/15/D/ST8/03969.

Conflicts of Interest: The authors declare no conflict of interest.

References

1. Noort, R.V. Review Titanium: The implant material of today. *J. Mater. Sci.* **1987**, *22*, 3801. [CrossRef]
2. Davidson, J.A.; Mishra, A.K.; Kovacs, P.; Poggie, R.A. New surface-hardened, low-modulus, corrosion-resistant Ti-13Nb-13Zr alloy for total hip arthroplasty. *Biomed. Mater. Eng.* **1994**, *4*, 231. [PubMed]
3. Lugowski, S.J.; Smith, D.C.; McHugh, A.D.; Loon, V.J. Release of metal ions from dental implant materials in vivo: Determination of Al, Co, Cr, Mo, Ni, V, and Ti in organ tissue. *Biomed. Mater. Res.* **1991**, *25*, 1443. [CrossRef] [PubMed]
4. Okazaki, Y.; Rao, S.; Ito, Y.; Tateishi, T. Corrosion resistance, mechanical properties, corrosion fatigue strength and cytocompatibility of new Ti alloys without Al and V. *Biomaterials* **1998**, *19*, 1197. [CrossRef]
5. Niinomi, M.; Hattori, T.; Morikawa, K.; Kasuga, T.; Suzuki, A.; Fukui, H.; Niwa, S. Development of low Rigidity β-type titanium alloy for biomedical applications. *Mater. Trans.* **2002**, *43*, 2970. [CrossRef]
6. Niinomi, M.; Nakai, M.; Hieda, J. Development of new metallic alloys for biomedical applications. *Acta Biomater.* **2012**, *8*, 3888. [CrossRef] [PubMed]
7. Zhang, L.B.; Wang, K.Z.; Xu, L.J.; Xiao, S.L.; Chen, Y.Y. Effect of Nb addition on microstructure, mechanical properties and castability of β-type Ti-Mo alloys. *T. Nonferr. Metal. Soc.* **2015**, *25*, 2214–2220. [CrossRef]
8. Cardoso, F.F.; Ferrandini, P.L.; Lopes, E.S.N.; Cremasco, A.; Caram, R. Ti-Mo alloys employed as biomaterials: Effects of composition and aging heat treatment on microstructure and mechanical behavior. *J. Mech. Behav. Biomed.* **2014**, *32*, 31–38. [CrossRef] [PubMed]
9. Zhang, W.D.; Liu, Y.; Wu, H.; Song, M.; Zhang, T.Y.; Lan, X.D.; Yao, T.H. Elastic modulus of phases in Ti-Mo alloys. *Mater. Charact.* **2015**, *106*, 302–307. [CrossRef]
10. Zhou, Y.L.; Niinomi, M. Ti-25Ta alloy with the best mechanical compatibility in Ti-Ta alloys for biomedical applications. *Mater. Sci. Eng. C* **2009**, *29*, 1061–1065. [CrossRef]
11. Liu, Y.; Li, K.; Wu, H.; Song, M.; Wang, W.; Li, N.; Tang, H. Synthesis of Ti-Ta alloys with dual structure by incomplete diffusion between elemental powders. *J. Mech. Behav. Biomed.* **2015**, *51*, 302–312. [CrossRef] [PubMed]
12. Mareci, D.; Chelariu, R.; Gordin, D.M.; Ungureanu, G.; Gloriant, T. Comparative corrosion study of Ti-Ta alloys for dental applications. *Acta Biomater.* **2009**, *5*, 3625–3639. [CrossRef] [PubMed]
13. Morgado, T.L.M.; Navas, H.; Brites, R. Wear study of Innovative Ti-Ta alloys. *Procedia Struct. Integrity* **2016**, *2*, 1266–1276. [CrossRef]
14. Zhou, Y.L.; Niinomi, M.; Akahori, T. Effects of Ta content on Young's modulus and tensile properties of binary Ti-Ta alloys for biomedical applications. *Mater. Sci. Eng. A* **2004**, *371*, 283–290. [CrossRef]
15. Sing, S.L.; Yeong, W.Y.; Wiria, F.E. Selective laser melting of titanium alloy with 50 wt% tantalum: Microstructure and mechanical properties. *J. Alloy. Compd.* **2016**, *660*, 461–470. [CrossRef]
16. Adamek, G. Tantalum foams prepared by the thermal dealloying process. *Int. J. Refract. Met. H.* **2017**, *65*, 88–93. [CrossRef]
17. Adamek, G. Tantalum-45S5Bioglass composite foams prepared in thermal dealloying process. *Int. J. Refract. Met. H.* **2019**, *81*, 58–62. [CrossRef]
18. Zadra, M. Mechanical alloying of titanium. *Mater. Sci. Eng. A* **2013**, *1*, 583. [CrossRef]
19. Adamek, G. Mechanical Alloying of Ti-20Ta-20Nb-(10 ÷ 20)Mg Alloys. *Acta Phys. Pol. A* **2014**, *126*, 871–874. [CrossRef]

materials

Article

Comparison of Mechanical Stability of Elastic Titanium, Nickel-Titanium, and Stainless Steel Nails Used in the Fixation of Diaphyseal Long Bone Fractures

Pei-Yuan Lee [1,2], Yen-Nien Chen [1,2,*], Jin-Jia Hu [1,3,*] and Chih-Han Chang [1]

1 Department of BioMedical Engineering, National Cheng Kung University, Tainan 70101, Taiwan; b1208b@ms26.hinet.net (P.-Y.L.); changbmencku@gmail.com (C.-H.C.)
2 Department of Orthopedics, Show Chwan Memorial Hospital, Changhua 50008, Taiwan
3 Medical Device Innovation Center, National Cheng Kung University, Tainan 70101, Taiwan
* Correspondence: yennien.chen@gmail.com (Y.-N.C.); jjhu@mail.ncku.edu.tw (J.-J.H.); Tel.: +886-06-2757575 (ext. 63431) (Y.-N.C); +886-06-2757575 (ext. 63421) (J.-J.H.)

Received: 5 October 2018; Accepted: 29 October 2018; Published: 1 November 2018

Abstract: Elastic nails made of the nickel-titanium shape memory alloy (Nitinol) have been reported to control bone modeling in animal studies. However, the mechanical stability of the Nitinol nail in the fixation of long bone fractures remains unclear. This study compared mechanical stability among nails made of three materials, namely Nitinol, titanium, and stainless steel, in the fixation of long bone fractures. These three materials had identical shapes (arc length: $\pi/2$ and radius: 260 mm). A cylindrical sawbone with a 10-mm gap and fixed with two C-shaped elastic nails was used to examine the stability of the nails. A finite element (FE) model was developed based on the sawbone model. The end cap for elastic nails was not used in the sawbone test but was considered based on a constraint equation in FE simulation. The results of stability tests appeared to depend on the presence or absence of the end cap. In the sawbone test, the titanium nail yielded a higher ultimate force against the applied load than did the stainless steel and Nitinol nails before the gap completely closed; the difference in linear stiffness between the nails was nonsignificant. In FE simulation, the titanium nail produced smaller gap shortening than did stainless steel and Nitinol nails without the end cap; the difference in gap shortening between the nails was minor with the end cap. The titanium elastic nail should be a better choice in managing diaphyseal long bone fractures when the end cap is not used. For Nitinol and stainless steel nails, the end cap should be used to stop the nail from dropping out and to stabilize the fractured bone.

Keywords: titanium; stainless; nitinol; diaphyseal fracture

---

1. Introduction

The elastic stable intramedullary nail (ESIN) is used for the fixation of diaphyseal long bone fractures in adolescents, children, and small people [1–3]. The major advantages of using the ESIN are the minimally invasive nature of the technique, the short operation time, and the preservation of the growth plate. Titanium and stainless steel are commonly used to fabricate elastic nails, and both of these metals result in adequate stabilization of the fractured bone [4–7]. However, complications, such as bursitis caused by the nail tip, delayed union, and nonunion, commonly occur in patients with body weights of >55 kg or patients aged >13 years [8,9]. Thus, development of an innovative technique or device for treating diaphyseal fractures is still required. In recent years, the nickel-titanium shape memory alloy (SMA) has received increasing attention, because it has been used as a functional intramedullary nail to apply control force during bone remodeling, and as a novel bone plate during osteotomy in animal studies [10–12].

Nitinol (Nickel Titanium Naval Ordnance Laboratory) is a type of SMA and has been widely used in many implant types, such as vascular stents and memory staples [13,14], as well as in medical instruments, such as the Nitinol rod used for treating scoliosis, tissue ablation devices, and endoscopes [15–17]. Nitinol is the most widely used SMA in medical applications because of its excellent biocompatibility, mechanical properties, and anticorrosion ability [18–20]. Furthermore, according to the American Society for Testing and Materials (ASTM F2063), the phase transformation temperature of Nitinol is close to the human body temperature. Nitinol is also widely used because of its unique shape memory effect and superelastic behavior. These two phenomena are attributed to the alloy's phase changes (Martensite phase ⇆ Austin phase). The difference between the shape memory effect and superelastic behavior is the starting temperature of phase transformation, which is dependent upon the proportion of nickel and titanium, as well as the thermomechanical parameters used during manufacture.

The Nitinol nail has been demonstrated to increase bone density and promote bone healing in animal models [10,11]. Furthermore, in human beings, Nitinol staples are used in the fixation of bone fractures for bones that bear relatively little weight, such as the carpal bones and phalanges, and these applications are associated with favorable outcomes [13,21]. However, the Nitinol nail is rarely used to fix long bone fractures of the lower limb, and the mechanical stability of the Nitinol nail in the fixation of long bone fractures has not been studied. The difference in mechanical stability between the currently used titanium and stainless steel elastic nails and the Nitinol nail in the fixation of long bone fractures, particularly in the lower limbs, must be clarified before clinical application. A previous finite element (FE) study reported the importance of the prebending degree of titanium nails in the fixation of long bone fractures [22]. A study reported that the effect of the end cap is crucial to mechanical stability when the prebending degree is not adequate to apply sufficient frictional force to stop the nail from sliding [23]. However, only the titanium nail was examined in these studies; the Nitinol and stainless steel nails were not considered. Therefore, in the present study, the mechanical stability of the Nitinol nail was compared with that of the currently used titanium and stainless steel nails in the fixation of diaphyseal long bone fractures by using the sawbone test and FE simulation. FE simulation is a numerical method used to solve complex nonlinear problems, and it has been used in many types of mechanical studies [24,25]. The present study hypothesized that there is no difference in mechanical stability between Nitinol, titanium, and stainless steel nails in the fixation of diaphyseal long bone fractures.

2. Materials and Methods

To investigate the effect of various nail materials, namely titanium, stainless steel, and Nitinol, on the stability of the fractured long bone fixed with two C-shaped nails, an experimental test and FE simulation were used.

2.1. Experimental Test

A cylindrical sawbone (Pacific Research Laboratories, Vashon, Washington, WA, USA) with a length of 260 mm, an outer diameter of 35 mm, and a wall thickness of 11 mm was used, and a 10-mm transverse gap was created at the middle of the sawbone. Subsequently, two prebent, C-shaped elastic nails were used to fix the fractured sawbone. The outer diameters of the nails was 3 mm [26]; this size is widely used in clinical settings. The nails had arc lengths of $\pi/2$ and radiuses of 260 mm (Figure 1). Holes were created using a 4-mm drill bit at the distal border to insert the nails, and the holes were located 45° to the long axis of the sawbone. A commercial titanium elastic nail (DePuy Synthes, Oberdorf, Switzerland) was used, and the stainless steel nail (316 L) and Nitinol nail (weight percentage of nickel and titanium: 56.1% and 43.9%, respectively) were manufactured by Nitinol Innovative Technology Co., Ltd. (Kaohsiung Taiwan). The curves of the nails made of different materials were set using a mold that was custom made by Nitinol Innovative Technology Co., Ltd. The commercial end cap that is typically used to prevent the nail from dropping out was not used in this experiment; thus, the nail end could be dropped out of the nail under loading (Figure 2a). A total of 18 samples, divided into three groups (6 samples in each group), were used in this experiment.

A material property test system (MTS 858; MTS Systems Corporation, Eden Prairie, MN, USA) was used to apply axial compression force to the fractured sawbone fixed with two C-shaped elastic nails (Figure 2a). The distal end of the sawbone was completely fixed. The loading rate was set to 1 mm/min, and loading was applied until the gap closed fully. The environmental temperature was set to 24°C by using an air conditioning system.

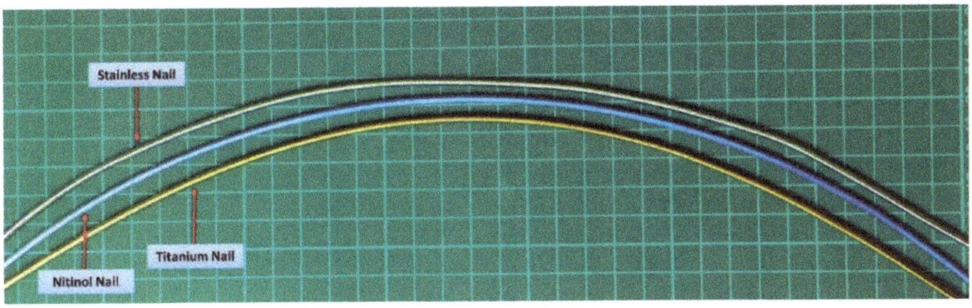

Figure 1. C-shaped nails used in this study.

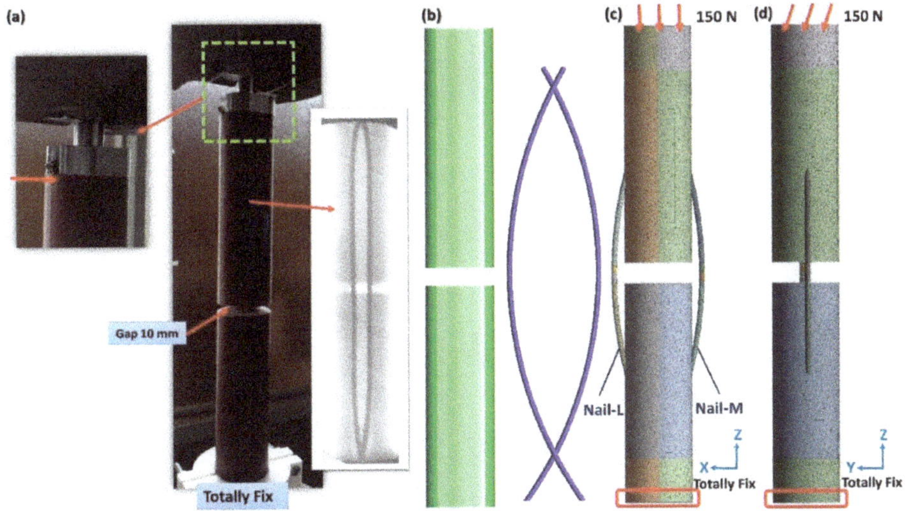

Figure 2. Sawbone model and experimental setup (**a**); The three-dimensional models of the nail and bone (**b**); Boundary conditions of axial compression (**c**) and bending (**d**).

2.2. FE Simulation

Based on the present sawbone nail model, a solid model was developed using SolidWorks 2014 (Dassault Systemes SolidWorks Corp., Waltham, MA, USA) to examine the internal responses of the nail and bone under loading. The outer diameter and prebending degrees of the nail were the same as those used in the experimental setting (Figure 2b). The nail close to the coordinate origin on the X axis was defined as Nail-M, and the other nail was defined as Nail-L. The three-dimensional (3D) model was then imported into ANSYS Workbench V17 (Swanson Analysis Systems, Inc., Houston, PA, USA) to generate the FE model. Ten-node quadratic tetrahedron elements (solid 187) were used to mesh the whole model. Compared with the linear element, quadratic elements can simulate the deformation of a bone and nail more precisely, and tetrahedron elements are the most powerful elements for meshing the complex geometry of a bone and nail. The element edges of the bone and nail were set to 2 and 1 mm, respectively, by using the command "sizing" in Workbench. The same command was used to set the element edge of inner surfaces that were in contact with the nail to 1.5 mm. A total of 151,554 nodes and 89,396 elements were used in the simulation. The contact behaviors between the nail and the canal, between the bone fragments (at the fracture site), and between the nails were all set to frictional surface-to-surface contact behaviors (target 174 and contact 170 in ANSYS). The frictional coefficients of the bone-to-bone, nail-to-bone, and nail-to-nail were set to 0.45, 0.3, and 0.2, respectively [27,28]. The use of the end cap was considered; however, 3D modeling of the end cap was not conducted. The effect of using the end cap to bond the distal surfaces of nails to surrounding bones was simulated using a constraint equation. The distal end of the nail was set to free if the end cap was not used.

The elastic modulus and Poisson's ratio of the bone were 12 GPa and 0.3 (according to the manufacturer), respectively. Titanium and stainless steel were set as plastic behavior, and the bilinear hardening effect was then adopted for these two metals. Material parameters were used from the engineering database in ANSYS Workbench (Table 1). The superelastic material property of Nitinol was used in this simulation, and material parameters were defined by the curve fitting of the stress-strain relationship (Figure 3) observed during the tensile test of the Nitinol rod at room temperature (24 °C).

Table 1. Material properties of titanium and stainless steel used in this study.

Material	Elastic Modulus (MPa)	Poisson Ratio	Yield Strength (MPa)	Tangent Modulus (MPa)
Titanium	110,000	0.3	800	1250
Stainless	200,000	0.3	210	1800

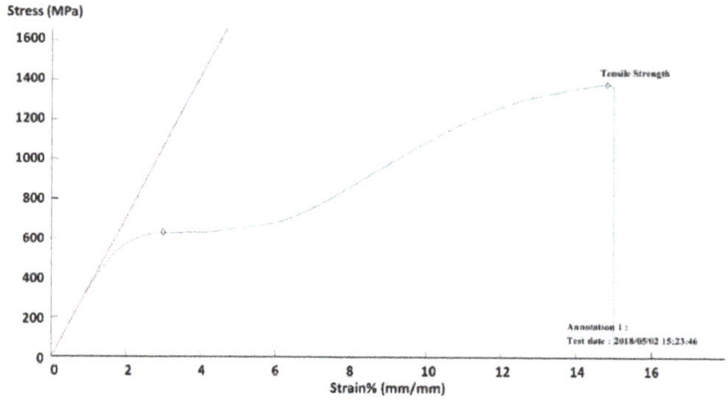

Figure 3. Relationship between the stress and strain of Nitinol rod used in this study.

Two types of loading conditions based on our previous study were used to examine the stability of the fractured bone fixed with nails fabricated from different materials [22]. To simulate the loading conditions of the lower limb in gait, a force of 150 N was applied to the superior surface of the bone model in the following two directions: vertical (axial compression) and 10° to the vertical line (bending) (Figure 2c,d). The distal end was completely fixed. The nails and bone exhibited initial partial volume interference; thus, volume interference was excluded by the program during the first step, and after the first step, the nails were constrained inside the intramedullary canal. The process of inserting the nails into the canal was simulated in this step, and a constraint force was developed from the canal and applied on the nails. The aforementioned loading of 150 N was applied on the bone in the subsequent loading step.

2.3. Incidence and Data Analysis

In the experimental compression test, the loading with the gap completely closed was recorded as "ultimate load," and the relationship between force and displacement at the initial linear stage was referred to as "linear stiffness" for comparison (Figure 4). The value of linear stiffness was calculated using linear regression. One-way ANOVA was used to test for significance between different nails and a P value of less than 0.05 was used. In FE simulation, gap shortening under compression was used as an index for stability. Under bending, the compression (shortened) side of the gap was used as an index.

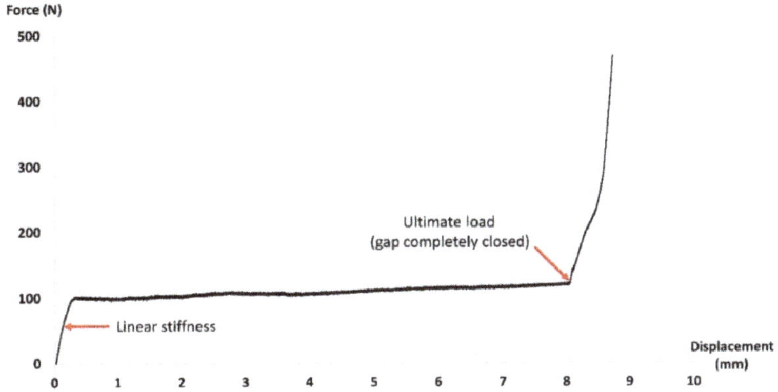

Figure 4. Relationship between the displacement and applied force of the sawbone with nails.

3. Results

3.1. Experimental Test

Experimental results indicated that compared with stainless steel and Nitinol nails, the titanium nail yielded a significantly higher ultimate load before the gap completely closed ($P < 0.05$); the difference in the ultimate load between stainless steel and Nitinol nails was not significant (Figure 5). Furthermore, no significant difference in linear stiffness was observed among titanium, stainless steel, and Nitinol nails. The average ultimate loads yielded by titanium, stainless steel, and Nitinol nails were 272 (SD: 50), 144 (SD: 24), and 111 (SD: 15) N, respectively. Additionally, the average linear stiffness values yielded by the titanium, stainless steel, and Nitinol nails were 199 (SD: 70), 264 (SD: 110), and 180 (SD: 64) N/mm, respectively.

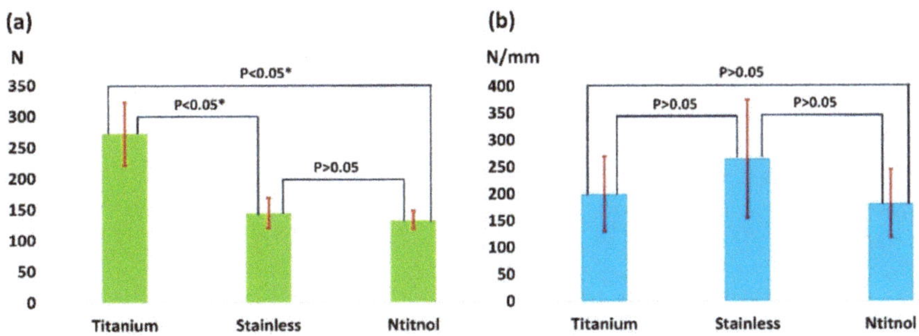

Figure 5. Ultimate load (**a**) and linear stiffness (**b**) of nails.

3.2. FE Simulation

FE simulation results indicated that the titanium nail could sustain the applied loading (150 N) without gap closing, either under axial compression or bending (Figure 6a,b), in the absence of the end cap. By contrast, stainless steel and Nitinol nails could not maintain the gap when 150 N of loading was fully applied in the absence of the endcap, and the gap closed completely during the process of load application. The gap fully closed at 82 and 68 N under axial compression, and at 90 and 86 N under bending for stainless steel and Nitinol nails, respectively (Figure 7). When the end cap was used, gap deformation was larger for the Nitinol nail than for titanium and stainless steel nails; the difference in gap deformation was not minor between titanium and stainless steel nails (Figure 7). The end cap could stop the gap from closing. In addition, the titanium nail developed a higher contact force between the nail and the bone than did stainless and Nitinol nails after the insertion of the nail (Table 2). The difference in the contact force between Nitinol and stainless steel nails was very minor.

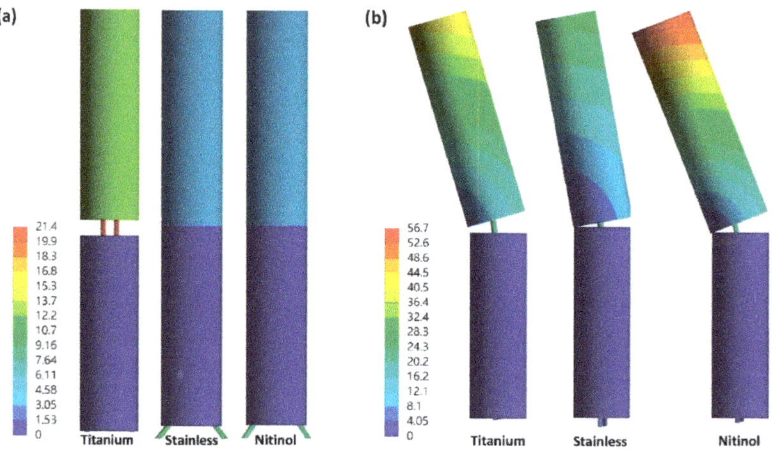

Figure 6. Total deformation (mm) of the fractured bone with nails and without end cap under compression (**a**) and bending (**b**).

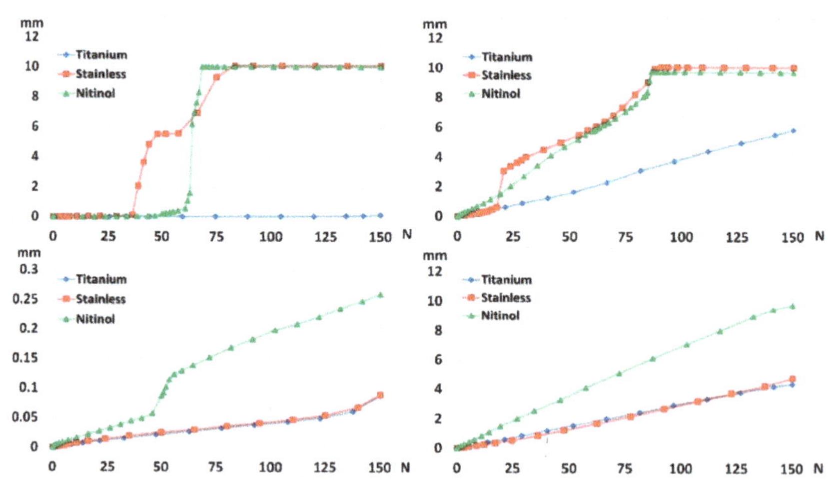

Figure 7. Gap shortening with (**bottom row**) and without (**top row**) end cap in compression (**left column**) and bending (**right column**).

Table 2. Initial contact force (N) between the canal and nails fabricated from various metals.

Nail	Titanium	Stainless	Nitinol
	Bone shaft		
Nail-M	107.7	34.2	37.2
Nail-L	108.5	31.5	36.7
	Inserting hole		
Nail-M	141	51	53
Nail-L	143.4	45.7	49.2

4. Discussion

In this study, the mechanical stability of the currently used titanium and stainless steel nails was compared with that of the Nitinol nail in the fixation of diaphyseal long bone fractures by using the sawbone test and FE simulation. The results of the sawbone test and FE simulation both indicated that the titanium nail provided the highest stability for the bone fracture, particularly when the end cap was not used. The Nitinol nail demonstrated the lowest stability. Although the Nitinol nail has facilitated a favorable outcome for bone union in animal studies [10,11], it provided lower mechanical stability than did titanium and stainless steel nails in the fixation of long bone fractures in the present study.

The titanium nail resulted in a higher ultimate load (sawbone test) and stability (FE simulation) than did stainless steel and Nitinol nails, because the yield stress of titanium (800 MPa) is considerably higher than that of stainless steel (210 MPa). Although the elastic modulus of titanium in the linear phase is only half of that of stainless steel, the linear elastic zone (before the yield stress) of titanium is much wider than that of stainless steel. Compared with the stainless steel nail, the titanium nail developed a higher contact force, thereby generating a higher frictional force against the applied load. The superelastic property of Nitinol could supply a consistent force when phase transformation was triggered, and the force did not increase with strain. Although the linear zone (before entering the martensite phase; 600 MPa) of Nitinol is slightly smaller than that of titanium, the elastic modulus of Nitinol is considerably lower than that of titanium. Compared with the titanium nail, the Nitinol nail developed a lower contact force between the nail and canal; consequently the Nitinol nail's ultimate force was also lower.

Titanium and stainless steel are commonly used metals for preparing ESINs. Nails composed of stainless steel and titanium may exhibit different responses under loading. Whether the titanium nail provides higher stability than the stainless steel nail remains controversial [29,30]. In the present study, the titanium nail resulted in lower gap deformation than did the stainless steel nail, which is consistent with the result reported by Perez [7,29]. Furthermore, compared with the stainless steel nail, the titanium nail provided a higher contact force between the nail and canal, resulting in a higher frictional force against the applied load. The prebending degree, which was not considered in Perez's simulation, was considered in the present simulation. The stainless steel nail provided higher structural stiffness than did the titanium nail in Kaiser's study [30]. Kaiser's result is in accordance with the results on linear stiffness in the present study. The average value of linear stiffness in this study was higher for the stainless steel nail than for the titanium nail, but the variance for each nail type was large. The large variance (70 and 110 N/mm in titanium and stainless steel, respectively) may explain why stainless steel nails have yielded higher stability in some studies, whereas other studies have reported that titanium nails yielded higher stability. Therefore, the ultimate load, which was first demonstrated in the present study, is another measure for determining the effect of nail properties on mechanical stability.

In previous FE simulation studies, the end cap was reported to be crucial to the stability of the titanium nail when the contact force between the nail and canal was not adequate to stop the nail from sliding against the inner aspect of the canal, particularly in nails with low prebending degrees [22,23]. In this study, the contact force developed by the titanium nail, which had an arc length of $\pi/2$ and a radius of 260 mm, was adequate to generate sufficient frictional force against the applied loading of 150 N and maintain the gap without the end cap. However, stainless steel and Nitinol nails could not develop sufficient contact and frictional forces to resist the applied load without the end cap, resulting in the closure of the gap and the dropping out of the nail. When the end cap was used, it provided adequate support to the nail end and prevented the nail from dropping out. Thus, the use of the end cap is recommended if the stainless steel or Nitinol nail is used for the management of diaphyseal long bone fractures.

Although the Nitinol nail demonstrated relatively low stability in this study, it remains a potential biomaterial for many medical devices because of its unique shape memory effect and superelastic material properties. Furthermore, compared with the elastic moduli of titanium (110 GPa) and stainless steel (210 GPa), the elastic modulus of Nitinol (30–70 GPa) is much closer to that of the cortical bone. Hence, the stress shielding effect caused by a material with a higher elastic modulus than that of the bone can be decreased by reducing the elastic moduli of implants. The shape memory effect enables the implant to be delivered into the bone through a small tunnel, and then recovery to the original complex shape for supporting, such as kyphoplasty [31,32]. In addition, the shape memory effect can be used to generate and apply a compressive force to the fracture site of non-weight-bearing bones, such as metatarsal and carpal bones [33,34]. Absolute stiffness is not necessary for non-weight-bearing bones, and force stimulation can contribute to bone healing [35].

The elastic modulus of Nitinol used in this study was only approximately 30 GPa, which was lower than that reported in the literatures [36–38]. In Taiwan, the available source of the Nitinol alloy is rarer than those of the commonly used titanium and stainless steel alloys. In the future, the stability of the Nitinol nail may be increased if the elastic modulus of the Nitinol alloy is increased.

This study involved some limitations that should be addressed. First, the simulation results are approximate solutions, and some small errors likely occurred during the transformation of the complex geometry into simple elements. Second, the FE bone model was simplified as a cylinder for comparison with the sawbone. The real contact between the nail and canal is more complex than that in the present model. Third, the effect of temperature on the material properties of Nitinol was not considered in the FE simulation. Fourth, the solid model of the end cap was not modeled, and the contribution of the end cap was replaced by a constraint equation in the simulation. Fifth, the difference of frictional coefficient between the materials was not considered in the simulation. Sixth, the Nitinol with a higher elastic modulus is unavailable, hence the stability may be underestimated. We have substituted the elastic modulus of the Nitinol with a higher value in the simulation and it indeed increased the stability.

5. Conclusions

Nitinol has received substantial attention in recent years, particularly for its applications in orthopedic implants. Thus, the differences in mechanical behavior between Nitinol and conventional metals should be clarified before this material is further applied clinically, particularly in weight-bearing limbs. The present study is the first to demonstrate the mechanical stability of titanium, stainless steel, and Nitinol nails in the fixation of diaphyseal long bone fractures. According to the results of this study, the C-shaped titanium elastic nail should be the top choice for the management of diaphyseal long bone fractures when the end cap is not used. For Nitinol and stainless steel nails, use of the end cap is suggested to prevent the nail from dropping out and to stabilize the fractured bone. Additionally, loading with the gap completely closed (the ultimate load proposed in this study) is a key index—along with structural stiffness and gap deformation—for judging the stability of elastic stable intramedullary nails fabricated from various materials.

Author Contributions: Conceptualization, P.-Y.L.; Methodology, Y.-N.C.; Writing—review and editing, J.-J.H.; Supervision, C.-H.C.

Funding: This research was funded by Show Chwan Memorial Hospital, grant number NCKUSCMH10704.

Conflicts of Interest: The authors declare no conflict of interest.

References

1. Nielsen, E.; Bonsu, N.; Andras, L.M.; Goldstein, R.Y. The effect of canal fill on paediatric femur fractures treated with titanium elastic nails. *J. Child. Orthop.* **2018**, *12*, 15–19. [CrossRef] [PubMed]
2. Popkov, A.; Foster, P.; Gubin, A.; Borzunov, D.; Popkov, D. The use of flexible intramedullary nails in limb lengthening. *Expert Rev. Med. Devices* **2017**, *14*, 741–753. [CrossRef] [PubMed]
3. Heare, A.; Goral, D.; Belton, M.; Beebe, C.; Trizno, A.; Stoneback, J. Intramedullary implant choice and cost in the treatment of pediatric diaphyseal forearm fractures. *J. Orthop. Trauma* **2017**, *31*, e334–e338. [CrossRef] [PubMed]
4. Mohamed, A.; Rajeev, A.S. Clinical outcomes and complications of titanium versus stainless steel elastic nail in management of paediatric femoral fractures-a systematic review. *Eur. J. Orthop. Surg. Traumatol. Orthop. Traumatol.* **2017**, *27*, 157–167. [CrossRef] [PubMed]
5. Gyaneshwar, T.; Nitesh, R.; Sagar, T.; Pranav, K.; Rustagi, N. Treatment of pediatric femoral shaft fractures by stainless steel and titanium elastic nail system: A randomized comparative trial. *Chin. J. Traumatol. Zhonghua Chuang Shang Za Zhi* **2016**, *19*, 213–216. [CrossRef] [PubMed]
6. Wall, E.J.; Jain, V.; Vora, V.; Mehlman, C.T.; Crawford, A.H. Complications of titanium and stainless steel elastic nail fixation of pediatric femoral fractures. *J. Bone Joint Surg. Am. Vol.* **2008**, *90*, 1305–1313. [CrossRef] [PubMed]
7. Perez, A.; Mahar, A.; Negus, C.; Newton, P.; Impelluso, T. A computational evaluation of the effect of intramedullary nail material properties on the stabilization of simulated femoral shaft fractures. *Med. Eng. Phys.* **2008**, *30*, 755–760. [CrossRef] [PubMed]
8. Canavese, F.; Marengo, L.; Andreacchio, A.; Mansour, M.; Paonessa, M.; Rousset, M.; Samba, A.; Dimeglio, A. Complications of elastic stable intramedullary nailing of femoral shaft fractures in children weighing fifty kilograms (one hundred and ten pounds) and more. *Int. Orthop.* **2016**, *40*, 2627–2634. [CrossRef] [PubMed]
9. Sarkar, S.; Bandyopadhyay, R.; Mukherjee, A. Titanium elastic nail—Complications in the treatment of paediatric diaphyseal fracture of femur [corrected]. *Open Orthop. J.* **2013**, *7*, 12–17. [PubMed]
10. Muller, C.W.; Pfeifer, R.; Meier, K.; Decker, S.; Reifenrath, J.; Gosling, T.; Wesling, V.; Krettek, C.; Hurschler, C.; Kramer, M. A novel shape memory plate osteosynthesis for noninvasive modulation of fixation stiffness in a rabbit tibia osteotomy model. *BioMed Res. Int.* **2015**, *2015*, 652940. [CrossRef] [PubMed]
11. Kujala, S.; Ryhanen, J.; Jamsa, T.; Danilov, A.; Saaranen, J.; Pramila, A.; Tuukkanen, J. Bone modeling controlled by a nickel-titanium shape memory alloy intramedullary nail. *Biomaterials* **2002**, *23*, 2535–2543. [CrossRef]
12. Kapanen, A.; Ryhanen, J.; Danilov, A.; Tuukkanen, J. Effect of nickel-titanium shape memory metal alloy on bone formation. *Biomaterials* **2001**, *22*, 2475–2480. [CrossRef]
13. Schipper, O.N.; Ford, S.E.; Moody, P.W.; Van Doren, B.; Ellington, J.K. Radiographic results of nitinol compression staples for hindfoot and midfoot arthrodeses. *Foot Ankle Int.* **2018**, *39*, 172–179. [CrossRef] [PubMed]
14. Soderlund, C.; Linder, S.; Bergenzaun, P.E.; Grape, T.; Hakansson, H.O.; Kilander, A.; Lindell, G.; Ljungman, M.; Ohlin, B.; Nielsen, J.; et al. Nitinol versus steel partially covered self-expandable metal stent for malignant distal biliary obstruction: A randomized trial. *Endoscopy* **2014**, *46*, 941–948. [CrossRef] [PubMed]
15. Bechtold, C.; de Miranda, R.L.; Chluba, C.; Quandt, E. Fabrication of self-expandable niti thin film devices with micro-electrode array for bioelectric sensing, stimulation and ablation. *Biomed. Microdevices* **2016**, *18*, 106. [CrossRef] [PubMed]
16. Lukina, E.; Kollerov, M.; Meswania, J.; Wertheim, D.; Mason, P.; Wagstaff, P.; Laka, A.; Noordeen, H.; Yoon, W.W.; Blunn, G. Analysis of retrieved growth guidance sliding lsz-4d devices for early onset scoliosis and investigation of the use of nitinol rods for this system. *Spine* **2015**, *40*, 17–24. [CrossRef] [PubMed]
17. Bastijanic, J.M.; Etscheidt, J.; Sattiraju, M.; Bonsignore, C.; Kopchok, G.; White, R.; Sarac, T.P. Fatigue and in vivo validation of a peritoneum-lined self-expanding nitinol stent-graft. *J. Endovasc. Ther.* **2014**, *21*, 735–746. [CrossRef] [PubMed]
18. Park, C.H.; Tijing, L.D.; Pant, H.R.; Kim, C.S. Effect of laser polishing on the surface roughness and corrosion resistance of nitinol stents. *Bio-Med. Mater. Eng.* **2015**, *25*, 67–75.

19. Nam, H.G.; Yoo, C.M.; Baek, S.M.; Kim, H.K.; Shin, J.H.; Hwang, M.H.; Jo, G.E.; Kim, K.S.; Cho, J.H.; Lee, S.H.; et al. Enhancement of mechanical properties and testing of nitinol stents in cerebral aneurysm simulation models. *Artif. Organs* **2015**, *39*, E213–E226. [CrossRef] [PubMed]
20. Janas, A.; Milewski, K.; Buszman, P.P.; Nowakowski, P.; Jelonek, M.; Orlik, B.; Krauze, A.; Samborski, S.; Beaudry, D.; Lecelrc, G.; et al. Safety and biocompatibility of a novel self-expanding nitinol carotid stent with hybrid cell design in a porcine model of neointimal hyperplasia. *Kardiol. Polska* **2015**, *73*, 240–245. [CrossRef] [PubMed]
21. Winkel, R.; Schlageter, M. Nitinol shape memory staple for osteosynthesis of the scaphoid. *Oper. Orthop. Traumatol.* **2009**, *21*, 361–372. [CrossRef] [PubMed]
22. Chen, Y.N.; Lee, P.Y.; Chang, C.W.; Ho, Y.H.; Peng, Y.T.; Chang, C.H.; Li, C.T. Biomechanical investigation of titanium elastic nail prebending for treating diaphyseal long bone fractures. *Australas. Phys. Eng. Sci. Med.* **2017**, *40*, 115–126. [CrossRef] [PubMed]
23. Chen, Y.N.; Lee, P.Y.; Chang, C.H.; Chang, C.W.; Ho, Y.H.; Li, C.T.; Peng, Y.T. Computational comparison of tibial diaphyseal fractures fixed with various degrees of prebending of titanium elastic nails and with and without end caps. *Injury* **2016**, *47*, 2339–2346. [CrossRef] [PubMed]
24. Chang, C.W.; Chen, Y.N.; Li, C.T.; Chung, Y.H.; Chang, C.H.; Peng, Y.T. Role of screw proximity in the fixation of transverse patellar fractures with screws and a wire. *J. Orthop. Surg.* **2018**, *26*. [CrossRef] [PubMed]
25. Chen, Y.N.; Chang, C.W.; Lin, C.W.; Wang, C.W.; Peng, Y.T.; Chang, C.H.; Li, C.T. Numerical investigation of fracture impaction in proximal humeral fracture fixation with locking plate and intramedullary nail. *Int. Orthop.* **2017**, *41*, 1471–1480. [CrossRef] [PubMed]
26. Ligier, J.N.; Metaizeau, J.P.; Prevot, J.; Lascombes, P. Elastic stable intramedullary pinning of long bone shaft fractures in children. *Z. Kinderchir.* **1985**, *40*, 209–212. [CrossRef] [PubMed]
27. Nuno, N.; Groppetti, R.; Senin, N. Static coefficient of friction between stainless steel and pmma used in cemented hip and knee implants. *Clin. Biomech.* **2006**, *21*, 956–962. [CrossRef] [PubMed]
28. Shirazi-Adl, A.; Dammak, M.; Paiement, G. Experimental determination of friction characteristics at the trabecular bone/porous-coated metal interface in cementless implants. *J. Biomed. Mater. Res.* **1993**, *27*, 167–175. [CrossRef] [PubMed]
29. Mahar, A.T.; Lee, S.S.; Lalonde, F.D.; Impelluso, T.; Newton, P.O. Biomechanical comparison of stainless steel and titanium nails for fixation of simulated femoral fractures. *J. Pediatr. Orthop.* **2004**, *24*, 638–641. [CrossRef] [PubMed]
30. Kaiser, M.M.; Wessel, L.M.; Zachert, G.; Stratmann, C.; Eggert, R.; Gros, N.; Schulze-Hessing, M.; Kienast, B.; Rapp, M. Biomechanical analysis of a synthetic femur spiral fracture model: Influence of different materials on the stiffness in flexible intramedullary nailing. *Clin. Biomech.* **2011**, *26*, 592–597. [CrossRef] [PubMed]
31. Chen, B.; Zheng, Y.H.; Zheng, T.; Sun, C.H.; Lu, J.; Cao, P.; Zhou, J.H. The implantation of a nickel-titanium shape memory alloy ameliorates vertebral body compression fractures: A cadaveric study. *Int. J. Clin. Exp. Med.* **2015**, *8*, 16899–16906. [PubMed]
32. Wilson, D.C.; Connolly, R.J.; Zhu, Q.; Emery, J.L.; Kingwell, S.P.; Kitchel, S.; Cripton, P.A.; Wilson, D.R. An ex vivo biomechanical comparison of a novel vertebral compression fracture treatment system to kyphoplasty. *Clin. Biomech.* **2012**, *27*, 346–353. [CrossRef] [PubMed]
33. Pech, J.; Veigl, D.; Hromadka, R.; Dobias, J.; Zatrapa, T. [shape-memory osteosynthesis for trapeziometacarpal joint arthrodesis]. *Acta Chir. Orthop. Traumatol. Cechoslov.* **2014**, *81*, 335–339.
34. Lorea, P.; Schmitz, S.; Aschilian, M.; Chirila-Dobrea, A.; Petrea, A.D. The preliminary results of treatment of symptomatic carpal boss by wedge joint resection, radial bone grafting and arthrodesis with a shape memory staple. *J. Hand Surg. Eur. Vol.* **2008**, *33*, 174–178. [CrossRef] [PubMed]
35. Bottlang, M.; Lesser, M.; Koerber, J.; Doornink, J.; von Rechenberg, B.; Augat, P.; Fitzpatrick, D.C.; Madey, S.M.; Marsh, J.L. Far cortical locking can improve healing of fractures stabilized with locking plates. *J. Bone Joint Surg. Am. Vol.* **2010**, *92*, 1652–1660. [CrossRef] [PubMed]
36. Ko, C.; Yang, M.; Byun, T.; Lee, S.W. Design factors of femur fracture fixation plates made of shape memory alloy based on the taguchi method by finite element analysis. *Int. J. Numer. Methods Biomed. Eng.* **2018**, *34*, e2967. [CrossRef] [PubMed]

37. Gautam, A.; Rani, A.B.; Callejas, M.A.; Acharyya, S.G.; Acharyya, A.; Biswas, D.; Bhandari, V.; Sharma, P.; Naik, G.R. Shape memory alloy smart knee spacer to enhance knee functionality: Model design and finite element analysis. In Proceedings of the 2016 38th Annual International Conference of the IEEE Engineering in Medicine and Biology Society (EMBC), Orlando, FL, USA, 16–20 August 2016; pp. 6046–6049.
38. Jiang, X.; Huang, J.; Wang, Y.; Li, B.; Du, J.; Hao, P. Finite element analysis for the self-loosening behavior of the bolted joint with a superelastic shape memory alloy. *Materials* **2018**, *11*, 1592. [CrossRef] [PubMed]

materials

MDPI

Article

Surface Characteristics and Hydrophobicity of Ni-Ti Alloy through Magnetic Mixed Electrical Discharge Machining

C.C. Feng [1], L. Li [1,*], C.S. Zhang [2], G.M. Zheng [1], X. Bai [1] and Z.W. Niu [1]

1 School of Mechanical Engineering, Shandong University of Technology, Zibo 255000, China; zy94fcc@163.com (C.C.F.); zhengguangming@sdut.edu.cn (G.M.Z.); cllyli1109@163.com (X.B.); niuzongwei@sdut.edu.cn (Z.W.N.)

2 School of Mechanical Engineering, Zibo Vocational Institute, Zibo 255000, China; sdutwd@163.com

* Correspondence: sdutlili@163.com; Tel.: 086-135-8335-0888

Received: 6 December 2018; Accepted: 21 January 2019; Published: 26 January 2019

check for updates

Abstract: Nickel–titanium (Ni-Ti) alloy has been selected as stent material given its good biocompatibility. In this study, experimental research on this material was conducted using magnetic field-assisted electrical discharge machining (EDM). The surface topography of the machined workpiece was analyzed with a scanning electron microscope (SEM). Hydrophobicity was measured by using an optical contact angle measuring instrument. The roughness values of different positions on the surface were measured using a TR200 roughness instrument. Results showed that the composite structure of solidification bulge–crater–pore–particle can be prepared on the surface of the Ni-Ti alloy through magnetic mixed EDM using suitable processing parameters. Moreover, the contact angle of the surface reaches 138.2°.

Keywords: Ni-Ti alloy; surface characteristics; hydrophobic; magnetic mixed EDM

1. Introduction

The element composition of medical Ni-Ti alloys generally suggests a minimal difference between nickel and titanium contents and the presence of only trace impurities. Ni-Ti alloy has been used in biomedicine given its good shape-memory property, superelasticity, and excellent corrosion resistance [1–3]. It is also an ideal material for vascular stents considering its excellent in-body deformation and high strength. Among vascular stents, porous drug-eluting stents use surface micropores, which have the efficacy of drug-eluting stents and the long-term safety of bare metal stents for drug storage and release, thereby showing attractive prospects [4]. As a biological implant, stents must demonstrate good blood compatibility, which is closely related to the hydrophobicity of the material surface. The improvement of materials' hydrophobicity can reduce the adhesion and activation of platelets on the material surface [5,6]. Simultaneously, the good hydrophobicity of the surface can also achieve a slow release of the drug and prolong its action time.

Ni-Ti alloy has the characteristics of low thermal conductivity, high ductility, and high viscosity. The use of traditional machining methods causes high tool wear and work hardening, resulting in the workpiece's poor surface quality [7,8]. Electrical discharge machining (EDM) is based on electrical corrosion to corrode a material [9,10]. EDM's optimal advantage is that it can efficiently process any conductive materials, regardless of their mechanical properties. It is only related to the electrical conductivity and thermal properties of materials [11,12]. At present, few studies are available on preparing multi-scale microporous and hydrophobic surfaces on a metal substrate through EDM. Researchers have found that a surface has good hydrophobic properties when its structural characteristics are composed of craters, bulges, micropores, and particles [13,14]. A new process of preparing hydrophobic surfaces with micropores through EDM is proposed in the present study on the basis of previous studies that have shown that the surface of Ni-Ti alloy machined through EDM is mostly composed of craters and bulges [15].

When a gas-rich working medium flows through the processing area, part of the gas dissolves into the molten metal to form pores because additional magnetic mixing allows additional air to mix into the working medium. Many pores are found on the surface of scaffolds, thereby enabling these scaffolds to have a high drug loading and improved hydrophobicity. A magnetic stirring device was added to the EDM machine to improve the microporous characteristics of the processed Ni-Ti alloy surface. The effects of different parameters on the surface characteristics were investigated, and the effect of surface morphology on the surface hydrophobicity of Ni-Ti alloy was analyzed.

2. Experimental Procedure

2.1. Equipment and Materials

The experimental device was modified on a DM71 (DM71, Changde Ltd., Taizhou, China) precision EDM machine. The magnetic mixed device was installed on the machine table. The fluid container was placed on the magnetic mixed device, thus quickly stirring the working fluid. The workpiece was fixed onto the workpiece fixture, whereas the tool electrode was fixed onto the tool fixture and then connected with the spindle. Figure 1 illustrates the structural principle.

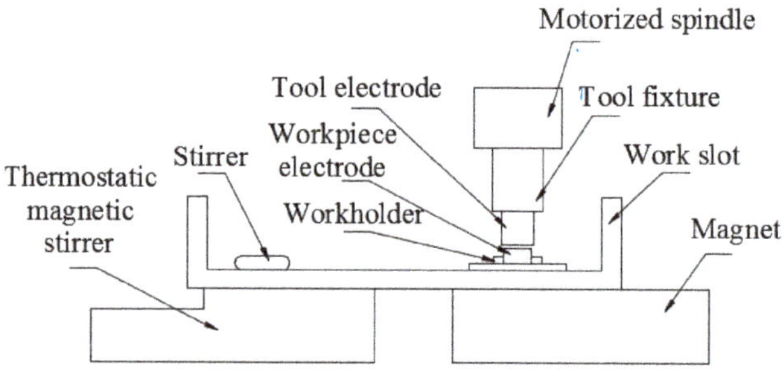

Figure 1. Schematic of the machine tool.

A special spark oil was selected as the working medium, and its index is listed in Table 1. The workpiece material was Ni-Ti alloy, and its chemical composition is presented in Table 2. The tool electrode material was red copper with the dimensions of Φ 16 mm × 150 mm. Table 3 displays the thermal and physical properties of copper and Ni-Ti alloy. In Table 3, both materials exhibit a minimal difference in density and melting point, but the thermal conductivity and specific heat capacity are much higher in copper than in Ni-Ti alloy. This observation indicates that Ni-Ti alloy has poor heat dissipation, and the same heat can persist on its surface for an extended period.

Table 1. Index of special electrical discharge machining (EDM) oil.

Index	Viscosity (40 °C)/($mm^2 \cdot s^{-1}$)	Density (20 °C)/($kg \cdot m^{-3}$)	Aromatic Content	Pour Point/°C
Typical data	2.20	0.79	≤0.08	−10

Table 2. Chemical composition of Ni-Ti alloy.

Chemical Composition	Ni	Ti	C	O	H
Content/%	≤50.9	>48	≤0.05	≤0.05	≤0.003

Table 3. Physical properties of materials.

Materials	Density/(g·cm^{-3})	Melting Point/°C	Thermal Conductivity/(W·m^{-1}·k^{-1})	Specific Heat/(J·kg^{-1}·°C)
Copper	8.96	1083	383.3	410
Ni-Ti	6.45	1310	10	0.32

2.2. Experimental Parameters

The magnetic stirring device had a power of 30 W and a rotating speed of 2600 r/min. The effects of peak current and pulse duration on surface microstructure, surface roughness, and static contact angle were investigated. The EDM parameters were as follows: the peak currents were 1.5, 4.5, and 9 A; the pulse duration was 15, 30, 60, and 90 μs; the voltage was 50 V; and the pulse interval was 5 μs.

2.3. Sample Testing and Characterization

The samples were ultrasonically cleaned for 8 min, followed by air drying, and were examined using a scanning electron microscope (SEM; Apreo, FEI Ltd., Hillsboro, OR, USA) and energy dispersive spectroscopy (EDS; Apreo, FEI Ltd., Hillsboro, OR, USA) to analyze the surface topography and elemental composition, respectively. The parameters of EDS were kv: 20, mag: 2000, takeoff: 36.1, live time(s): 26.7, and amp time (μs): 7.68. The EDS adopts area analysis for the machined surface. X-ray diffraction (XRD; D8 Venture, Bruker Ltd., Madison, WI, USA) analysis was used to analyze the surface compounds. The static contact angle of the surface was measured using a video optical contact angle measuring instrument (OCA15EC, Dataphysics Ltd., Stuttgart, Germany), and the volume of the water droplet was 5 μL. Five positions on the surface were selected to measure the contact angle, and the average value of the measurement results was obtained. A roughness meter (TR200, Times Group Ltd., Shandong, China) was used to measure roughness. Five workpiece positions were selected for measurement, and then the average value was determined. The samples were mounted, polished, ultrasonically cleaned, and etched for a subsurface microstructure observation using the SEM. The etchant consisted of 10 ml HF, 40 ml HNO_3, and 50 ml deionized water.

3. Results and Analysis

3.1. Effect of Parameters on Surface Morphology

Two sets of data were selected for the EDM of Ni-Ti alloy without being magnetically mixed to highlight clearly the role of magnetic mixing. The results are presented in Figure 2. Figure 2a shows very few holes on the surface when the current is 1.5 A and the pulse duration is 60 μs. No stomatal feature on the surface of Figure 2b is observed when the current is 9 A and the pulse duration is 30 μs.

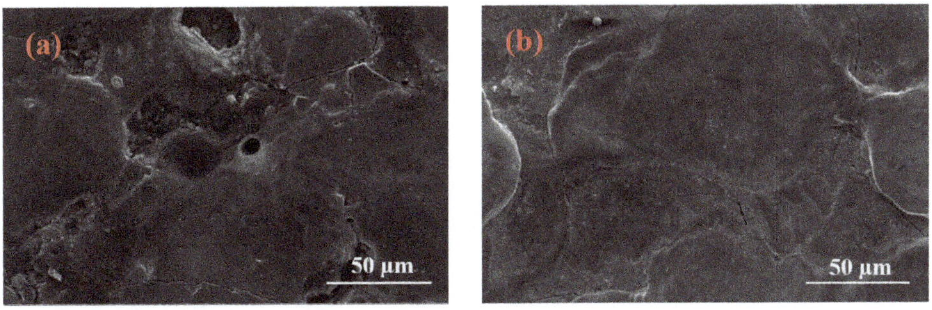

Figure 2. Surfaces machined by EDM without additional magnetic mixing. (**a**) $I = 1.5$ A, $T_{on} = 60$ μs, (**b**) $I = 9$ A, $T_{on} = 30$ μs.

The surface micrographs of Ni-Ti alloy after magnetic mixed EDM under different processing parameters are different. The surface contains a different number of pores under most processing parameters. The formation of pores is due to a large amount of heat generated during the EDM of Ni-Ti alloy, which causes the material to enter a molten state. At that time, gas dissolves into the molten metal. The molten metal quickly cools and solidifies with the cooling flow of the working fluid, and then some of the gas bubbles are expelled with the splash of the molten material. The unexpelled gas is trapped in the re-solidified material, resulting in micropores [16]. The gas is mainly produced in two steps. The first step occurs when the working fluid is quickly stirred using the magnetic stirring device, which enables air to enter the liquid and the working fluid becomes rich in gas. When the gas-carrying working fluid flows through the molten metal surface, the gas enters the molten metal. The other step is when the insulating medium is ionized to produce hydrions during the EDM's discharge process. The hydrions in the discharge channel are reduced on the negative polarity surface to produce hydrogen gas. Furthermore, a certain amount of working fluids evaporates at a high temperature to form water vapor, while the liquid produces oxygen due to pyrolysis at a high temperature. These gases also dissolve into the molten metal. Moreover, the velocity of the liquid passing through the processing area is 1.8 m/s. Rapid cooling prevents the gas from overflowing in large quantities from molten metals. Thus, the stomatal characteristics on the surface increase. However, the formation of pore morphology is due to a large amount of gas mixed by magnetic stirring because, in the absence of magnetic stirring, and even with the same process parameters, only an individual pore was found on the surface [17]. Figure 2 shows that, when magnetic mixing is not used, only an individual pore is obtained, or no pore exists on the surface.

Figure 3a–c indicates that the surface of Figure 3a contains many shallow pores. The result is that the parameter of the peak current is 1.5 A and the pulse on time is 15 μs. A short discharge time indicates a short time of gas entering the molten metal, resulting in a minimal gas entry. In addition, the discharge pulse energy is small, and the pressure is inadequate to remove numerous materials. Thus, additional molten materials are re-solidified on the surface. The surface forms numerous shallow pores, and several solidification bulges are connected by craters. When the pulse duration increases to 30 μs, the number of pores does not increase significantly, whereas the craters between the solidification bulges are clearly enlarged to form gaps, and several spherical fragments are found in the gap. This phenomenon occurs because, when the peak current is 1.5 A, the pulse energy increases with the pulse duration. The gap between the solidification bulges expands and occupies the position of the pores, and the pores do not increase considerably. Spherical debris is formed by re-solidifying the material on the surface of the workpiece when it meets the flowing working fluid after gasification. When the pulse duration is further extended to 60 μs, the number of pores does not increase, but the gap between the solidification bulges is further enlarged. Moreover, the spherical debris in the gap develops into a "coral reef" structure. This phenomenon occurs due to the continuous increase in pulse duration, which expands the discharge channel and leads to low discharge energy, and most of the molten materials remain on the surface, rather than being splashed. The continuously expanding gap and "coral reef" morphology further occupy the position of the pores. Therefore, although the pulse duration increases, the number of pores does not increase considerably. The "coral reef" is formed by the re-solidified materials with random irregular debris, and the debris is re-solidified from the splashed molten material or the vaporized material through rapid dielectric quenching [18].

Figure 3d–f illustrates that, when the pulse duration was 15 μs, the SEM found 14 holes on its surface. These holes are due to the short discharge time and the minimal amount of gas to enter the molten metal. In comparison with Figure 3a, the discharge pulse energy is larger, and the molten material is removed. Thus, the number of shallow pores on its surface reduces. When the pulse duration increases to 30 μs, the material removal rate increases with the continuous increment in pulse discharge energy, and the shallow pores on the surface are invisible. Furthermore, given the increase in pulse duration, the gas has additional time to enter the molten metal, and the depth of the pore increases slightly. When the pulse duration further increases to 60 μs, the increase in pulse duration leads to additional gas that enters the molten metal, thereby forming deep pore characteristics on the surface.

Figure 3g–i demonstrates that craters and pores are nonexistent on the surface when the peak current is 9 A and the pulse duration is 15 μs. This result is due to the small pulse duration that leads to minimal gas entering the molten metal. Simultaneously, considering the high discharge pulse energy, the molten metal has been considerably removed before forming pores. Thus, no pore characteristic is observed on the surface. This trend can be observed from the changes in Figure 3a,d. When pulse duration increases to 30 μs, many deep pores are found on the surface, thereby increasing time and the quantity of gas that enters the molten metal due to the increase in pulse duration. In comparison with Figure 3e, the discharge pulse energy increases, resulting in the removal of the shallow surface pore characteristics. Thus, the surface pore presents a deep state. When the pulse width increases to 60 μs, additional gas enters the molten metal. The shallow pores on the surface are removed considerably, and the depth of the pores on the surface is deeper than that of Figure 3h considering the large discharge pulse energy.

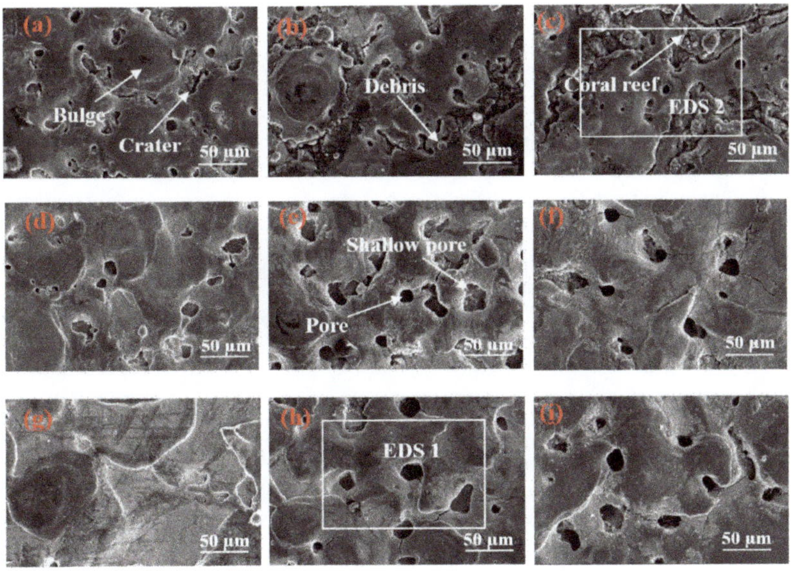

Figure 3. Surface morphology of surface at different processing parameters. (**a**) $I = 1.5$ A, $T_{on} = 15$ μs; (**b**) $I = 1.5$ A, $T_{on} = 30$ μs; (**c**) $I = 1.5$ A, $T_{on} = 60$ μs; (**d**) $I = 4.5$ A, $T_{on} = 15$ μs; (**e**) $I = 4.5$ A, $T_{on} = 30$ μs; (**f**) $I = 4.5$ A, $T_{on} = 60$ μs; (**g**) $I = 9$ A, $T_{on} = 15$ μs; (**h**) $I = 9$ A, $T_{on} = 30$ μs; (**i**) $I = 9$ A, $T_{on} = 60$ μs.

3.2. *Effect of Surface Topography on Contact Angle*

Wettability is an important property of a solid surface. The wettability of a material surface is typically measured by the static contact angle θ. That is, $\theta > 90°$ are hydrophobic surfaces, whereas $\theta < 90°$ are hydrophilic surfaces. The intrinsic contact angle of Ni-Ti alloy is 70°, and the surface exhibits a hydrophilic characteristic. An appropriate surface morphology must be constructed to improve the contact angle of Ni-Ti alloy and obtain improved hydrophobicity.

Figure 4 shows a remarkable difference in the contact angles of different surface morphologies. When the pulse duration is 60 μs and the peak current is 1.5 and 4.5 A, the surface contact angles of the two parameters are greater than 130°, and the size of the solidification bulges on their surfaces is large. The solidification bulges exhibited in Figure 4c are more regular and denser than those of Figure f. Therefore, when the droplets come into contact with the surface, the bulges form support points to lift the droplets, and the gap between the bulges closes considerable air. The difference in pressure prevents the droplet from fully entering the gap between the solidification bulges, thereby resulting in the formation of an "air cushion" effect, which reduces the contact between the droplet and the solid [19–21]. Furthermore, Figure 4c displays that the "coral reef" structure between the solidification bulges is not in a single dimension. In addition, the droplets that enter the gap form the secondary "air cushion" effect on the "coral reef" structure, and the surface of the material has changed from hydrophilicity to hydrophobicity. The contact angle presented in Figure 4f is slightly smaller because no clear demarcation line is found between the solidification bulges on the surface, and no special morphology between the bulges is observed. However, a certain number of pores are found on the surface. In addition, Figure 3f illustrates that the pore size of the surface is small and has a certain depth. Thus, when the droplet is on its surface, the bulges form support points to lift the droplets, and the droplet cannot enter into the micropore completely given the difference in pressure. The "air cushion" effect is formed and transformed into a hydrophobic surface.

The surface morphology depicted in Figure 4a,b is similar to that of Figure 4c, but the size of the solidification bulges increases with the pulse duration. The solidification bulges in Figure 4a,b are smaller. When a droplet is acted upon by gravity, the contact area is small and the pressure is high. Thus, small solidification bulges can pierce the droplet and lead to a large area in contact with the droplet and therefore a small contact angle.

Figure 4d,e shows that the contact angle of the surface with several shallow pores is below 90°; this surface is not hydrophobic. This condition is due to several shallow pores on their surfaces, thereby enabling droplets to easily enter the shallow pores on this type of surface given gravity and capillary forces [22]. Therefore, the contact area between the droplet and the solid increases, and catching sufficient air to form the pressure difference is not feasible; thus, no "air cushion" effect is observed [19–21]. The contact angle is small considering the large contact surface between the droplet and the solid. However, given the change in the surface microstructure, the contact angle remains slightly improved in comparison with the intrinsic contact angle. In addition, the surface morphology depicted in Figure 4d changes minimally, that is, neither the large size of the solidification bulges nor the formation of a large number of shallow pores; therefore, the contact angle increases slightly. When the peak current is 4.5 A, the pulse duration is 15 and 30 μs, and the surface morphology contains numerous shallow pores. Thus, their contact angles are nearly the same, and the surface remains hydrophilic. However, when the pulse width is 60 μs, many deep pores are formed, and the contact angle is considerably increased by the "air cushion" effect.

When the peak current and pulse duration are 9 A and 15 μs, respectively, Figure 4g shows the formation of large solidification bulges, but these bulges are only disorderly overlapping. No boundaries are found between these bulges, and no pores are formed on the surface. When the droplet is on its surface, the height of the convex overlap is different, thus leading to the different heights of the bulges that support the droplets, and the surface is only slightly hydrophobic. With the increase in the pulse duration, the surface morphology also changes considerably. Figure 4h,i indicates similar surface morphologies, but their contact angles differ remarkably because, when the pulse duration is 30 μs, the surface morphology contains additional shallow pores. Moreover, the droplets will be in contact with additional solids, and the contact angle will be smaller than the one displayed in Figure 4g. When the pulse duration increases to 60 μs, the depth of the pores deepens, and the local "air cushion" effect occurs. The contact angle is larger than the one presented in Figure 4g.

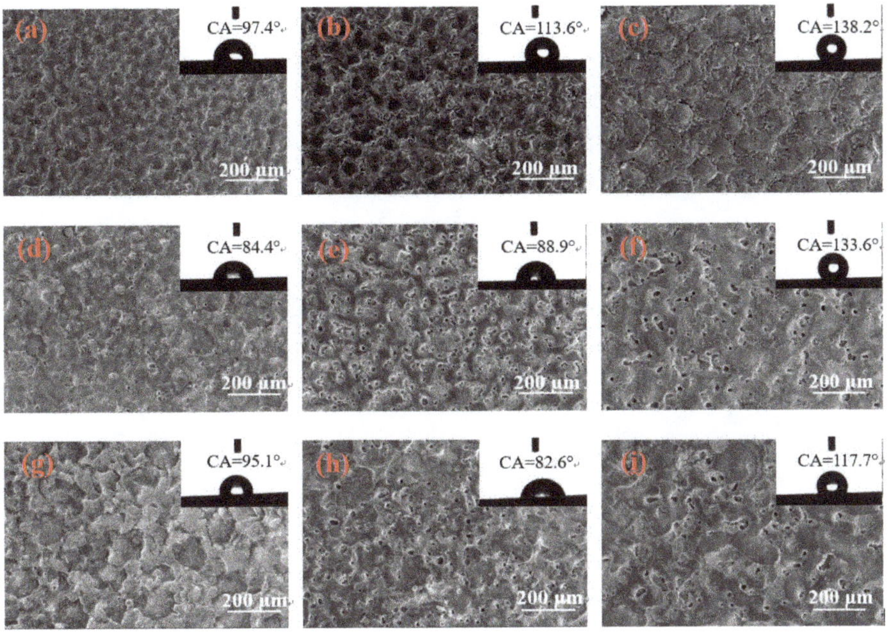

Figure 4. Surface morphology and contact angle under various processing parameters. (**a**) I = 1.5 A, T_{on} = 15 μs; (**b**) I = 1.5 A, T_{on} = 30 μs; (**c**) I = 1.5 A, T_{on} = 60 μs; (**d**) I = 4.5 A, T_{on} = 15 μs; (**e**) I = 4.5 A, T_{on} = 30 μs; (**f**) I = 4.5 A, T_{on} = 60 μs; (**g**) I = 9 A, T_{on} = 15 μs; (**h**) I = 9 A, T_{on} = 30 μs; (**i**) I = 9 A, T_{on} = 60 μs.

Figure 3 shows that the number of pores on the surface increases when the peak currents are 4.5 and 9 A given the increase in pulse duration. Figure 4 illustrates that, when the peak currents are 1.5 and 4.5 A, the hydrophobicity of the surface increases with the pulse duration. Considering that the peak current is less than 1.5 A, the discharge becomes atypical, and the processing efficiency is extremely low. Only the pulse duration is considered to continuously increase to observe the change in surface characteristics. When the pulse duration is increased to 90 μs, the results obtained are depicted in Figure 5. Figure 5b demonstrates a high multiple surface topography under the peak current of 1.5 A.

Figure 5 shows that, when the pulse duration increases to 90 μs, the contact angles of the three currents are reduced. When the peak current and pulse duration are 1.5 A and 90 μs, respectively, the size is slightly larger in the solidification bulges than with the pulse duration of 60 μs. However, Figure 5b displays no "coral reef" structure on the surface. Therefore, a large solidification bulge indicates a large contact area. Furthermore, considering that no "coral reef" structure is found on the surface, the secondary "air cushion" effect cannot be formed, thereby reducing the contact angle. When the peak current is 4.5 A, additional craters are found on the surface. The surface morphology is poor, and the air cushion effect cannot be formed. Thus, the contact angle decreases considerably. When the peak current is 9 A, irregular discharge traces are found on the surface and the number of pores decreases. Thus, the "air cushion" effect is reduced, and the contact angle is also reduced.

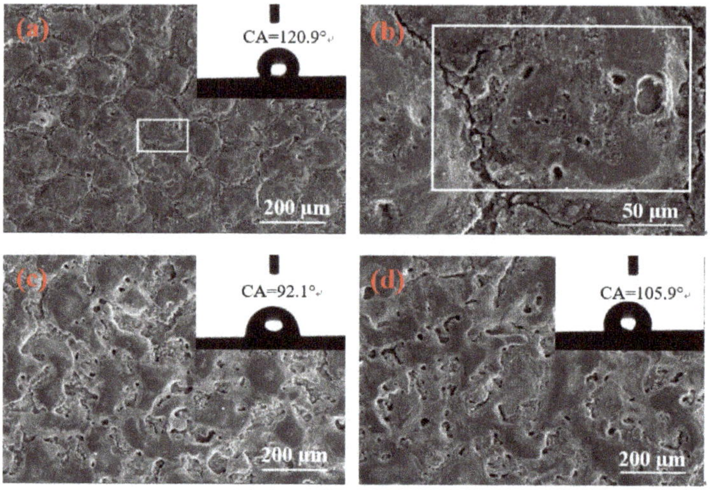

Figure 5. Surface topography and contact angle under different peak current conditions with pulse duration of 90 μs. (**a**,**b**) I = 1.5 A, (**c**) I = 4.5 A, (**d**) I = 9 A.

3.3. Element Analysis and XRD Analysis of Workpiece Surface

The analysis presented in Section 3.2 suggests that the contact angles of different surface morphologies are considerably different. Thus, the two samples with the largest difference in contact angles were selected for surface element analysis. The analysis area is illustrated in the white box in Figure 3c,h, and the results are presented in Figure 6.

The very high temperature of plasma instantly melted and vaporized the material during the discharge process. The copper electrode and Ni-Ti workpiece were eroded, and complex chemical reactions between the vaporized gas and the molten pool were observed. Therefore, the existence of Cu was found on the surface of the workpiece. As a result of magnetic mixing, the high cooling rate of liquid dielectric resulted in Cu diffusion on the surface, rather than it being flushed away. The presence of O can be attributed to decomposing water in the discharge gap. Figure 6 shows that different surface topographies have various C contents, and the surface morphology exhibited in Figure 6b contains "coral reef" with the highest C contents on the surface. This result is due to its processing parameters, I = 1.5 A and T_{on} = 60 μs, and to the large pulse duration, thereby leading to expanding discharge channels. The corrosion products are not thrown out in time but are rather re-solidified on the surface. Simultaneously, given the poor thermal conductivity and extensive pulse width of Ni-Ti alloy, the working fluid decomposes considerable C for the highly active Ti to absorb the considerable C in the process of re-solidification. Thus, the C content on the surface is high. Figure 6a,b indicates that the Ni content is different under various processing parameters because Ni has a lower evaporation temperature than Ti. Moreover, with the increase in the pulse duration, additional Ni is removed from the surface by the liquid dielectric. In addition, TiC substances are found on the surface of Figure 6c,d, which will further decrease the content of Ni. Thus, the Ni content decreases.

Ni-Ti alloy has good biocompatibility and corrosion resistance considering the natural titanium film found on its surface. However, the natural titanium film cannot withstand the long-term erosion of human physiological body fluid and blood. Once the natural titanium film breaks down, nickel ions will precipitate, and excessive nickel ions will cause anaphylaxis and tissue necrosis [23]. The TiC prepared on the surface of Ni-Ti alloy can effectively improve its biocompatibility and mechanical strength, and inhibit nickel ion precipitation [24]. Figure 6c,d presents the XRD of two characteristic surfaces. TiC strengthening phases are found on the machined surface, thereby confirming that C on the workpiece surface exists in the form of a compound. The positions of TiC diffraction peaks in the two graphs are the same but the intensity is different, indicating that the content of TiC varies.

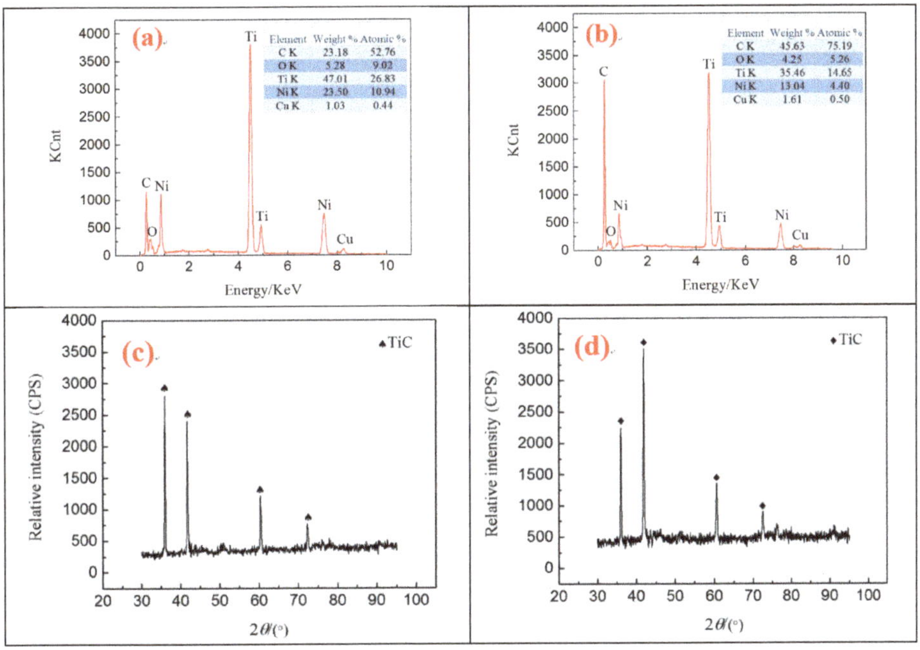

Figure 6. Surface element analysis and XRD analysis under different processing parameters. (**a**) $I = 9$ A, $T_{on} = 30$ μs; (**b**) $I = 1.5$ A, $T_{on} = 60$ μs; (**c**) $I = 9$ A, $T_{on} = 30$ μs; (**d**) $I = 1.5$ A, $T_{on} = 60$ μs.

3.4. Subsurface Microstructure

Two specimens with the largest difference in contact angle were selected for the cross-section analysis. The results are shown in Figure 7.

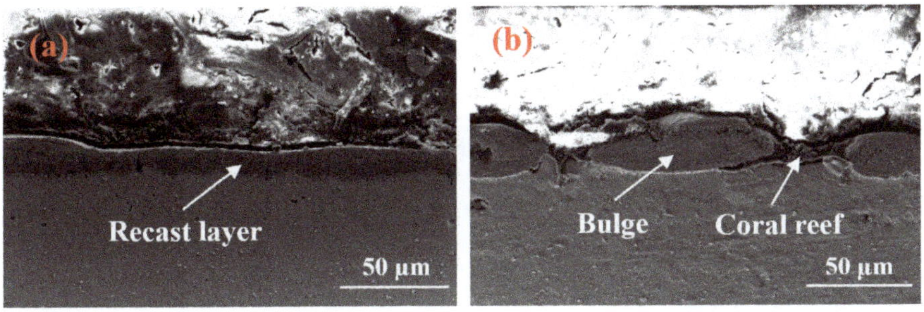

Figure 7. Cross-sectional view of recast layer. (**a**) $I = 9$ A, $T_{on} = 30$ μs; (**b**) $I = 1.5$ A, $T_{on} = 60$ μs.

The recast layer depicted in Figure 7b is non-uniform, but this layer is thicker than the one shown in Figure 7a because Figure 7b exhibits a large pulse duration. Moreover, when the duration of discharge is extended, melting isothermals penetrate further into the material interior, thereby resulting in the molten zone's further extension into the material and a large recast layer thickness. The surface topography is displayed in Figure 3c. This layer is composed of solidification bulges, "coral reef," micropores, and craters. Thus, the thick part of the recast layer is the solidification bulge, and the thinner part is the "coral reef." The contact angle can be considerably increased because this unique structure can seal considerable air and form an "air cushion" effect. Figure 7a indicates that numerous materials are removed given the high pulse discharge energy, and the cross-section morphology is uniform and cannot form the "air cushion" effect.

3.5. Relationship of Surface Roughness and Contact Angle

Figure 8a illustrates the influence of processing parameters on the contact angle. When the peak currents are 1.5 and 4.5 A, the contact angle increases first and then decreases with the increase in pulse duration. According to the analysis in Section 3.2, this phenomenon occurs because the surface morphology has dramatically changed with the increase in pulse duration. When the pulse duration is below 60 μs, the size of the solidification bulges increases with the pulse duration, but the surface that forms the "air cushion" effect has not formed or is insufficiently large. This phenomenon leads to a large contact area between liquids and solids so that the contact angle increases are insufficiently large. Although the size of the solidification bulges is large when the pulse duration increases to 90 μs, the contact angle decreases differently with the decrease or absence of the "air cushion" effect. When the peak current is 9 A, the contact angle varies irregularly with the increase in pulse duration. However, when the pulse durations are 60 and 90 μs, the law is consistent with the law of other parameters because the suitable surface can form an "air cushion" effect, which helps enhance the contact angle. Within the range of parameters used in this study, the optimum parameters for preparing a hydrophobic surface are I = 1.5 A, T_{on} = 60 μs, and V = 50 V. However, considering the number of pores on the surface to facilitate drug loading, the parameters are I = 4.5 A, T_{on} = 60 μs, and V = 50 V.

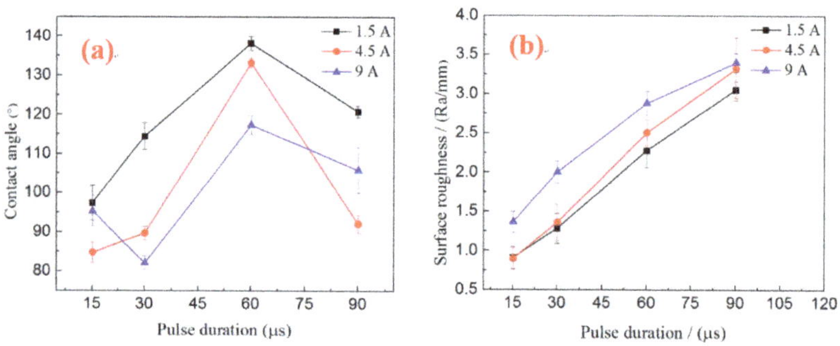

Figure 8. Effect of machining parameters on contact angle and surface roughness. (**a**) Influence of parameters on contact angle; (**b**) Influence of parameters on roughness.

Figure 8b depicts the effect of processing parameters on surface roughness. The pulse energy enlarges with the increase in the pulse duration and peak current. Thus, surface roughness increases accordingly. Several researchers have suggested that a contact angle increases with roughness [25]. However, Figure 8a,b shows that, when the pulse duration is between 60 and 90 μs, the surface roughness increases with the pulse duration, but the contact angle decreases with the increase in the pulse duration. Under the same pulse duration, the peak current and the roughness of the surface are large, but the contact angle is not necessarily large. The experimental results show that surface contact angles are not entirely increasing with surface roughness. Therefore, only suitable surface roughness and surface topography can achieve improved hydrophobicity.

4. Conclusions

This study investigated the processing of Ni-Ti alloy through magnetic mixed EDM with different parameters and carried out an analysis of its surface properties. The key results are summarized as follows.

(1) When the peak currents are 4.5 and 9 A and the pulse durations are 30 and 60 μs, the porous surface can be prepared on the surface of Ni-Ti alloy through magnetic mixed EDM. The composite structure of solidification bulge–crater–pore–particle can be prepared on the surface of Ni-Ti alloy through magnetic mixed EDM with suitable processing parameters.

(2) The processing parameters considerably affect the surface morphology. Different processing parameters can obtain various surface microstructures, and the surface microstructures directly affect the hydrophobicity of the surface. When the peak current and pulse duration are 1.5 A and 60 μs, respectively, the surface is composed of many large-sized solidification bulges and "coral reef" structures, and the contact angle of the surface can reach 138.2°. However, considering the number of pores on the surface to facilitate drug loading, the optimal parameters of this study are I = 4.5 A, T_{on} = 60 μs, and V = 50 V.

(3) The C content on the surface of the workpiece increases rapidly with the increase in the pulse width and the decrease in the peak current.

Author Contributions: Conceptualization, C.C.F. and L.L.; methodology, C.C.F. and L.L.; software, C.C.F. and C.S.Z.; validation, C.C.F., L.L. and G.M.Z.; formal analysis, C.C.F. and L.L.; investigation, X.B.; resources, Z.W.N.; data curation, C.C.F.; writing—original draft preparation, C.C.F.; writing—review and editing, C.C.F. and L.L.; visualization, C.S.Z.; supervision, L.L. and G.M.Z.; project administration, L.L. and X.B.; funding acquisition, L.L.

Funding: This research was funded by Natural Science Foundation of Shandong Province (ZR2018MEE025).

Conflicts of Interest: The authors declare no conflict of interest.

References

1. Jani, J.M.; Leary, M.; Subic, A.; Gibson, M.A. A review of shape memory alloy research, applications and opportunities. *Mater. Des.* **2014**, *56*, 1078–1113. [CrossRef]
2. Sun, L.; Huang, W.M.; Ding, Z.; Zhao, Y.; Wang, C.C.; Purnawali, H.; Tang, C. Stimulus-responsive shape memory materials: A review. *Mater. Des.* **2012**, *33*, 577–640. [CrossRef]
3. Henderson, E.; Buis, A. Nitinol for prosthetic and orthotic applications. *J. Mater. Eng. Perform.* **2011**, *20*, 663–665. [CrossRef]
4. Park, C.; Kim, S.; Kim, H.E.; Jang, T.S. Mechanically stable tantalum coating on a nano-roughened NiTi stent for enhanced radiopacity and biocompatibility. *Surf. Coat. Technol.* **2016**, *305*, 139–145. [CrossRef]
5. Moradi, S.; Hadjesfandiari, N.; Toosi, S.F.; Kizhakkedathu, J.N.; Hatzikiriakos, S.G. Effect of extreme wettability on platelet adhesion on metallic implants: From super hydrophilicity to super hydrophobicity. *ACS Appl. Mater. Interfaces* **2016**, *8*, 17631–17641. [CrossRef] [PubMed]
6. Sanli, M.; Victoria, L.; Arun, K.; Popat, K.C. Effect of super hydrophobicity/super hydrophilicity of titanium surface on hemocompatibility. *Front. Bioeng. Biotechnol.* **2016**. [CrossRef]
7. Li, W.; Wang, Z.; Wang, D.; Zhang, Z.; Li, D.; Qiu, R.; Maple, C. Super hydrophobic dual micro-and nanostructures fabricated by direct laser interference lithography. *Opt. Eng.* **2014**, *53*, 34–109.
8. Tepe, G.; Schnorr, B.; Albrecht, T.; Brechtel, K.; Claussen, C.D.; Scheller, B.; Speck, U.; Zeller, T. Angioplasty of femoral-popliteal arteries with drug-coatedballoons: 5-year follow-up of the thunder trial. *JACC Cardiovasc. Interv.* **2015**, *8*, 102–108. [CrossRef]
9. Weinert, K.; Petzoldt, V.; Kötter, D. Turning and drilling of niti shape memory alloys. *CIRP Ann.* **2004**, *53*, 65–68. [CrossRef]
10. Biermann, D.; Kahleyss, F.; Surmann, T. Micromilling of niti shape-memory alloys with ball nose cutters. *Adv. Manuf. Proc.* **2009**, *24*, 1266–1273. [CrossRef]
11. Xu, Y.D.; Zhu, Z.Y.; Xu, T.Z.; Abadikhah, H.; Wang, J.W.; Xu, X.; Agathopoulos, S. Fabrication and characterization of robust hydrophobic lotus leaf-like surface on Si_3N_4, porous membrane via, polymer-derived SiNCO inorganic nanoparticle modification. *Ceram. Int.* **2018**. [CrossRef]

12. Kietzig, A.M.; Mirvakili, M.N.; Kamal, S.; Englezos, P.; Hatzikiriakos, S.G. Laser-patterned super-hydrophobic pure metallic substrates: Cassie to wenzel wetting transitions. *J. Adhes. Sci. Technol.* **2011**, *25*, 2789–2809.
13. Jeong, C.; Choi, C.H. Single-step direct fabrication of pillar-on-pore hybrid nanostructures in anodizing aluminum for superior super hydrophobic efficiency. *ACS Appl. Mater. Interfaces* **2012**, *4*, 842–848. [CrossRef] [PubMed]
14. Wan, Y.L.; Lian, Z.X.; Yu, H.D.; Xu, J. Experimental study on super-hydrophobic copper surface fabricated by HS-WEDM. *Appl. Mech. Mater.* **2014**, *687–691*, 4367–4371. [CrossRef]
15. Hsieh, S.F.; Lin, M.H.; Chen, S.L.; Qu, S.F.; Huang, T.S.; Zhou, X.Q. Surface modification and machining of tini/tinb-based alloys by electrical discharge machining. *Inter. J. Adv. Manuf. Technol.* **2016**, *86*, 1475–1485. [CrossRef]
16. Liu, J.F.; Guo, Y.B. Process capability of Wire-EDM of NiTi shape memory alloy at main cut and trim cut modes. *Proc. Manuf.* **2015**, *1*, 904–914. [CrossRef]
17. Alidoosti, A.; Ghafari-Nazari, A.; Moztarzadeh, F.; Jalali, N.; Moztarzadeh, S.; Mozafari, M. Electrical discharge machining characteristics of nickel–titanium shape memory alloy based on full factorial design. *J. Intell. Mater. Syst. Struct.* **2013**, *24*, 1546–1556. [CrossRef]
18. Li, L.; Wei, X.T.; Li, Z.Y. Surface integrity evolution and machining efficiency analysis of w-edm of nickel-based alloy. *Appl. Surf. Sci.* **2014**, *313*, 138–143. [CrossRef]
19. Wan, Y.L.; Xu, L.N.; Yu, H.D. Fabrication of superhydrophobic aluminium alloy surface by twice nanosecond laser scanning. *Micro Nano Lett.* **2017**, *13*, 469–472. [CrossRef]
20. Liu, W.Y.; Luo, Y.T.; Sun, L.Y.; Wu, R.M.; Jiang, H.Y.; Liu, Y.J. Fabrication of the superhydrophobic surface on aluminum alloy by anodizing and polymeric coating. *Appl. Surf. Sci.* **2013**, *264*, 872–878. [CrossRef]
21. Wan, Y.L.; Liu, Z.G.; Dong, B.; Xu, L.N.; Xu, J.K.; Yu, H.D.; Yu, Z.J. Orthogonal experiment on super-hydrophobic aluminum surface by high-speed wire electrical discharge machining. In Proceedings of the 2016 IEEE International Conference on Mechatronics and Automation, Harbin, China, 7–10 August 2016.
22. Wan, Y.L.; Lian, Z.X.; Lou, J.; Yu, H.D. Fabrication and Coupling Mechanism of Super-hydrophobical Aluminium Alloy Surface. *CHN Surf. Eng.* **2014**. [CrossRef]
23. Shanaghi, A.; Chu, P.K. Enhancement of mechanical properties and corrosion resistance of NiTi alloy by carbon plasma immersion ion implantation. *Surf. Coat. Technol.* **2018**. [CrossRef]
24. Poon, R.W.; Yeung, K.W.; Liu, X.Y.; Chu, P.K.; Chung, C.Y.; Lu, W.W.; Cheung, K.M.C.; Chan, D. Carbon plasma immersion ion implantation of nickel-titanium shape memory alloys. *Biomaterial* **2005**, *26*, 2265–2272. [CrossRef] [PubMed]
25. Xu, J.; Zhang, L.S.; Wan, Y.L.; Xu, L.Z.; Huang, Y. Experimental study of super-hydrophobic brass surface obtained by WEDM. *J. Changchun Univ. Sci. Technol.* **2015**, 3–40.

materials

MDPI

Article

Evaluation of the Shape Memory Effect by Micro-Compression Testing of Single Crystalline Ti-27Nb Ni-Free Alloy

Takashi Nagoshi [1,*], Takahisa Yasuda [2,3], Nao Otaki [2], Masaki Tahara [2], Hideki Hosoda [2] and Masato Sone [2,3]

1 National Institute of Advanced Industrial Science and Technology, 1-2-1 Namiki, Tsukuba, Ibaraki 305-8564, Japan
2 Institute of Innovative Research, Tokyo Institute of Technology, Yokohama 226-8503, Japan; yasuda.t.aee@gmail.com (T.Y.); otaki.8pn.nao@jp.nipponsteel.com (N.O.); tahara.m.aa@m.titech.ac.jp (M.T.); hosoda.h.aa@m.titech.ac.jp (H.H.); sone.m.aa@m.titech.ac.jp (M.S.)
3 CREST, Japan Science and Technology Agency, Yokohama 226-8503, Japan
* Correspondence: nagoshi-t@aist.go.jp; Tel.: +81-29-861-8286

Received: 26 November 2019; Accepted: 24 December 2019; Published: 25 December 2019

Abstract: In this work, micro-compression tests are performed at various temperatures with Ti-27Nb (at.%) single crystalline pillars to investigate anisotropic deformation behavior, including the shape memory effect. In non-tapered single-crystal pillars with loading directions parallel to [001], [011], and [111], transformation strain and stress show orientation dependence. [001]-oriented micropillars with aspect ratios of 2 and 1.5 demonstrate temperature-dependent transformation stress during micro-compression at various temperatures. Although more stress is required to induce martensite transformation in the pillar with the lower aspect ratio, the temperature dependence of ~1.8 MPa/K observed in both pillars is in good agreement with that of bulk Ti-27Nb.

Keywords: shape memory alloy; temperature variable micro-compression test; single crystal; biomedical alloy

1. Introduction

Shape memory is a property of materials in which materials can recover their shape even after plastic deformation. The unique properties of shape memory alloys originate from martensitic transformation. Shape recovery in these alloys occurs with unloading or heating [1]. The shape memory effect has been widely utilized in sensing, actuation, energy conversion, and smart devices [2,3]. Shape memory alloys are used in bone plates, coils, clips, and blood vessel stents owing to their superior shape memory properties, which include super-elasticity and corrosion resistance. Ti–Ni shape memory alloys are widely employed as biomedical materials. On the other hand, development of Ni-free biomedical shape memory alloys is important for biomedical applications due to concerns about Ni hypersensitivity [4,5].

Recent research has shown that β-type Ti–Nb binary alloys have attractive properties for a wide range of biomedical applications [6,7]. Ti–Nb alloys retain the β phase at room temperature when they contain sufficient amounts of the β-stabilizing element Nb. The shape memory effect in these materials originates in martensitic transformation from the body-centered cubic β phase to C-centered orthorhombic α″ martensite, which can be transformed back to the β phase by unloading or heating. The shape memory properties of Ti–Nb-based alloys have been investigated in various studies, and the effects of alloying [8,9], thermomechanical treatment [10,11], and interacting deformations [12,13] on shape memory have been evaluated. However, the anisotropic features of their mechanical properties have not been well documented due to the difficulty of single crystal fabrication. For a better understanding of the nature of Ti–Nb shape memory alloys, it is very important to investigate their mechanical properties using single crystalline pillars. In addition, the mechanical behavior of a sample at the microscale is different from that of the bulk material because the strength of a metal increases as sample size decreases [14,15]. The effect of sample size on the shape memory of alloys has also been a topic of interest for applications in microelectromechanical systems (MEMS). Recent experimental studies have revealed a wide range of influences that sample size has on the shape memory effect [16–18]. In the present work, the shape memory effect in Ti–27Nb single crystals at scales relevant for MEMS applications was investigated using micro-compression testing at various temperatures.

2. Materials and Methods

Ti–27Nb (at.%) alloy ingots were prepared by arc melting with a copper crucible and tungsten electrode in an argon atmosphere. Ingots were solution treated at 1173 K for 1 h in an argon atmosphere followed by water quenching, and the oxidized surface layer that formed by water quenching was removed by chemical etching at 333 K with a solution of HF:HNO_3:H_2O = 7:8:10 in volume. Then, the ingots were mechanically thinned to a thickness of 0.1 mm. Compression micropillars were fabricated at the edges of thin plates secured in a custom-built aluminum holder which could be easily transferred for SEM, focused ion beam (FIB) modification, and experiments with a mechanical testing machine. Non-tapered pillars were formed by ion irradiation vertical to the pillar axis using a FB2001 FIB (Hitachi, Tokyo, Japan). Details of the fabrication method can be found elsewhere [19,20]. Two square sections were milled from each pillar by ion irradiation normal to the plate, leaving a rectangle pillar on the plate edge. The corners of the pillar were then removed by irradiation at 45° angles from the plate to form a smaller square pillar. To obtain a smooth surface, damaged areas on each face of the pillar were removed by irradiation with a low-current beam. The final dimensions of the pillars fabricated for micro-compression tests were $10 \times 10 \times 20$ μm^3, and additional pillars $9 \times 9 \times 13.5$ μm^3 in size were fabricated for the temperature-variable micro-compression tests. Measurements of the crystal orientations of the pillars were performed before and after fabrication with a S-4300SE SEM (Hitachi, Tokyo, Japan) equipped with an e-Flash electron back-scatter diffraction (EBSD) detector (Bruker AXS GmbH, Karlsruhe, Germany) to obtain the desired orientations of [111], [011], and [001].

Micro-compression tests were conducted with a test machine developed by our group [19,20]. A schematic diagram of the machine is shown in Figure 1a. The specimen holder attached to the X-Y-Z stage was carefully manipulated to align each pillar with a flat-ended indenter, which was viewed with a charge-coupled device (CCD) camera (WRAYMER, Osaka, Japan). A piezoelectric actuator equipped with an indenter and a load cell was used to apply displacement force at a rate of 0.05 μm/s. Applied displacement and load were recorded with an AC/DC converter every 33 ms. In situ optical images were obtained with the CCD camera, as shown in Figure 1b. For temperature-variable micro-compression, the sample holder equipped with a thermocouple was sandwiched between two aluminum nitride heaters. Repeated loading and unloading cycles were performed and the temperature was increased by 10 K for each cycle. A sufficient waiting period during each cycle allowed the temperature to stabilize until the highest testing temperature of 393 K was reached.

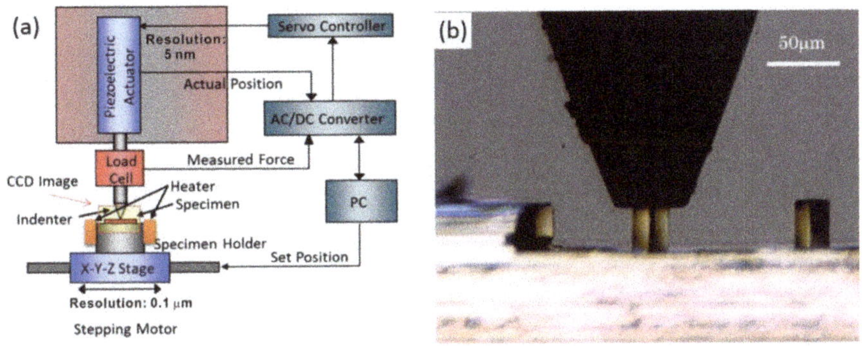

Figure 1. (**a**) Schematic illustration of the temperature-variable micro-compression setup and (**b**) image captured with a charge-coupled device (CCD) camera during in situ compression.

3. Results and Discussion

The compositions of the solution-treated samples were determined with inert gas fusion and inductively coupled plasma analysis with a measurement error of less than 0.3% for each analysis. The material compositions are shown in Table 1. The Nb concentration was very close to the target amount and only low concentrations of other elements were observed. The martensitic transformation start temperature, M_s, at this Nb concentration was estimated to be ~230 K in a previous study [21]. Thus, martensitic transformation and recovery were expected at all temperatures used for the compression tests, including room temperature.

Table 1. Compositions of the Ti–27Nb samples.

Elements	Ti	Nb	Fe	Ni	Cu	Mo	W	Au	O	H	N
Measured composition (at.%)	Bal.	26.9	0.0519	0.0988	0.141	0.0302	0.0221	0.0118	0.119	0.36	0.0936

SEM images of the micro-compression samples with different orientations collected before and after compressive deformation are shown in Figure 2. The pillars had smooth surfaces and were not tapered due to the fabrication method with FIB milling from vertical to pillar axis. The engineering stress–strain curves at orientations of [111], [011], and [001] parallel to the loading axis are shown in Figure 3. The deformation behavior of the pillars differed significantly depending on their crystal orientation. Transformation strain during martensitic transformation from the β to the α″ phase was calculated from changes in the lattice constants. Based on a previous study by Kim et al. [21], the lattice constants of the β phase (a_0) and the orthorhombic α″ phase (a', b', c') were calculated from Nb concentration of samples and were found to be a_0 = 0.3290 nm, a' = 0.3225 nm, b' = 0.4770 nm, and c' = 0.4615 nm. Considering the correspondence between the β and α″ phase lattices, the maximum possible transformation strains during compression at orientations of [111], [011], and [001] were calculated to be 1.2%, 0.8%, and 2.0%, respectively. Plastic strain exceeded the maximum possible transformation strain during deformation of the [011]-oriented pillar, which implied that martensitic transformation did not occur or was mixed with deformation by slip activation. SEM observation following deformation of the [011]-oriented pillar did not reveal any trace of martensitic transformation. The [111]-oriented pillar yielded with martensitic transformation at transformation start stress, σM and deformed at stress for activation of dislocation slip (σSlip). Martensite recovered at transformation recover stress (σRM) with unloading. Less than 1% of the strain was generated from martensitic transformation and further recovered. Low σM was observed in the [001]-oriented pillar, which was unloaded before slip deformation could occur. The transformation strain was about 1%, and partial recovery was observed. The transformation strain should be smaller than the maximum value during pillar compression because transformation is limited to the body of the pillar. This is illustrated clearly in Figure 2d. The middle of the pillar was transformed and traces of the post-transformation planes are visible at each end. Taking the non-transformed areas at the top and bottom of the pillar into consideration, the calculated transformation strain and the observed strain during martensitic transformation are in good agreement.

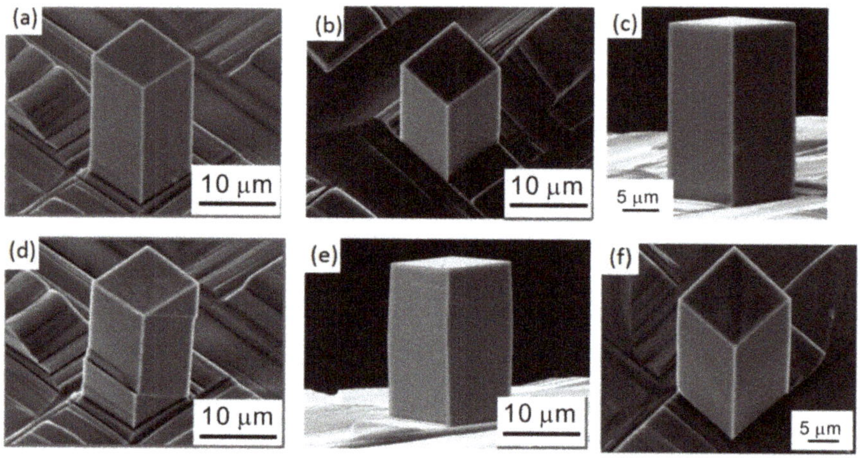

Figure 2. SEM images of the pillars before (**a**–**c**) and after (**d**–**f**) the micro-compression test. Orientations of the specimens: (**a**,**d**) [111]; (**b**,**e**) [011]; and (**c**,**f**) [001].

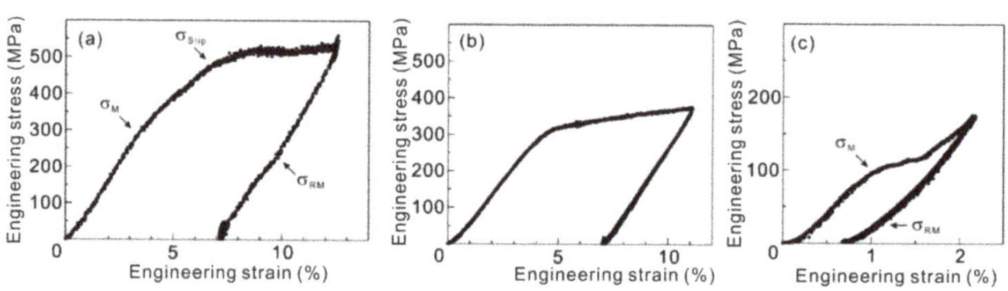

Figure 3. Engineering stress–strain curves of the single crystalline pillars with orientations (**a**) [111], (**b**) [011], and (**c**) [001].

Temperature-variable micro-compression tests were performed with the [001]-oriented sample, which clearly yielded to transformation at a low stress of ~100 MPa. Transformation stress, σ (MPa), was recorded as the stress at initial yielding in each loading and unloading cycle. Transformation stress has been plotted according to testing temperature in Figure 4. Transformation stress was found to be higher in pillars with an aspect ratio of 1.5, which may have been due to the larger impact of lateral stress at the tops of the pillars from the contacting indenter. However, the relationship between transformation stress and testing temperature at both aspect ratios was almost identical. The stress required for martensitic transformation reflected the thermodynamic stability of the β phase, which is undisputedly thought to obey the well-known Clausius-Clapeyron relationship, i.e.:

$$\frac{\Delta\sigma}{\Delta T} = -\frac{\rho\Delta S}{\varepsilon_M} = -\frac{\rho\Delta H}{T \times \varepsilon_M} \tag{1}$$

Here, ρ is the density of the transformation phase, ΔS and ΔH are the changes in entropy and enthalpy during transformation, respectively, and ε_M is the maximum transformation strain. These parameters are constant in a given alloy system with the exception of transformation strain, which depends on the direction of stress relative to the crystal orientation. Al-Zain et al. [22] have reported the dependence of transformation stress on temperature in the Ti–27Nb alloy as 2 MPa/K. This value is slightly larger than the value of 1.8 MPa/K with a root mean square error (RMSE) of 0.3 obtained in this work for compression of the [001]-oriented single crystal. However, considering the difference in transformation strain, the dependence of transformation strain on temperature observed in the present study is in good agreement with the bulk value. The difference in transformation strain would also be responsible for the orientation-dependent transformation stress observed during single-crystal pillar compression. Higher transformation stress in the [111] pillar was due to the lower transformation strain in the [111] orientation.

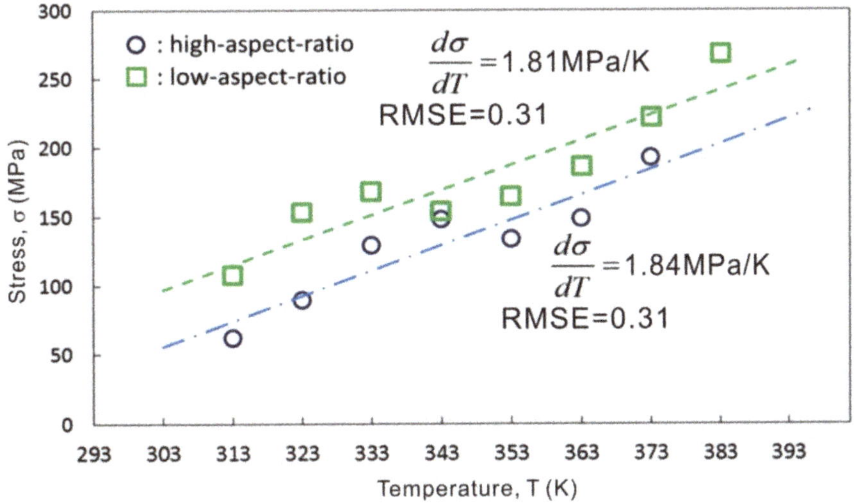

Figure 4. Dependence of transformation stress on temperature during micro-compression. Legend: RMSE, root mean square error.

4. Conclusions

In this study, we investigated the shape memory effect of a Ti–27Nb biomedical alloy. Non-tapered single-crystal micropillars were fabricated by FIB and underwent compression testing with a custom-designed, variable temperature micro-compression system. Anisotropic deformation behavior was observed during compression at room temperature in pillars with orientations of [001], [011], and [111]. Changes in strain and stress with martensitic transformation were explained well by the calculated transformation strains for the given loading directions. Taking into account the specimen orientation, the dependence of transformation strain on temperature was in good agreement with the bulk value. The shape memory effect in Ti–27Nb at scales relevant for MEMS applications was consistent with its bulk properties.

Author Contributions: Conceptualization, H.H., M.S. and T.N.; methodology, T.N.; software, T.N.; validation, T.N., M.T. and M.S.; formal analysis, T.N. and H.H.; investigation, T.Y.; resources, N.O. and M.T.; data curation, T.N., M.T. and H.H.; writing—original draft preparation, T.N.; writing—review and editing, T.N.; visualization, T.N.; supervision, T.N., H.H. and M.S.; project administration, M.S.; funding acquisition, M.S. All authors have read and agreed to the published version of the manuscript.

Funding: This work is supported by CREST Project operated by the Japan science and Technology Agency (JST) and by the Grant-in-Aid for Scientific Research (S) (JSPS KAKENHI Grant number 26220907).

Conflicts of Interest: The authors declare no conflict of interest.

References

1. Otsuka, K.; Wayman, C.M. *Shape Memory Materials*; Cambridge University Press: Cambridge, UK, 1998.
2. Liu, C. *Foundations of MEMS*; Pearson Prentice Hall: Upper Saddle River, NJ, USA, 2006.
3. Worden, K.; Bullough, W.A.; Haywood, J. *Smart Technologies*; World Scientific: River Edge, NJ, USA, 2003.
4. Biesienkierski, A.; Wang, J.; Gepreel, A.-H.M.; Wen, C. A new look at biomedical Ti-based shape memory alloys. *Acta Biomater.* **2012**, *8*, 1661–1669. [CrossRef] [PubMed]
5. Es-Souni, M.; Es-Souni, M.; Fischer-Brandies, H. Assessing the biocompatibility of NiTi shape memory alloys used for medical applications. *Anal. Bioanal. Chem.* **2005**, *381*, 557–567. [CrossRef] [PubMed]
6. Grosdidier, T.; Philippe, M.J. Deformation induced martensite and superelasticity in a β-metastable titanium alloy. *Mater. Sci. Eng. A* **2000**, *291*, 218–223. [CrossRef]
7. Wang, W.; Soper, S.A. *Bio-MEMS: Technologies and Applications*; CRC Press: Boca Raton, FL, USA, 2007.
8. Al-Zain, Y.; Kim, H.Y.; Hosoda, H.; Nam, T.H.; Miyazaki, S. Shape memory properties of Ti–Nb–Mo biomedical alloys. *Acta Mater.* **2010**, *58*, 4212–4223. [CrossRef]
9. Ijaz, M.F.; Kim, H.Y.; Hosoda, H.; Miyazaki, S. Effect of Sn addition on stress hysteresis and superelastic properties of a Ti–15Nb–3Mo alloy. *Scr. Mater.* **2014**, *72*, 29–32. [CrossRef]
10. Hosoda, H.; Kinoshita, Y.; Fukui, Y.; Inamura, T.; Wakashima, K.; Kim, H.Y.; Miyazaki, S. Effects of short time heat treatment on superelastic properties of a Ti–Nb–Al biomedical shape memory alloy. *Mater. Sci. Eng. A* **2006**, *438*, 870–874. [CrossRef]
11. Al-Zain, Y.; Kim, H.Y.; Koyano, T.; Hosoda, H.; Nam, T.H.; Miyazaki, S. Anomalous temperature dependence of the superelastic behavior of Ti–Nb–Mo alloys. *Acta Mater.* **2011**, *59*, 1464–1473. [CrossRef]
12. Zhan, H.; Zeng, W.; Wang, G.; Kent, D.; Dargusch, M. On the deformation mechanisms and strain rate sensitivity of a metastable β Ti–Nb alloy. *Scr. Mater.* **2015**, *107*, 34–37. [CrossRef]
13. Tobe, H.; Kim, H.Y.; Inamura, T.; Hosoda, H.; Miyazaki, S. Origin of {3 3 2} twinning in metastable β-Ti alloys. *Acta Mater.* **2014**, *64*, 345–355. [CrossRef]
14. Uchic, M.D.; Dimiduk, D.M.; Florando, J.N.; Nix, W.D. Sample dimensions influence strength and crystal plasticity. *Science* **2004**, *305*, 986–989. [CrossRef] [PubMed]
15. Greer, J.R.; Oliver, W.C.; Nix, W.D. Size dependence of mechanical properties of gold at the micron scale in the absence of strain gradients. *Acta Mater.* **2005**, *53*, 1821–1830. [CrossRef]
16. Norfleet, D.M.; Sarosi, P.M.; Manchiraju, S.; Wagner, M.X.; Uchic, M.D.; Anderson, P.M.; Mills, M.J. Transformation-induced plasticity during pseudoelastic deformation in Ni–Ti microcrystals. *Acta Mater.* **2009**, *57*, 3549–3561. [CrossRef]
17. Chen, Y.; Schuh, C.A. Size effects in shape memory alloy microwires. *Acta Mater.* **2011**, *59*, 537–553. [CrossRef]
18. Frick, C.P.; Orso, S.; Arzt, E. Loss of pseudoelasticity in nickel–titanium sub-micron compression pillars. *Acta Mater.* **2007**, *55*, 3845–3855. [CrossRef]
19. Nagoshi, T.; Shibata, A.; Todaka, Y.; Sato, T.; Sone, M. Mechanical behavior of a microsized pillar fabricated from ultrafine-grained ferrite evaluated by a microcompression test. *Acta Mater.* **2014**, *73*, 12–18. [CrossRef]
20. Nagoshi, T.; Chang, T.F.M.; Sone, M. Mechanical Property Evaluation of Electrodeposited Nanocrystalline Metals by Micro-testing. In *Electroplating of Nanostructures*; Aliofkhazraei, M., Ed.; IntechOpen: London, UK, 2015; pp. 211–240.
21. Kim, H.Y.; Ikehara, Y.; Kim, J.I.; Hosoda, H.; Miyazaki, S. Martensitic transformation, shape memory effect and superelasticity of Ti–Nb binary alloys. *Acta Mater.* **2006**, *54*, 2419–2429. [CrossRef]
22. Al-Zain, Y.; Kim, H.Y.; Koyano, T.; Hosoda, H.; Miyazaki, S. A comparative study on the effects of the ω and α phases on the temperature dependence of shape memory behavior of a Ti–27Nb alloy. *Scr. Mater.* **2015**, *103*, 37–40. [CrossRef]

materials

MDPI

Article

Mechanical Alloying and Electrical Current-Assisted Sintering Adopted for In Situ Ti-TiB Metal Matrix Composite Processing

Andrzej Miklaszewski and Mieczyslaw Jurczyk *

Institute of Materials Science and Engineering, Poznan University of Technology, Jana Pawla II 24, 61-138 Poznan, Poland; andrzej.miklaszewski@put.poznan.pl

* Correspondence: mieczyslaw.jurczyk@put.poznan.pl; Tel.: +48-61-665-3508

Received: 22 January 2019; Accepted: 19 February 2019; Published: 21 February 2019

check for updates

Abstract: In this work, mechanical alloying and electrical current-assisted sintering was adopted for in situ metal matrix composite material processing. Applied at the initial powder stage, mechanical alloying enables a homogeneous distribution of the starting elements in the proposed precursor powder blends. The accompanying precursor preparation and the structurally confirmed size reduction allow obtainment of a nanoscale range for the objects to be sintered. The nano precursors aggregated in the micro-sized particle objects, subjected to electrical current-assisted sintering, characterize the metal matrix composite sinters with high uniformity, proper densification, and compaction response, as well as maintaining a nanoscale whose occurrence was confirmed by the appearance of the highly dispersed reinforcement phase in the examined Ti-TiB material example. The structural analysis of the sinters confirms the metal matrix composite arrangement and provides an additional quantitive data overview for the comparison of the processing conditions. The mechanical alloying examined in this work and the electrical current-assisted sintering approach allow in situ metal matrix composite structures to create their properties by careful control of the processing steps.

Keywords: mechanical alloying; nanoprecursor; electric pulse-assisted sintering; metal matrix composites

---

1. Introduction

For the growing expectations and challenges of modern engineering, Metal Matrix Composite (MMC) materials appear to remain a popular choice. The interest and significant attention accompanying the application of this material group always grows rapidly with the newly introduced general relations or the enhancement of the processing approach. The fields of present and prospective applications of the MMC materials are a result of their structure and their component distribution finds its source mainly in the high mechanical properties also available for high-temperature regimes. The properties of lightweight and corrosion-resistant MMC materials are good not only for military armors and automobile or aircraft engine parts [1–3] but also for biomaterials and hard tissue applications [4]. The investigations conducted in this work focused on the influence of the starting precursors (obtained by mechanical alloying) and current-assisted sintering provide the results that allow some new generalized conclusions for the MMC materials. The final composite material microstructure, through its above mentioned properties with their mutual relations, show an evident relation with the applied processing method.

The starting precursor compositions proposed in this work, based on titanium and boron powders, following the application of mechanical alloying (MA) and current-assisted sintering, allow the obtainment of a nearly fully compacted sinter composite structures, composed of a titanium matrix and a highly dispersed in situ nucleated titanium boride reinforcement phase. For the above type of MMC structures, other research considering different approaches investigated in literature and industrial practice [5–9] has not clearly indicated the relation between the processing method and the starting material, thus limiting the new possibilities of creation of the material properties.

For the proposed example of Ti-TiB metal matrix composite, the function of the phases remains strictly defined. Titanium, distinguished by its specific properties [10], shows a broad possible spectrum of applications. Considered as a matrix phase, it remains challenging during the processing due to its high reactivity with the interstitial elements such as O_2, N_2, C, or H_2 contained in crucibles, the atmosphere or the material of the tools [11]. The selected reinforcement TiB phase appears as the best choice compared to other phases such as TiC, TiN, TiB_2, SiC, B_4C, Al_2O_3 or Si_3N_4 [12]. With its excellent mechanical properties, the thermal expansion coefficient close to that of the the matrix, and balanced thermochemical stability it is distinguished as the best in such a phase. For Ti + TiB composites, a significant improvement of properties could usually be obtained at the expense of the fracture toughness and ductility, especially at high volume fractions of the reinforcement phase [13], where interlocking may occur. The amount of the reinforcement phase should be precisely controlled knowing that its higher share in the composite structure may affect the phase morphology and size [14]. Research conducted by other investigators [15] shows that an average size of the TiB whiskers remains tightly connected with the Ti content in the initial mixtures. The same research also shows that the TiB whiskers in the composites synthesized from the mixtures containing low content of Ti powders have a high tendency to agglomerate. Besides the above, it is noteworthy that in general, the reinforcement phase could appear differently in a composite structure. For example, it could be added and mixed ex situ [16], transformed from another phase [17], or nucleated in situ [18]. The way that the reinforcement phase appears in the composite matrix classifies the applied approach directly by the required processing stages and the undertaken preparation steps or specific actions with its economic profitability and, finally, the material properties. The most important property for the MMC's materials, however, remains the size of the precipitation phase, its amount, and the way of its distribution with the interface region sharpness as well as the overall features of the composite's high density and the microstructure homogeneity. The processing approach that allows manipulating the above-mentioned features remains the best choice for recommendation to undertake a possibly quick industrial implementation path or to conduct breakthrough research in the field of interest.

The data analyzed and concluded to date have shown that much work has been done. However, sufficient attention has not yet been devoted to considering the starting material characteristics including the morphology and the size dimensions. The composition, purity, and morphology of the starting materials as essential densification factors regarding the process control, have not yet been analyzed in terms of the evaluation of the crystallite size range and their influence on the final structure and the obtained relation of properties. As will be confirmed in this work, the nanoscale could be preserved at the final product stage, after a properly defined material processing. Involving earlier data [19], the dimensions of the starting materials may influence the processing temperatures and the time periods as well as the resultant material properties. The obtained values of porosity and density for the compacts or the final sinters are also important and at the same time dependent on the selected processing path, different precursor material composition, or the starting dimension [18]. The density also has a distinct effect on the fracture toughness [12], which is why the presence of pores in any form should be avoided.

The 'top-down' mechanical milling approach applied for the required size transition of the starting materials by itself remains the source of additional effects. The pulverized or ripped reactor edge and vial material pollution as well as the aggregation effect that appears as a result of the powder (fragile or ductile) character remain commonly occurring problems in mechanically milled materials [20]. On the other hand, increased defect density, reduction of the particle size and a possible atomic penetration into the substitutional sites of the lattice with uniform dispersion of the starting components allow simplifying the diffusion processes also by their occurrence at the interfacial regions. For the nanoscale range structure with its inherent relation, the relevant material preparation and processing steps should be adhered to. High temperatures, long time periods of processing, or starting material contamination may often cause a grain growth-, phase-, or porosity- aggregation effect, as well as reduced properties in the final products. The electric current-assisted sintering method with its advantages [21,22] and proper processing control has the potential of overcoming the said limitations.

In this work, the innovative path of MMC material processing by mechanical alloying and current-assisted sintering was adopted. The analysis was carried out on the example mixtures of different nanoscale-sized precursor powder Ti-B compositions with homogenously distributed reactants to evaluate their influence on the final sintered composite properties. General relations were analyzed and conclusions drawn for the in situ MMC's materials group based on the analyzed example. Uniform reinforcement phase dispersion was analyzed for the approach proposed in this work with its size in the microstructure and high densification and compaction response that contribute to the improvement of the composite properties.

2. Materials and Methods

Commercially pure titanium and boron powders (Alfa Aesar, Karlsruhe, Germany) with an average particle size of 45 and 80 μm respectively and a purity level of 99.5% were used for the precursor mixtures preparation. The starting powder morphology and the size were analyzed by the Scanning Electron Microscopy SEM (Vega-Tescan, Brno, Czech Republic) technique for control purposes. Three compositions of precursor blends proposed for the preparation of sinters with different weights of the starting elements were gathered and shown in Table 1 for higher data clarity.

Table 1. Basic information about the starting element powders with the proposed precursor blends compositions and their indications.

Element	Average Particle Size [μm]	Purity Level [%]	Blend Composition 1 [wt.%]	Blend Composition 2 [wt.%]	Blend Composition 3 [wt.%]
Titanium	50	99.5	98	95	90
Boron	80	99.5	2	5	10
Symbol of the obtained after the MA process precursor composition			2%	5%	10%

The starting powders were weighed and mixed in the proposed proportions in the glove box (LabMaster 130, MBraun, Garching, Germany) under high purity argon atmosphere and poured into the hardened tool steel vial. Then, the blends were processed for 48 h, by mechanical alloying (MA) in a Spex Mixer Mill (SPEX SamplePrep, Metuchen, NJ, USA) with the ball-to-powder ratio (BPR) of 15:1. The obtained precursor compositions were structurally analyzed with Cu Kα radiation λ = 1.5406 Å (Panalytical Empyrean-Netherlands) in the range 30−80° to estimate the lattice parameters, the stress and the crystallite size after MA. The Willamson–Hall analysis was applied, where the size and strain induced broadening is deconvoluted by considering the peak width as a function of 2Θ [23] as:

$$\mathrm{B}\cos\Theta = \frac{0.9\lambda}{\mathrm{D}} + 4\varepsilon\sin\Theta \tag{1}$$

where β is the full-width at half-maximum (FWHM) of the diffraction peak, Θ is the position of peak maximum, λ is the X-ray wavelength (λ = 0.15406 nm), D is crystallite size, and ε is the lattice strain. The linear plot course equation obtained for the proposed Uniform Deformation Model approach allows estimating the stress and the crystallite size of the analyzed precursor powder blend compositions.

In order to investigate the obtained precursor powder blends, the aggregated particle size, its morphology (according to GOST 25849 standard) and its distribution profiles, the SEM technique was repeatedly used.

For the sinters preparation, the obtained precursor compositions after synthesis were placed in a non-conductive ceramic die between two graphite punches with the initially applied force of 1 kN under vacuum conditions of 4 Pa according to the processing diagram shown in Figure 1 referred to as isolative die setup.

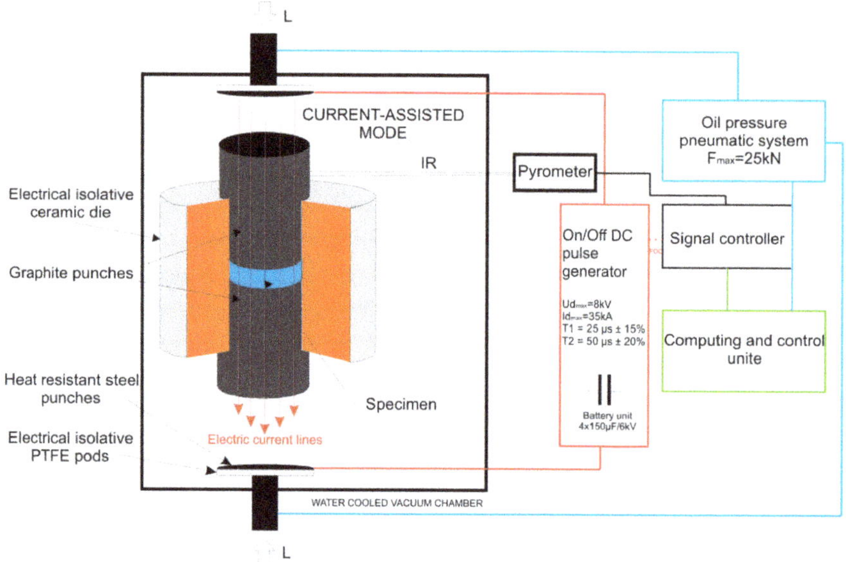

Figure 1. Schematic arrangement of the isolative die setup for a Pulse Plasma Sintering Press module.

The Pulse Plasma Sintering (PPS) module equipment, designed and made by Elbit (Koszyce Wielkie, Poland), allows generating current pulses within a maximum stroke of 60 kA by a discharge of a battery unit capacitor (Elbit, Koszyce Wielkie, Poland), charged to a voltage of maximum 8 kV. The specific pulse duration and the frequency realize the sintering process that is automatically controlled by the measurement of the return temperature whose fluctuation is kept at ±35 °C. The voltage, the pulse frequency, the force, the temperature, and the real-time vacuum measurement remain under constant control during the process performed by a two-way software unit (Elbit, Koszyce Wielkie, Poland) with the latency of 0.1 s. Based on earlier research [18,19,24], the PPS process parameters were proposed for the preparation of sinters gathered and shown in Figure 2. The sample processing was realized using the 'save/load' mode function available in the upgraded software version (Prasa 1.26, Elbit, Koszyce Wielkie, Poland). For each proposed sintering temperature regime, 3 samples of the diameter of Φ10 mm were prepared.

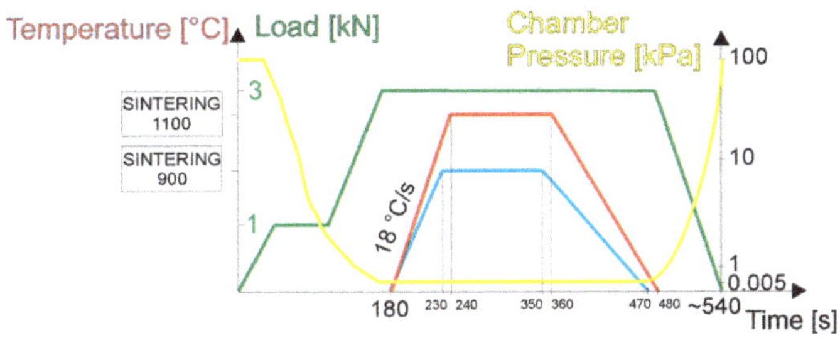

Figure 2. The Pulse Plasma Sintering processing parameter setup using the 'save/ load' operational function, presented in the form of a graph.

A parameter setup including the die material, its geometry and size with the assumed punch distance and the material amount was loaded by the software operational module and was comparatively controlled for all the samples in terms of specific curves of the temperature relation. For further examination, the obtained samples after Pulse Plasma Sintering were ground and polished to a 1-μm finish.

The X-ray diffraction (XRD, Malvern Panalytical, Almelo, Netherlands) method was used for the sample structural characterization. The collected data were refined to estimate the structural parameters and variables such as the background, the profile coefficients, the lattice parameters and the linear absorption coefficients. The instrumental broadening effect for the collected data was eliminated by subtracting the full width at half-maximum (βo) of a standard (Si) sample from β of the respective Bragg peaks. The lattice parameter estimation as well as phase quantitative analysis were based on the Rietveld profile fitting method realized on the Maud software. The applied approach involved a simulation of the diffraction pattern based on the analyzed structural model for:

- Ti(α) (COD 9008517),
- TiB (COD 9008946)

The calculated pattern of the model structure was fitted to the observed one by minimizing the sum of the squares and after a refinement using the Marquardt least squares algorithm. For clarity, residual pattern indicators of the modeled data were revealed:

- Rwp—weighted pattern residual indicator
- Rexp—expected residual indicator
- GOD—goodness of fit

The relative density of the obtained composite sinters was measured by the Archimedes drainage method in order to reveal the microstructure compactness. The porosity of the samples was estimated based on the observation of the optical microscopy images on the non-etched surfaces, realized in the bright and dark field mode on the Olympus (GX51, Olympus, Shinjuku, Tokio, Japan) unit, fitted with Plan Fluorite Objectives and a material solution plug-in dedicated software (Olympus, Shinjuku, Tokio, Japan). The microhardness measurements were carried out on the samples to determine the average hardness level by a Vickers (Innovatest Nexus 4302, INNOVATEST Europe BV, Maastricht, The Netherlands) tester within the 10 indents at an applied load of 300 g and loading time of 10 s. For microstructural observations, the samples were etched in the Kroll's reagent and then characterized by the SEM technique.

3. Results

3.1. Precursor Powder Blend Stadium

The starting substrate elementary titanium and boron powders used for the blend composition preparation were characterized for the purposes of the processing control. The morphology and average particle size were determined from the SEM microphotographs shown in Figure 3 where the powder preparation stadium was presented schematically.

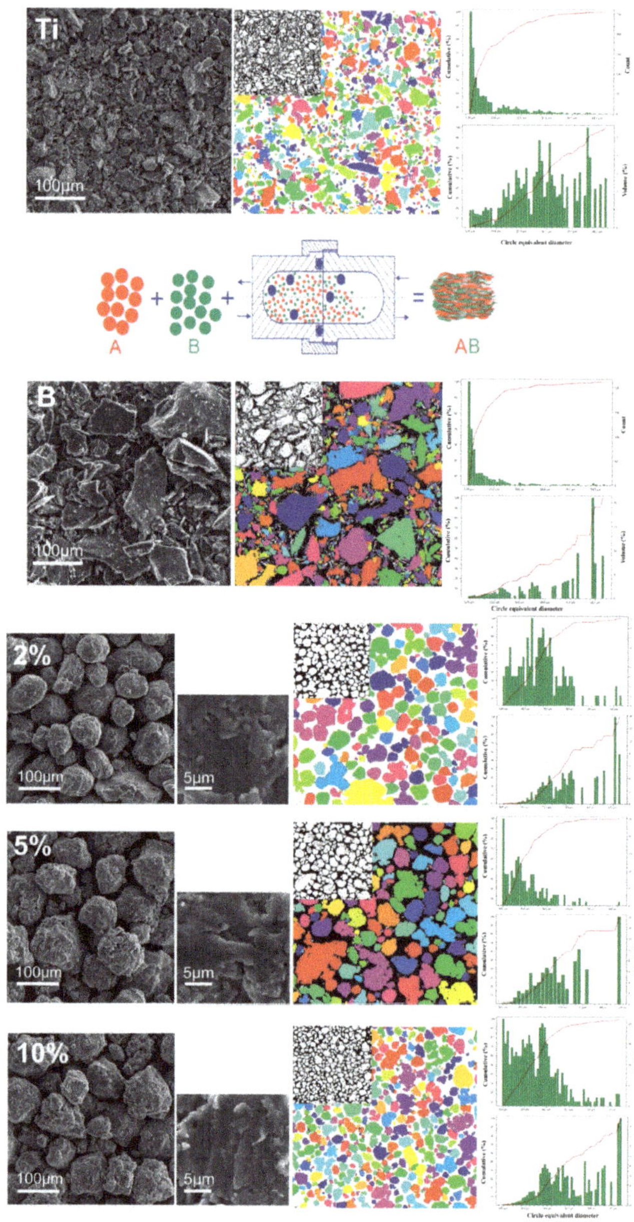

Figure 3. SEM microphotographs showing the powder blends processing stadium from the starting elementary titanium–Ti and boron–B powders to the 48 h mechanically alloyed precursor mixtures with different amounts of boron in the starting composition and their particle size distribution profiles in volumetric and count scales.

The average powder particle size, measured and confirmed with catalogue data, as shown in the attached size distribution profiles, is additionally characterized by an angular shape of the plate-like morphology with a higher size scatter for elementary boron and for a sharp-edged and flat surfaced angular-type titanium. After 48 h, the mechanically alloyed precursor powders, as shown on the SEM microphotographs, are characterized by spheroidal-type particle morphology and a growing angularity along with the boron amount. The additional particle diameter distribution profiles of the obtained precursors in the volumetric and count scales show an ongoing stadium of synthesis dependent on the boron amount that influences the powder average particle size and its scatter—from aggregation to fracture behavior. The profiles manifest a powder fragmentation stadium that occurs during processing and its relation with the boron amount during the synthesis also investigated and confirmed elsewhere [4,19]. The highly magnified SEM microphotographs show a powder particle aggregate structure after MA within smaller visible crystallite parts, also confirmed structurally.

The XRD spectra correlated and shown in Figure 4 show a strong relation between the structural response from the precursor material composition and its processing time.

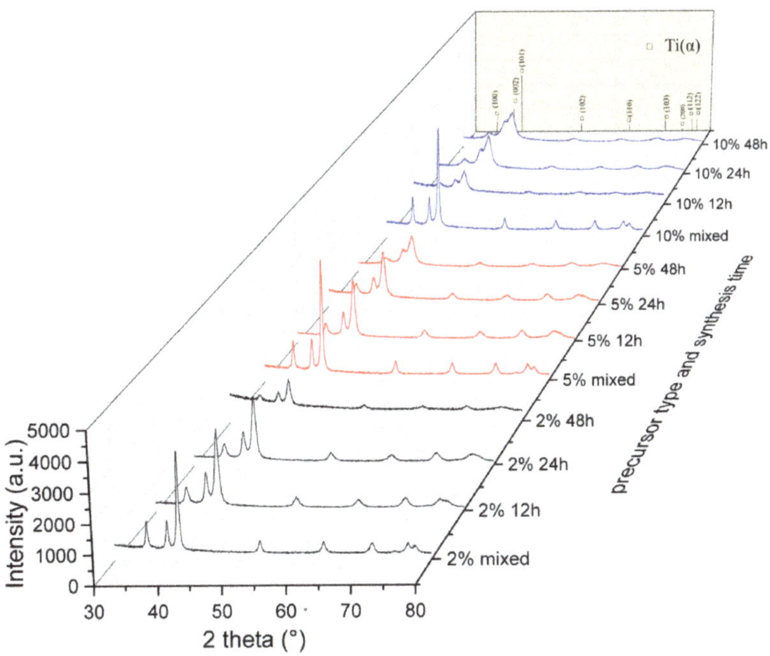

Figure 4. X-ray diffraction spectra of the precursors during mechanical alloying.

The structural analysis of the obtained precursor compositions after mechanical alloying confirms the broadening of the growing peaks and their reduction due to the prolonged processing and the amount of boron. The starting boron amount influences the powder fragmentation characteristics during processing, depending on the strengthening mechanism of the reactants. After 48 h of synthesis, the precursors exhibited fragmentation, homogenization, and structure disintegration. The applied Willamson–Hall Uniform Deformation Model (UDM) approach allows an estimation (from the linear plots) of the crystallite size and the microstrain level related to the starting precursor composition after 48 h of MA. The functions in Figure 5 show a different course of the relation of the largest calculated crystallite size and the smallest microstrains for the 2% precursor sample to the smallest calculated crystallite size and the largest microstrains for the 10% precursor sample.

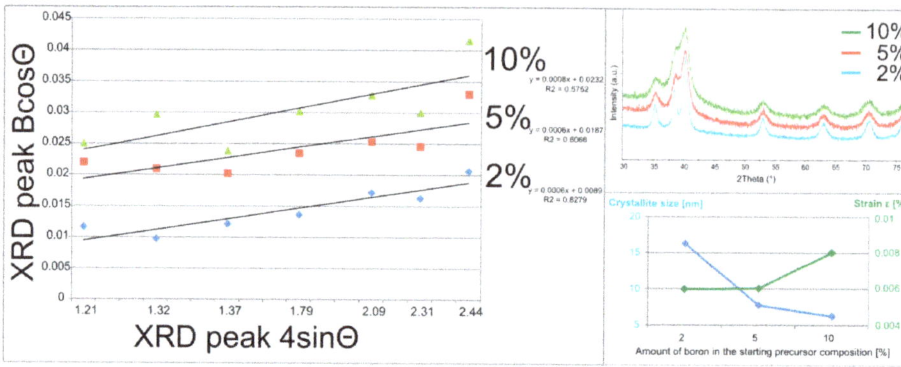

Figure 5. The crystallite size and the strain estimation based on the Willamson–Hall Uniform Deformation Model approach, realized on the XRD spectra of the obtained (after 48 h) mechanically alloyed precursor compositions with their linear plot course.

The nanoscale range confirmed in the obtained precursor compositions, dependent on the boron amount, indicates the same synthesis time, and advanced fragmentation stadium for the higher boron addition also confirmed in the synthesis time dependence in the previously conducted research [4,19]. The growing boron addition for the analyzed precursor powder examples, as the results confirm, increases the speed of precursors fragmentation, homogenization and structure disintegration reactions.

3.2. Sintered Composite Material Stadium

The precursor powder compositions, sintered in different temperature regimes by the PPS method, allowed obtaining metal matrix composite sinters with low porosity and high dispersion of the reinforcement phase in the nanoscale range.

The XRD spectra of the obtained sinters in Figure 6, confirm the composite structure in all the examined samples. The growing amount of boron in the starting precursor powder compositions (as reflected by the presented data for the obtained sinters by the change of ratio of the intensity of the peaks) increases the titanium boride precipitation phase with a simultaneous reduction of the peaks of the titanium matrix phase. The composite phase amount (PA) values estimated from the Rietveld analysis and presented in Table 2, also confirm the differences between the considered processing variants.

Table 2. Crystallographic data of the composite sinters obtained for different PPS temperature regimes (900, 1100 °C) and starting precursor compositions (2%, 5%, 10%), with the estimation of their phase amount and residual pattern indicators from the Rietveld approach analysis.

Treatment Temperature	Sample	Ti(α)			TiB				Rexp [%]	Rwp [%]	GOD
		a [Å]	c [Å]	PA [%]	a [Å]	b [Å]	c [Å]	PA [%]			
900	2%	2.9641(5)	4.7134(8)	89.96	4.5532(70)	6.0850(11)	3.0529(57)	10.04	8.67	19.67	2.27
	5%	2.9629(7)	4.7226(13)	68.82	4.5510(23)	6.0953(33)	3.0581(18)	31.18	7.04	19.73	2.80
	10%	2.9608(5)	4.7313(11)	42.63	4.5524(9)	6.1146(16)	3.0487(7)	57.36	9.06	14.10	1.56
1100	2%	2.9637(3)	4.7137(5)	90.65	4.5608(23)	6.1161(34)	3.0617(19)	9.35	9.22	15.55	1.69
	5%	2.9663(6)	4.7195(11)	70.58	4.5557(22)	6.1169(38)	3.0481(18)	29.42	7.29	19.18	2.63
	10%	2.9692(6)	4.7423(14)	40.59	4.5586(8)	6.1341(15)	3.0488(6)	59.41	9.57	16.66	1.74

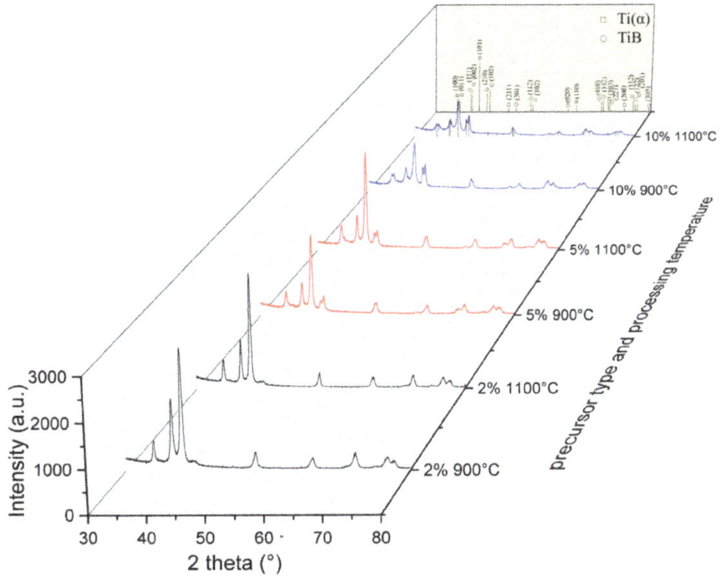

Figure 6. X-ray diffraction spectra of the obtained composite structures analyzed for different Pulse Plasma Sintering (PPS) temperature regimes and starting precursor compositions.

The crystallographic phase data for the titanium matrix also confirm the value of the c-axis parameter growing with the increase in the amount of boron in the precursor powder composition. The above relation remains directly connected with the structure boron intake, possible in accordance with the phase diagram which, for the maximum solubility, remains lower than 0.2 at. % for the hexagonal Ti(α) [25].

The microstructural analysis of sinters shown in Figure 7 confirms the obtained composite structure that consists of a titanium matrix and an in situ nucleated highly dispersed titanium boride reinforcement phase. The higher amount of boron in the starting precursor compositions, as seen in the microphotographs in Figure 7, increases the share of the reinforcement phase and leads to its visible coarsening in the obtained composite microstructure. The above growth at the same time corresponds to the reinforcement phase interlocking with the visible high dispersion in lower sintering temperature regimes. The grain size of the reinforcement phase, with the whisker-type morphology, falls in the nano to submicrometer scale range. Microstructurally compared different sinter treatment temperatures (900 and 1100 °C), also exhibit the precipitation phase clustering behavior in the higher proposed sintering regime for the obtained composite microstructure. The above supposition has its source in the interlocked area transformation. The locally occurring highly dispersed concentration regions of the precipitation phase remain an easy-to-transform microstructure area due to its energy-related favorability. The highest amount of boron in the starting precursor powders may also be recognized in the sinters microstructure by the locally occurring spatial boron diffusion centers described as regular in all-directions, spatial centers of the growth of the reinforcement phase.

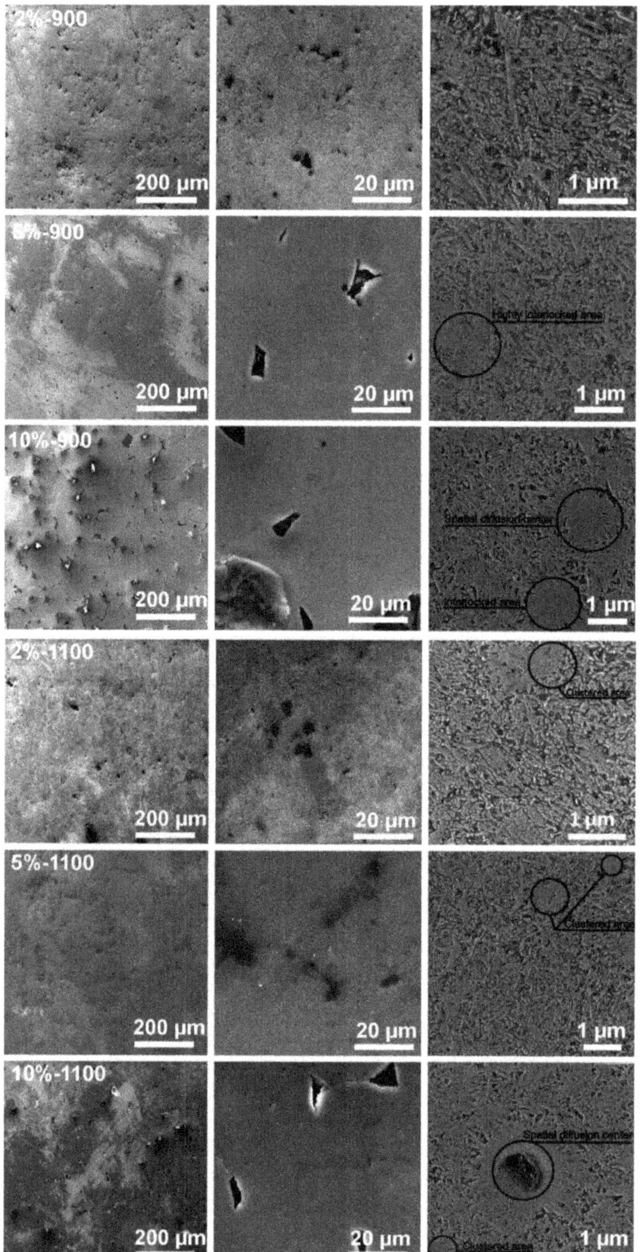

Figure 7. SEM microphotographs of the sinters obtained at different processing temperatures and from different starting precursor compositions (etched and non-etched surfaces).

The examined, low magnified images allow relating the starting precursor particle morphology, the size and composition with its processing temperature to the obtained porosity and microstructure of the sinters. The porosity depends on the amount of boron as well as on the PPS processing temperature and the occurring shapes and its distribution remains strongly related to the above. The higher boron amount in the starting precursors, revealed larger and less regular empty spaces in the obtained sinters (see Figure 7) thus influencing its compactness characteristics-see Figure 8.

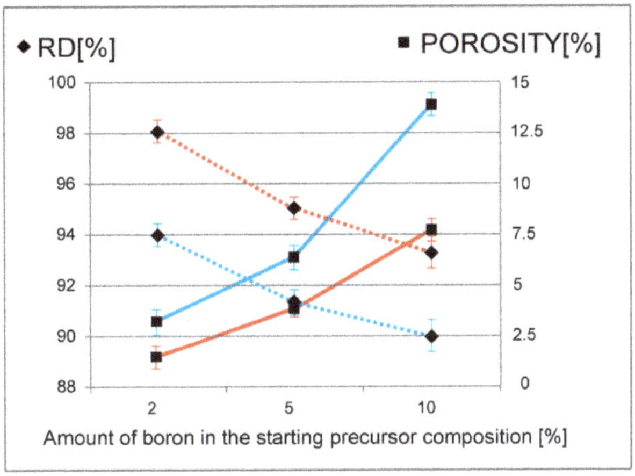

Figure 8. Relative density (RD–dotted lines) and porosity (full lines) analysis of the obtained composite sinters for different PPS temperature regimes (blue—900 °C, red—1100 °C) and ranges of the starting precursor compositions (2%, 5%, 10%).

The observed relation indicates a consistency/boundary regions for the sintered aggregated particles that could be modified by such diffusion moderators as temperature, composition, substrates' size range, and lattice defects density. Higher-boron-amount precursors posses a higher potential for the reinforcement phase formation (composition imbalance + centers of the reaction between the substrates + diffusion kinetics), however, its growing amount blocks the movement of the atoms decreasing the speed of transformation and leading to its possible local concentration as well as microstructure locking. Higher-sintering-temperature regime decreases the appearance of the smaller voids in favor of the larger ones with their shape change to more regular during densification. The energy excess, as the microphotographs show, remains rapidly consumed by the matrix to remodel the microstructure. The amount of the reinforcement phase for the higher sintering temperature does not increase evidently, however, the characteristics of the compactness differs significantly (see Figure 8). The porosity change may thus become reduced by up to 50% by the processing temperature. The above-mentioned size, morphology, and composition of the starting precursor influence the compactness characteristics, the microstructure and the mutual relation between the final phase and its size range.

The densification and consolidation of the sinters proceeds under local plasma discharge conditions that occur at the particle grain boundaries as well as creates additional convection heating waves that spread in the whole material and remain mostly reflected in the ceramic die material protected by the boron nitride layer. The porosity in the material may change its share due to the starting precursor morphological relation, concentration of intrinsic structural defects, dimensions of the crystallite and diffusion mechanism energy relation dependent on the material composition as well as its processing conditions.

The performed microhardness test results analysis presented in Figure 9 for the obtained composite sinters, shows the preferred material response, dependent on the amount ratio of the precipitation phase. The above-mentioned high microhardness test results could be easily adjusted by controlling the starting precursor composition. The compared PPS temperature variant in the microhardness tests shows a slight difference among the obtained results, distinguishing, however, the high-temperature red side (1100 °C) followed by an increase in other fundamental properties of the sinters.

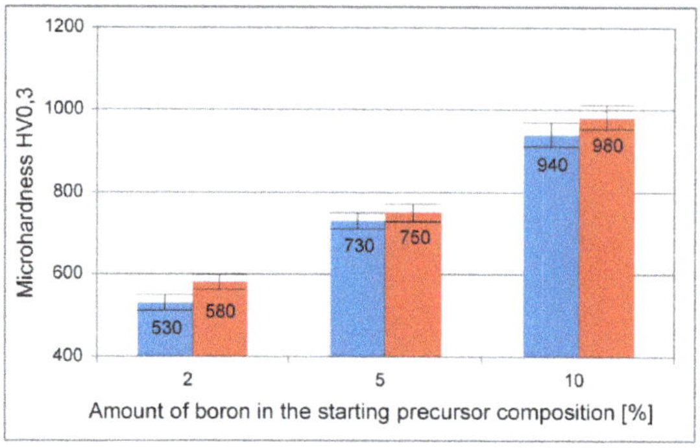

Figure 9. Microhardness analysis of the obtained composite sinters conducted for different PPS temperature regimes (blue—900 °C, red—1100 °C) and starting precursor compositions (2%, 5%, 10%).

4. Discussion

As the present results confirm, the starting powder morphology, its average size and composition analyzed in the mechanical alloying processing stage, remain responsible for the final fundamental properties of the compacts or sinters. For example, the analyzed porosity share and its dispersion in the composite sinters microstructure obtained after the PPS procedure and visualized in Figure 7, justifies the attention focused on the morphological features of the particles to be sintered. The precursor powder compositions obtained after MA, characterized by different average sizes of the aggregates, scattering, as well as assumed morphology, remains influenced by the processing fragmentation mechanism. For the above, the growing boron amount in the starting powder blends composition, for which the cold welding and fracturing stages overlap, manifests itself by the crystallite size reduction with the simultaneous increasing in the microstrains. The differences in the observed porosity and relative density among the sinters distinguish the starting spheroidal morphology, higher processing temperature, and a lower boron addition to the starting precursor composition.

The obtained MMC sinters microstructure, shows relevant differences in the range of the analyzed variable amount of boron in the starting precursor compositions and the temperature processing setup. The amount of the precipitation phase, for example, corresponds not only to the variable amount of boron but also, as earlier data confirm [18], to its size promoted by its higher dispersion in the structure. According to the earlier presented data [18], micro-type precursor materials with the same compositions and processed by the same methods show nearly half the amount of the precipitation phase. The submicron and nano-sized whisker-type precipitation morphology that characterizes the microstructure of the sinters obtained in this work, for the growing amount of the reinforcement phase, distinguishes the occurrence of the interlocked centers. High dispersion of the precipitation phase, ensured by the crystallite size reduction and homogeneous elements distribution in the starting precursor compositions, finally leads to the obtainment of sinters of high uniformity. The presence of the interlocked area verified in the microstructure of all sinters, which most likely transform to the clustered area for the obtained processing temperature, remain responsible for the size growth of the reinforcement phase. Nonetheless, as other researchers have confirmed using the TEM and HRTEM methods, the in situ ceramic reinforcements, formed in the titanium or its alloys [26,27], generally exhibit a clean interface that promotes strong bonding between the phases allowing an obtainment of high properties of the composites. High microhardness test results could easily be adjusted by controlling the starting precursor composition as well as its size ratio as the previously presented results may confirm [18].

5. Conclusions

In this work, mechanical alloying and pulse plasma sintering methods were adopted for the processing of metal matrix composite materials. The applied approach allows for the observation and analyzing of the characteristics of the starting powder materials, their transformation, and their influence on the final properties of the composite materials. The following conclusions from the performed research can be drawn:

- the mechanical alloying approach allows a homogeneous distribution of the elements in the prepared precursor powder blends;
- the mechanical alloying approach allows obtaining the nanoscale range in the prepared precursor powder blends;
- the sinters obtained after PPS processing of the MA precursor powders are characterized by a composite microstructure with a highly dispersed reinforcement phase, whose size falls in the submicron to nanoscale range;
- the obtained in situ metal matrix composite sinters are characterized by low porosity, yet heavily dependent on the starting material composition characteristics and the proposed processing variants; and
- the final material properties can be adjusted not only by the composition, the processing approach or the temperature dependence but also by the starting material morphology, its size ratio, and proper distribution of the components.

Author Contributions: Conceptualization, A.M.; Methodology, A.M.; Software, A.M.; Validation, A.M.; Formal Analysis, A.M.; Investigation, A.M.; Resources, A.M.; Data Curation, A.M.; Writing-Original Draft Preparation, A.M.; Writing-Review & Editing, A.M.; Visualization, A.M.; Supervision, A.M. and M.J.; Project Administration, A.M.; Funding Acquisition, A.M. and M.J.

Funding: This research was funded by the National Science Centre decision No. DEC-2014/13/N/ST8/00601.

Acknowledgments: Special thanks to polish Tescan Company agent Uni-Export Instruments Polska for SEM analysis.

Conflicts of Interest: The authors declare no conflict of interest. The funders had no role in the design of the study; in the collection, analyses, or interpretation of data; in the writing of the manuscript, and in the decision to publish the results.

References

1. Saito, T. The automotive application of discontinuously reinforced TiB-Ti composites. *JOM* **2004**, *56*, 33–36. [CrossRef]
2. Kainer, K.U. *Basics of Metal Matrix Composites*; WILEY-VCH Verlag GmbH & Co. KGaA: Weinheim, Geramny, 2006; ISBN 3527313605.
3. Montgomery, J.S.; Wells, M.G.H.; Roopchand, B.; Ogilvy, J.W. Low-cost titanium armors for combat vehicles. *JOM* **1997**, *49*, 45–47. [CrossRef]
4. Miklaszewski, A.; Jurczyk, M.U.; Kaczmarek, M.; Paszel-Jaworska, A.; Romaniuk, A.; Lipińska, N.; Żurawski, J.; Urbaniak, P.; Jurczyk, M. Nanoscale size effect in in situ titanium based composites with cell viability and cytocompatibility studies. *Mater. Sci. Eng. C* **2017**, *73*, 525–536. [CrossRef]
5. Selvakumar, M.; Chandrasekar, P.; Mohanraj, M.; Ravisankar, B.; Balaraju, J.N. Role of powder metallurgical processing and TiB reinforcement on mechanical response of Ti-TiB composites. *Mater. Lett.* **2015**, *144*, 58–61. [CrossRef]
6. Tjong, S.C.; Mai, Y.W. Processing-structure-property aspects of particulate- and whisker-reinforced titanium matrix composites. *Compos. Sci. Technol.* **2008**, *68*, 583–601. [CrossRef]
7. Luan, J.H.; Jiao, Z.B.; Chen, G.; Liu, C.T. Effects of boron additions and solutionizing treatments on microstructures and ductility of forged Ti-6Al-4V alloys. *J. Alloys Compd.* **2015**, *624*, 170–178. [CrossRef]
8. Feng, H.; Zhou, Y.; Jia, D.; Meng, Q. Microstructure and mechanical properties of in situ TiB reinforced titanium matrix composites based on Ti-FeMo-B prepared by spark plasma sintering. *Compos. Sci. Technol.* **2004**, *64*, 2495–2500. [CrossRef]

9. Zhang, Z.; Shen, X.-B.; Wen, S.; Luo, J.; Lee, S.; Wang, F. In situ reaction synthesis of Ti–TiB composites containing high volume fraction of TiB by spark plasma sintering process. *J. Alloys Compd.* **2010**, *503*, 145–150. [CrossRef]
10. Lütjering, G.; Williams, J.C. *Titanium: Engineering Materials and Processes*; Springer: New York, NY, USA, 2007.
11. Gil, F.J.; Aparicio, C.; Planell, J.A. Effect of Oxygen Content on Grain Growth Kinetics of Titanium. *J. Mater. Synth. Process.* **2002**, *10*, 263–266. [CrossRef]
12. Gorsse, S.; Chaminade, J.P.; Le Petitcorps, Y. In situ preparation of titanium base composites reinforced by TiB single crystals using a powder metallurgy technique. *Compos. Part A Appl. Sci. Manuf.* **1998**, *29*, 1229–1234. [CrossRef]
13. Huang, L.J.; Geng, L.; Wang, B.; Xu, H.Y.; Kaveendran, B. Effects of extrusion and heat treatment on the microstructure and tensile properties of in situ TiBw/Ti6Al4V composite with a network architecture. *Compos. Part A Appl. Sci. Manuf.* **2012**, *43*, 486–491. [CrossRef]
14. Zhang, J.; Ke, W.; Ji, W.; Fan, Z.; Wang, W.; Fu, Z. Microstructure and properties of in situ titanium boride (TiB)/titanium (TI) composites. *Mater. Sci. Eng. A* **2015**, *648*, 158–163. [CrossRef]
15. Wei, S.; Zhang, Z.H.; Wang, F.C.; Shen, X.B.; Cai, H.N.; Lee, S.K.; Wang, L. Effect of Ti content and sintering temperature on the microstructures and mechanical properties of TiB reinforced titanium composites synthesized by SPS process. *Mater. Sci. Eng. A* **2013**, *560*, 249–255. [CrossRef]
16. Attar, H.; Bönisch, M.; Calin, M.; Zhang, L.C.; Scudino, S.; Eckert, J. Selective laser melting of in situ titanium-titanium boride composites: Processing, microstructure and mechanical properties. *Acta Mater.* **2014**, *76*, 13–22. [CrossRef]
17. Sahay, S.S.; Ravichandran, K.S.; Atri, R.; Chen, B.; Rubin, J. Evolution of microstructure and phases in in situ processed Ti-TiB composites containing high volume fractions of TiB whiskers. *J. Mater. Res.* **1999**, *14*, 4214–4223. [CrossRef]
18. Miklaszewski, A. Ultrafast densification and microstructure evolution of in situ Ti/TiB metal matrix composite obtained by PPS approach. *Int. J. Refract. Met. Hard Mater.* **2017**, *65*, 34–38. [CrossRef]
19. Miklaszewski, A. Effect of starting material character and its sintering temperature on microstructure and mechanical properties of super hard Ti/TiB metal matrix composites. *Int. J. Refract. Met. Hard Mater.* **2015**, *53*, 56–60. [CrossRef]
20. Zadra, M. Mechanical alloying of titanium. *Mater. Sci. Eng. A* **2013**, *583*, 105–113. [CrossRef]
21. Yurlova, M.S.; Demenyuk, V.D.; Lebedeva, L.Y.; Dudina, D.V.; Grigoryev, E.G.; Olevsky, E.A. Electric pulse consolidation: An alternative to spark plasma sintering. *J. Mater. Sci.* **2014**, *49*, 952–985. [CrossRef]
22. Groza, J.R.; Zavaliangos, A. Sintering activation by external electrical field. *Mater. Sci. Eng. A* **2000**, *287*, 171–177. [CrossRef]
23. Suryanarayana, C.; Norton, M.G. Practical Aspects of X-ray Diffraction. In *X-ray Diffraction: A Practical Approach*; Springer Science + Buisness Media: New York, NY, USA, 1998; ISBN 978-1-4899-0148-4.
24. Miklaszewski, A.; Garbiec, D.; Niespodziana, K. Sintering behavior and microstructure evolution in cp-titanium processed by spark plasma sintering. *Adv. Powder Technol.* **2017**, *29*, 50–57. [CrossRef]
25. Kessler, H.D.; Rostoker, W.; Van Thyne, R.J. *Titanium Phase Diagrams*; WADC Tech. Rep. 52-335 Ti; Statement A: Approved for public release; Wright-Patterson Air Force Base: Dayton, OH, USA, 1953.
26. Feng, H.; Zhou, Y.; Jia, D.; Meng, Q.; Rao, J. Growth mechanism of in situ TiB whiskers in spark plasma sintered TiB/Ti metal matrix composites. *Cryst. Growth Des.* **2006**, *6*, 1626–1630. [CrossRef]
27. Feng, H.; Zhou, Y.; Jia, D.; Meng, Q. Stacking faults formation mechanism of in situ synthesized TiB whiskers. *Scr. Mater.* **2006**, *55*, 667–670. [CrossRef]

materials

MDPI

Article

Influence of Heat Treatments on Microstructure and Mechanical Properties of Ti–26Nb Alloy Elaborated In Situ by Laser Additive Manufacturing with Ti and Nb Mixed Powder

Jing Wei [1], Hongji Sun [1], Dechuang Zhang [2,*], Lunjun Gong [2], Jianguo Lin [1,*] and Cuie Wen [3]

1 School of Materials Science and Engineering, Xiangtan University, Xiangtan 411105, China; weijing19930624@163.com (J.W.); sun_hong_ji@163.com (H.S.)

2 Key Laboratory of Materials Design and Preparation Technology of Hunan Province, Xiangtan University, Xiangtan 411105, China; ljgong@xtu.edu.cn

3 School of Aerospace, Mechanical and Manufacturing Engineering, RMIT University, Melbourne 3083, Australia; cuie.wen@rmit.edu.au

* Correspondence: dczhang@xtu.edu.cn (D.Z.); lin_j_g@xtu.edu.cn (J.L.); Tel.: +86-731-5829-8119 (D.Z. & J.L.)

Received: 25 November 2018; Accepted: 18 December 2018; Published: 25 December 2018

check for updates

Abstract: In the present work, a Ti–26Nb alloy was elaborated in situ by laser additive manufacturing (LAM) with Ti and Nb mixed powders. The alloys were annealed at temperatures ranging from 650 °C to 925 °C, and the effects of the annealing temperature on the microstructure and mechanical properties were investigated. It has been found that the microstructure of the as-deposited alloy obtained in the present conditions is characterized by columnar prior β grains with a relatively strong <001> fiber texture in the build direction. The as-deposited alloy exhibits extremely high strength, and its ultimate tensile strength and yield strength are about 799 MPa and 768 MPa, respectively. The annealing temperature has significant effects on the microstructure and mechanical properties of the alloys. Annealing treatment can promote the dissolution of unmelted Nb particles and eliminate the micro-segregation of Nb at the elliptical-shaped grain boundaries, while increasing the grain size of the alloy. With an increase in annealing temperature, the strength of the alloy decreases but the ductility increases. The alloy annealed at 850 °C exhibits a balance of strength and ductility.

Keywords: heat treatment; in situ alloying; laser additive manufacturing; mechanical properties; microstructure; Ti–Nb alloy

---

1. Introduction

Beta (β) titanium (Ti) alloys containing non-cytotoxic elements such as niobium (Nb), zirconium (Zr), tantalum (Ta), tin (Sn), and molybdenum (Mo) have been intensively studied for applications in biomedical domains [1–5]. Among these, Ti–Nb alloys have attracted particular attention due to their high strength (600 MPa) and their very low elastic modulus (50 GPa), close to that of cortical bone (30 GPa), which was observed to reduce the stress-shielding phenomenon [6]. These alloys are considered to be potential substitute materials for conventional materials such as Ti–Ni or Ti–6Al–4V in order to prevent the release of toxic nickel (Ni), aluminium (Al), or vanadium (V) ions into the human body [7–11].

Recently, laser additive manufacturing (LAM), based on coaxed powder melting and rapid solidification through layer-upon-layer deposition, has attracted a great deal of attention in the fabrication of fully dense near net-shaped metallic components [12–15]. It has several significant advantages over traditional manufacturing methods, such as a wide range of material forms, a simple manufacturing process for complex parts, short production cycle, raw material saving, excellent performance, etc. So it is an effective method for the personal customization of biological implants.

Generally, the powders for LAM are obtained from a prealloyed material that is converted into powder by plasma arc or gas atomization to achieve spherical particles with a particle size between 10 and 110 μm. In the case of Ti–Nb alloys, their high melting temperatures make it very complicated to produce powders in this way. So the Ti–Nb alloy powders that can be applied to direct laser metal-forming are still rare or even commercially unavailable. Recently, some researchers attempted to fabricate Ti–Nb alloys using selective laser melting (SLM) with mixtures of elemental powders of pure Ti and Nb. For example, Fischer et al. [16] fabricated Ti–26Nb (at.%) using SLM with elemental Ti and Nb mixed powders with high energy levels, and the alloy exhibited a homogeneous and dense microstructure with a β structure. Wang et al. [17] also fabricated Ti–Nb alloys with different Nb content using the same method and investigated the effects of Nb content on the microstructure and mechanical properties; they found that SLM could be used for in situ fabrication of Ti–Nb bone implants with tailored mechanical and biomedical properties by adjusting the level of Nb.

It has been reported that, compared to the microstructure achieved with conventional casting and forging, Ti alloys prepared using LAM usually exhibit a quite different microstructure due to repeated rapid solidification and rapid annealing during laser forming. Moreover, defects can be found in the as-deposited part, such as inconsistency in the structure and stability of mechanical properties, residual stress, and pores, which can weaken the mechanical properties of the alloys. Post-heat treatment can have an important influence on the microstructure and properties of the alloys [18–20]. However, microstructure transformation after heat treatment of as-deposited Ti–Nb alloys has rarely been investigated. Therefore, a deeper understanding of the microstructural evolution of Ti–Nb formed by LAM during heat treatment would allow improvement of the mechanical mechanical of these alloys.

In the present work, a Ti–26at.%Nb alloy was prepared by directly melting a mixture of elemental Ti and Nb powders under a laser beam. The alloy was annealed at temperatures ranging from 650 °C to 925 °C and the influence of the annealing temperature on the microstructure evolution and mechanical properties of the alloy was investigated.

2. Experimental Methods

2.1. Material Manufacturing

Commercial gas-atomized pure Ti powder (purity 99.99%, Changsha Tianjiu Metal Materials Co. Ltd., Changsha, China) with a particle size of 45–105 μm and pure Nb powder (purity 99.99%, Changsha Tianjiu Metal Materials Co. Ltd., Changsha, China) with a particle size of 48–75 μm were used as the raw materials. The two powders were mixed in the weight ratio Ti:Nb = 59.5:40.5 (namely 74:26 in atomic ratio) by ball milling for 1 h. Figure 1 shows the morphologies of the Ti and Nb powders and the mixtures. It is clear that the Ti powder has regular spheres with the presence of satellites (Figure 1a), whereas the Nb has irregular shapes (Figure 1b). After mixing, the Nb particles were uniformly distributed around the Ti particles (see Figure 1c). The melting point of Nb (2477 °C) is much higher than that of Ti (1668 °C). To facilitate better melting of the Nb and its faster diffusion in Ti, Nb powder with a smaller particle size than the Ti powder was used in the present work.

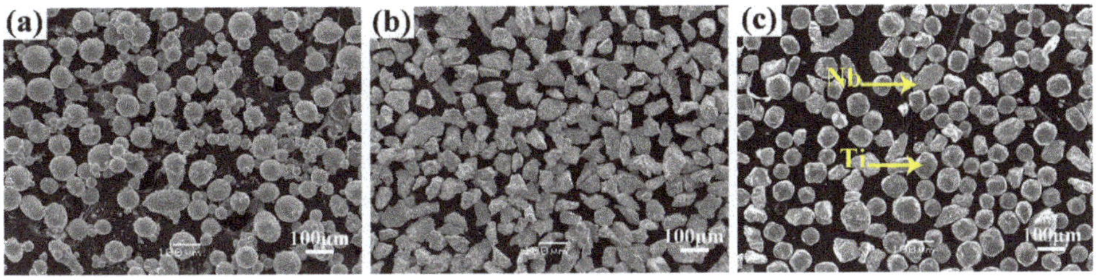

Figure 1. Scanning electron microscope (SEM) images of feedstock powders: (**a**) pure Ti; (**b**) pure Nb; and (**c**) Ti–26Nb powder mixture.

A square block sample of the obtained Ti–26Nb alloy with dimensions of 400 mm × 350 mm × 230 mm (see Figure 2) was fabricated by a LAM process on a YLS-4000-CL machine (IPG photonics corporation, Oxford, MA, USA), in which a fiber laser was produced by the IPG photonics corporation and the powder was fed in coaxial feeding mode with argon as the carrier gas. SD, LD, and BD represent the scanning direction, lateral direction, and build direction, respectively. The laser-deposition processing parameters were: laser nominal output power 750 W; laser beam diameter 2.5 mm; scanning speed 480 mm/min; and powder feed rate 2.2 g/min.

Figure 2. Bulk Ti–26Nb sample fabricated by laser additive manufacturing (LAM).

2.2. Heat Treatment

The samples for heat treatment were sectioned along the BD with dimensions of 25 mm in length (BD), 10 mm in width (SD), and 1.2 mm in thickness (LD). The samples were vacuum encapsulated in quartz tubes and annealed at temperatures ranging from 650 °C to 925 °C for a duration of 0.5 h followed by water quenching, as listed in Table 1.

Table 1. Details of heat treatment for LAM Ti–26Nb alloys.

Sample No.	Heat Treatment
1	650 °C/0.5 h + WQ
2	750 °C/0.5 h + WQ
3	850 °C/0.5 h + WQ
4	925 °C/0.5 h + WQ

Note: WQ is water quenching.

2.3. *Microstructure Characterization*

The microstructures of the alloy samples before and after heat treatment were characterized by optical microscopy (OM; BX51M, Olympus, Tokyo, Japan), scanning electron microscopy (SEM; MIRA3 LMU, Tescan, Brno, Czech), and X-ray diffraction (XRD; D/max 2500, Rigaku, Tokyo, Japan). Grain size and grain orientation were determined by electron backscatter diffraction (EBSD; HKL, Oxford, UK). Several regions in the samples were chosen for the EBSD analysis, and average grain size and grain orientations were determined by statistical analysis.

2.4. *Mechanical Property Testing*

Tensile tests of the Ti–26Nb alloy were carried out on an Instron 5569 universal testing machine (Instron, Boston, MA, USA). The samples for the tests were cut from the middle section of the block prepared by LAM with a gauge section of 1 mm × 2.5 mm × 8 mm, the geometry of which is schematically shown in Figure 3. All specimens are shown along the BD. Three samples were measured for each condition in order to reduce measurement error.

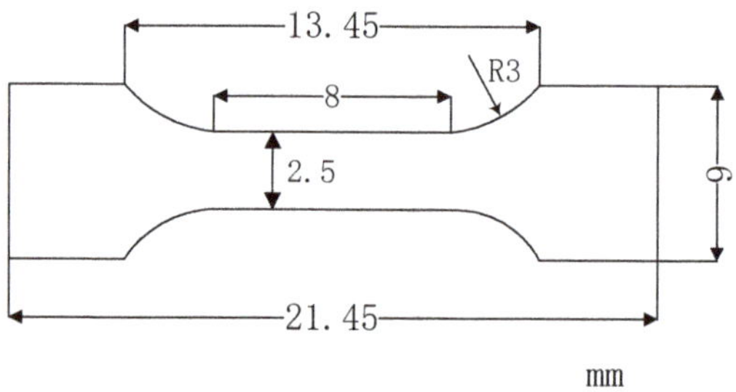

Figure 3. Schematic geometry of tensile specimen.

For comparison, another Ti–26Nb alloy was prepared using a conventional arc melting method with pure Ti and Nb as the raw materials. The ingot was hot-rolled by 90% in thickness, then tensile testing of the mechanical properties according to the above method was also carried out in the same conditions.

3. Results and Discussion

3.1. *Phase Composition*

Figure 4 shows the X-ray diffraction (XRD) patterns of the Ti–26Nb alloy fabricated by LAM and after annealing at different temperatures for 0.5 h. It can be seen that the as-deposited alloy shows a single β phase with a body-centered cubic (bcc) structure. After annealing at 650 °C for 0.5 h, diffraction peaks from an α phase could be seen in the XRD pattern of the alloy, implying that an α phase precipitated in the alloy. As the annealing temperature increased to 925 °C, the peaks from the α phase completely disappeared. By careful comparison, it can be seen that the diffraction peaks of the β phase on the XRD pattern of the alloy after annealing have slightly shifted to the left relative to the as-deposited sample (as shown in Figure 4). This implies that more Nb atoms have dissolved in the Ti lattice. This is because the radius of the solute atom Nb is slightly larger than that of the Ti atom, and thus a higher concentration of Nb leads to an increase in the lattice constant of the β phase.

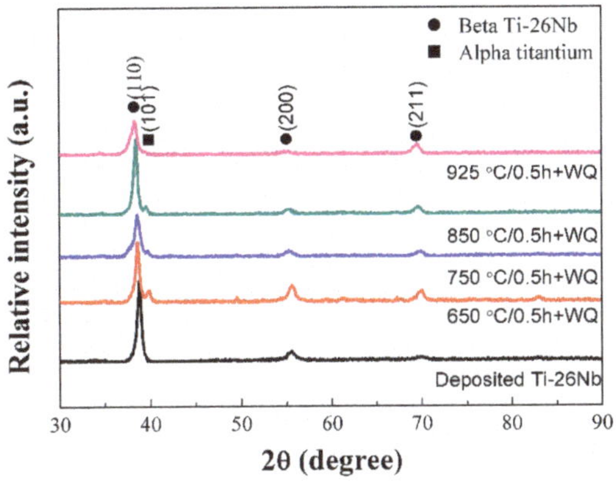

Figure 4. X-ray diffraction (XRD) patterns of LAM-processed Ti–26Nb alloys after annealing treatment at different temperatures.

3.2. Microstructure

Figure 5 illustrates the microstructure of the as-deposited Ti–26Nb alloy. It is clear that multilayer deposits with uniform thickness and regular distribution have been formed in the as-deposited Ti–26Nb alloy. The prior β phase with columnar grains oriented more or less in the BD, which penetrate the multilayer cladding layer, can be observed (see Figure 5a). The special formation mechanism of columnar grains has been clearly explained in previous work [21]. Some unmelted Nb can be observed in the Ti–Nb alloy due to its relatively high melting temperature of 2477 °C, as shown in Figure 5a. The results indicate that the applied energy density of the LAM process is sufficient to completely melt the Ti powder, but some of the larger Nb particles are only partially melted. Similar behavior was also found in a previous report [22]. Moreover, some pores can be seen on the surface of the sample, which may have been caused by ball formation and gas inclusion generated during the melting and remelting process [23].

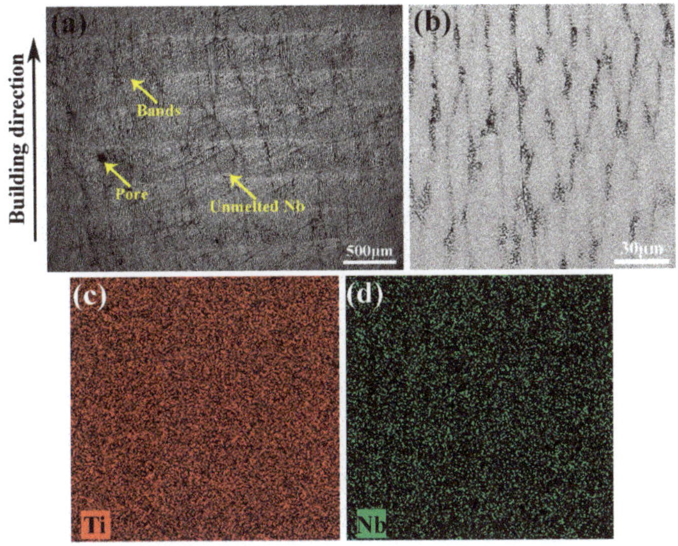

Figure 5. (**a**) and (**b**) Microstructures of as-deposited Ti–26Nb alloy; (**c**) and (**d**) elemental mapping images of as-deposited Ti–26Nb of (**b**).

A magnified image reveals that some ultra-fine elliptical-shaped grains (dendritic grains) with dimensions of ~10 μm in width and ~80 μm in length were formed in the representative regions (see Figure 5b). This may be attributed to the fact that the solidified layer acts as a substrate for the solidification of the melt, leading to the formation of elliptical-shaped grains perpendicular to the solidification front [24]. Moreover, the elliptical grains grew in a wavelike fashion and their boundaries are poor in Nb, as seen in Figure 5c,d.

Figure 6 shows the distributions of Ti and Nb elements in the molten pool of Ti-26Nb alloy. It can be seen that the Ti and Nb content at the melt pool boundaries (MPBs) is identical to that inside of the molten pool, implying that the solidification of the molten pool is completed in a short time, which is very different from conventional casting. As a result, the Nb particles do not have enough time to sink to the bottom of the molten pool and thus remain.

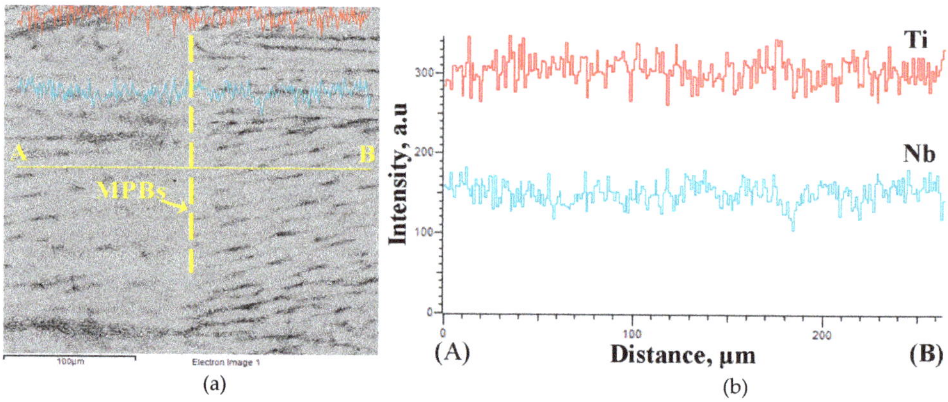

Figure 6. (**a**) Energy-dispersive X-ray spectroscopy (EDS) line scanning of melt pool in lateral plane of Ti–26Nb alloy (**b**) corresponding line scanning results of Ti and Nb elements.

To investigate the effects of the heat treatment on the microstructure of the alloy fabricated by LAM, the as-deposited alloy samples were annealed at temperatures ranging from 650 °C to 925 °C. Figure 7 illustrates the microstructures of the samples after annealing at the different temperatures.

For the sample annealed at 650 °C for 0.5 h, elliptical grains began to grow and their size distribution became more uniform (as shown in Figure 7a) as compared to as-deposited sample. Close observation reveals that the some fine acicular secondary α_S phases precipitated in the areas poor in Nb atoms in the alloys annealed at 650 °C and 750 °C, as shown in the inset of Figure 7c,d). As the annealed temperature increased to 850 °C, the boundaries of the ellipses became unclear, implying the dissolution of the dendritic grains due to the diffusion of Nb atoms. Moreover, almost all the α_S phases disappeared in the alloys as the annealing temperature increased over 850 °C, and the alloy that annealed at 925 °C exhibited an even microstructure with all β phases (see Figure 7e,f). This is consistented with the XRD analysis.

Figure 8 illustrates the Nb concentration in the elliptical-shaped grains and at the grain boundaries as obtained by energy-dispersive X-ray spectroscopy (EDS) analysis. It can be seen that, for the as-deposited alloy, the Nb content at the grain boundaries area is much lower than in the elliptical grains. After annealing at 650 °C, the Nb content at the grain boundaries increased to become close to that inside the grains, implying the diffusion of Nb from inside to the grain boundaries during the annealing treatment. As the annealing temperature further increased, the Nb content both inside and at the boundaries of the grains slightly increased, implying the dissolution of the unmelted Nb particles.

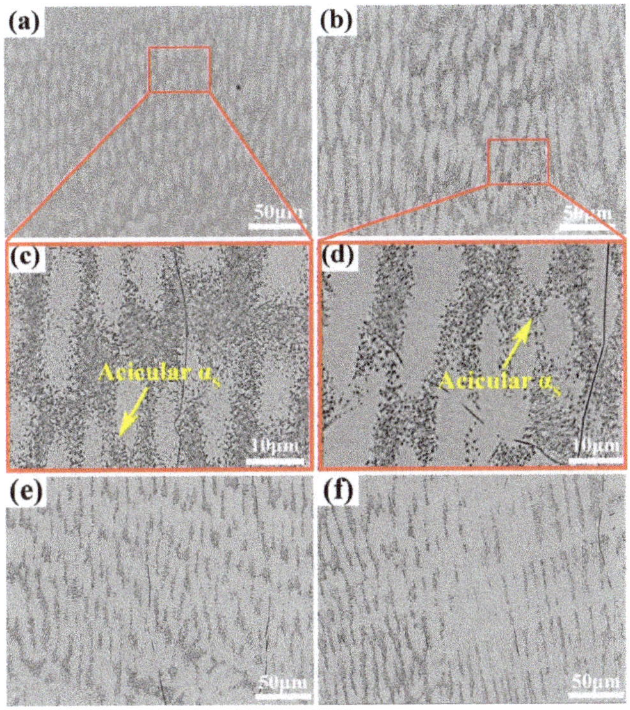

Figure 7. Microstructures of annealed samples at different temperatures for 0.5 h: (**a**) 650 °C; (**b**) 750 °C; (**c**) and (**d**) corresponding high-magnification morphologies of (**a**) and (**b**), respectively; (**e**) 850 °C; (**f**) 925 °C.

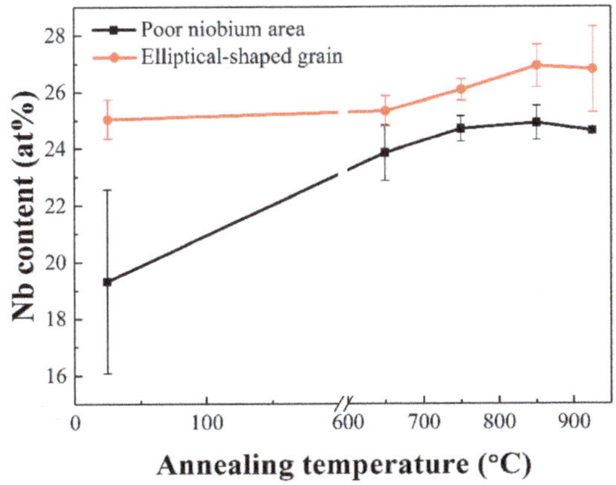

Figure 8. Comparison of Nb content in different regions of Ti–26Nb alloys after annealing at different temperatures.

3.3. Grain Orientation and Grain Size

As mentioned above, the microstructure of the as-deposited Ti–26Nb alloy exhibited columnar β grains with fine dendrites inside oriented more or less in the BD. The preferred direction of the dendrite growth in the Ti–Nb alloy can be ascribed to the largest thermal gradient parallel to BD during the LAM process, leading to the formation of this texture [25]. The texture was further verified by EBSD orientation maps, as seen in Figure 9a. It is clear that the coarse columnar β grains of the as-deposited alloy exhibits a relatively strong <100> fiber texture. The annealing treatment tended to make the grain orientation uniform, as shown in Figure 9b,c.

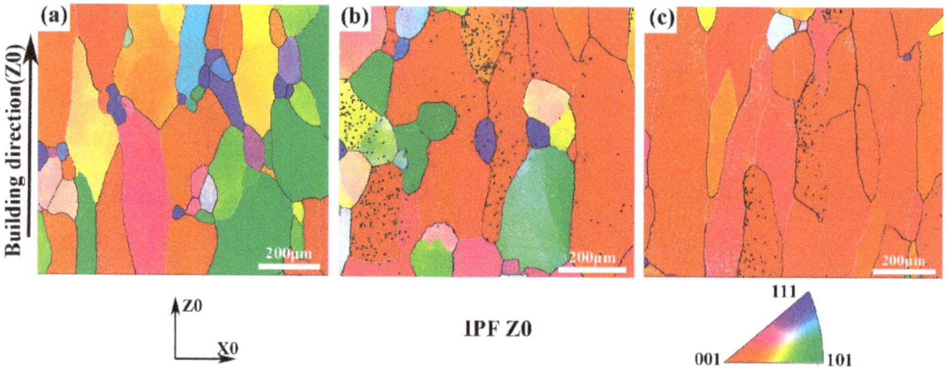

Figure 9. IPFZ electron backscatter diffraction (EBSD) map presented with respect to z-direction showing multiple grain orientations: (**a**) as-deposited Ti–26Nb alloy; (**b**) after annealing at 650 °C; (**c**) after annealing at 850 °C.

Further confirmation of the preferred orientation is illustrated by the corresponding EBSD pole figure in Figure 10. It can be seen that the as-deposited alloy exhibits the strongest <001> texture. After annealing treatment at 650 °C, the <001> texture of the alloy was weakened. However, with the annealing temperature further increasing to 850 °C, the texture became stronger.

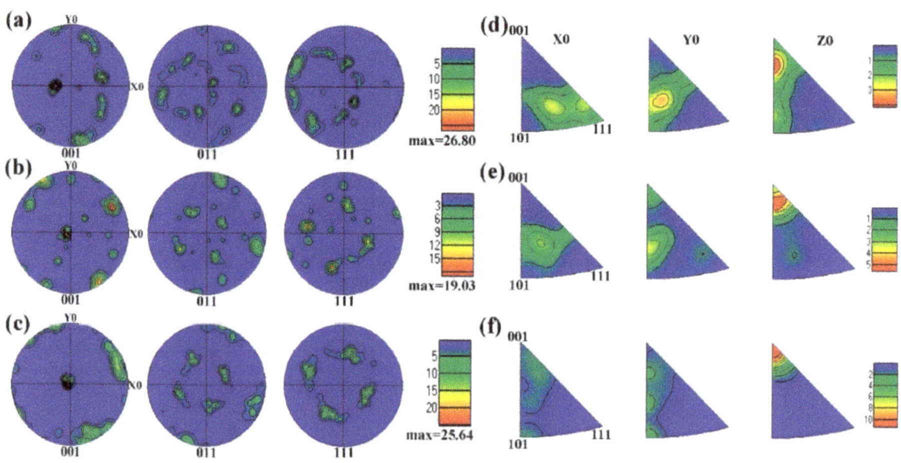

Figure 10. EBSD pole figure and inverse pole figure of Ti-26Nb samples: (**a**) and (**d**) as-deposited Ti–26Nb alloy; (**b**) and (**e**) after annealing at 650 °C; (**c**) and (**f**) after annealing at 850 °C.

The effects of annealing temperature on the columnar grain size of the alloy were also investigated. Figure 11 shows the grain size distribution of the alloy before and after annealing at different temperatures. It can be seen that the average grain size of the as-deposited alloy is about 143.3 μm.

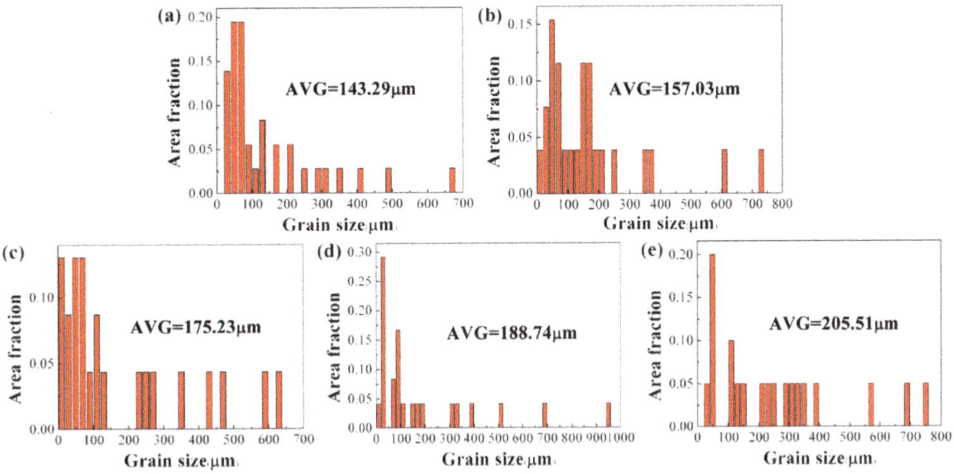

Figure 11. Grain size distributions of specimens: (**a**) as-deposited Ti–26Nb alloy; (**b**) after annealing at 650 °C; (**c**) after annealing at 750 °C; (**d**) after annealing at 850 °C; (**e**) after annealing at 925 °C.

While annealing treatment leads to clear growth in the grain size of the columnar β grains, with an increase in annealing temperature the size of the columnar β grains increases. The alloy after annealing at 925 °C exhibited the largest grain size, which reached 205.5 μm. The growth in grain size caused by the heat treatment may have decreased the strength of the alloy.

3.4. Mechanical Properties

The stress-strain curves of Ti–26Nb alloys after annealing at different temperatures (see Figure 12). The mechanical properties of the alloy before and after annealing at different temperatures were evaluated by Figure 12 and the results are shown in Figure 13 and Table 2. It can be seen that the as-deposited Ti–26Nb sample shows relatively high strength, with the ultimate tensile strength (UTS) and yield strength (YS) about 799 MPa and 768 MPa, respectively, which are much higher than those of the as-rolled Ti–26Nb alloy (see Table 2). The fine-grained structures present inside the β grains produced by the rapid solidification process may be responsible for the high strength [26].

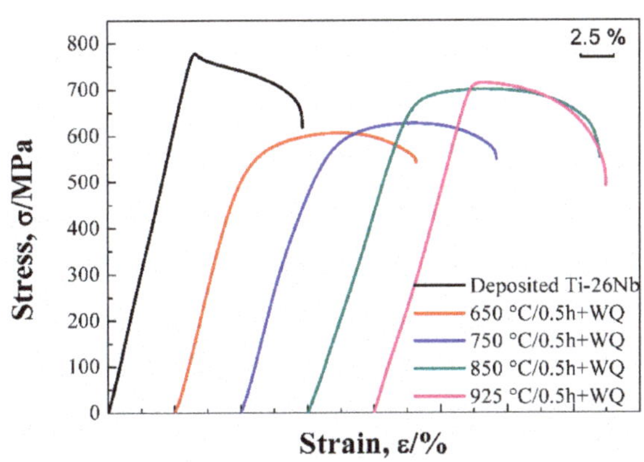

Figure 12. Stress-strain curves of Ti–26Nb alloys after annealing at different temperatures.

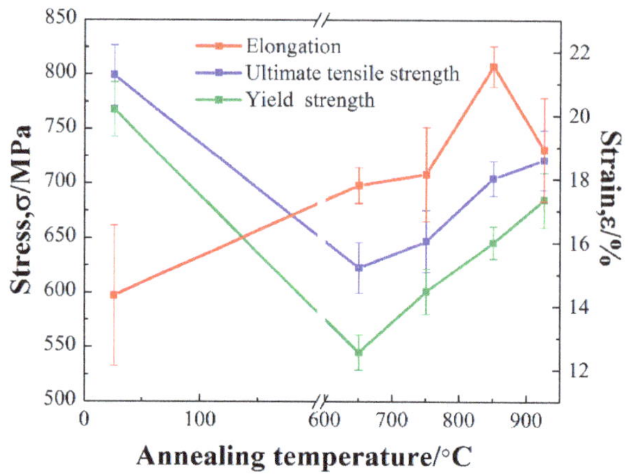

Figure 13. Mechanical properties of Ti–26Nb alloys after annealing at different temperatures.

Table 2. Mechanical properties of Ti–Nb alloys fabricated by LAM and casting.

State of Materials	Ultimate Tensile Strength/MPa	Yield Strength/MPa	Elongation/%
As–deposited	799 ± 27	768 ± 25	14.3 ± 2.2
650 °C/0.5 h + WQ	623 ± 23	545 ± 16	17.8 ± 0.6
750 °C/0.5 h + WQ	647 ± 28	601 ± 21	18.2 ± 1.5
850 °C/0.5 h + WQ	704 ± 16	646 ± 15	21.6 ± 0.6
925 °C/0.5 h + WQ	722 ± 27	685 ± 25	18.9 ± 1.6
Hot–rolled of cast Ti–26Nb	428 ± 15	415 ± 12	18.2 ± 0.8

The heat treatment decreased the strength but increased the ductility of the alloy, depending on the annealing temperature. It can be seen that, after annealing at 650 °C for 0.5 h, the UTS and YS of the alloy reduced by about 176 and 223 MPa, respectively, in comparison with the as-deposited alloy due to the growth of the prior columnar β grains and the dissolution of the fine dendrites [26]. It should be noted that, as the annealing temperature further increased, the strength of the alloys gradually increased despite the growth of the grains. The UTS and YS of the alloy annealed at 925 °C were about 722 MPa and 685 MPa, respectively. The increase in strength of the alloy annealed at a relatively high temperature may be attributable to the solid solution strengthening the effect of the Nb solute atoms. As mentioned above, annealing treatment can lead to the dissolution of unmelted Nb atoms into the Ti lattice and, with an increase in annealing temperature, the content of Nb in the β increases. This may cause more severe lattice distortion, which can increase the resistance of the dislocation motion and consequently promote the strength of the alloy.

With regard to the ductility of the alloy, the as-deposited alloy exhibited relatively low ductility with elongation of 14.3%. This may be attributable to the presence of a large amount of residual stress in the as-built condition and defects such as pores, micro-segregation, and unmelted Nb particles in the as-deposited alloy. After annealing at 650 °C, the ductility of the alloy slightly increased due to the elimination of residual stress [27] and microstructural homogenization. However, the precipitation of secondary α phases at the grain boundaries may have deteriorated the ductility of the alloy. As the annealing temperature increased to 850 °C, most unmelted Nb particles dissolved into the β phase, which may have increased its stability, and thus no precipitation of secondary α phases occurred. This led to an increase in the ductility of the alloy. As a result, the alloy annealed at 850 °C exhibited relatively high ductility with elongation of 21.6%. However, with the annealing temperature further increasing to 925 °C, the grains of the β phase became coarse, which may have decreased the ductility of the alloy, as its elongation decreased to 18.9%.

4. Conclusions

In this work, a Ti–26Nb alloy was successfully prepared from a mixture of titanium and niobium powders by laser additive manufacturing. The as-deposited alloy was annealed at different temperatures ranging from 650 °C to 925 °C, then the microstructure evolution and tensile mechanical properties were evaluated. The main conclusions can be summarized as follows:

(1) The microstructure of the as-deposited Ti–26Nb alloy was characterized by prior columnar β grains with a relatively strong <001> fiber texture due to a large temperature gradient and remelting penetration in the build direction. Defects such as pores, unmelted Nb particles, and micro-segregation at the grain boundaries of the elliptical-shaped grains could be observed in the as-deposited Ti–26Nb alloy. Its ultimate tensile strength and yield strength were about 799 MPa and 768 MPa, respectively, much higher than those of hot-rolled Ti–26Nb (428 MPa and 415 MPa, respectively).

(2) The heat treatment had an important influence on the microstructures of the as-deposited alloy, depending on the annealing temperature. After annealing at 650 °C for 0.5 h, many fine acicular secondary α_S phases precipitated at the boundaries of the elliptical-shaped grains with poor Nb content. With the annealing temperature increasing, the unmelted Nb particles dissolved into the Ti lattice, leading to an increase in Nb concentration in the β matrix, and the composition of the alloy tended to become uniform by the diffusion of Nb atoms. Furthermore, the annealing treatment led to growth in the size of the grains.

(3) Heat treatment decreased the strength but increased the ductility of the alloy. After annealing at 650 °C for 0.5 h, the prior columnar β grain growth and fine dendrite dissolution resulted in the ultimate tensile strength and yield strength decreasing to 623 MPa and 543 MPa, respectively. The increase in annealing temperature slightly increased the strength of the alloy. The solution strengthening due to the dissolution of the unmelted Nb particles together with the coarsening of the β grains may be responsible for the changing trend of the mechanical properties of the alloy with annealing temperature. The alloy annealed at 850 °C for 0.5 h exhibited a good balance of strength and ductility.

Author Contributions: J.W., J.L. and D.Z. designed experiments; J.W. and H.S. carried out experiments; J.L., J.W. and L.G. analyzed experimental results. J.L., D.Z. and C.W. assisted with Illumine sequencing. J.W. wrote the manuscript. J.L. and C.W. revised the paper.

Funding: This research was financially supported by the National Natural Science Foundation of China (11402220, 11872053), Scientific Research Fund of Hunan Provincial Educational Department (2016JC2005) and Hunan Provincial Natural Science Foundation of China (2018JJ4053).

Conflicts of Interest: The authors declare no conflict of interest.

References

1. Laheurte, P.; Prima, F.; Eberhardt, A.; Gloriant, T.; Wary, M.; Patoor, E. Mechanical properties of low modulus titanium alloys designed from the electronic approach. *J. Mech. Behav. Biomed. Mater.* **2010**, *3*, 565–573. [CrossRef] [PubMed]
2. Zhang, D.C.; Mao, Y.F.; Yan, M.; Li, J.J.; Su, E.L.; Li, Y.L. Superelastic behavior of a β-type titanium alloy. *J. Mech. Behav. Biomed. Mater.* **2013**, *20*, 29–35. [CrossRef] [PubMed]
3. Banerjeea, D. Perspectives on titanium science and technology. *Acta Mater.* **2013**, *61*, 844–879. [CrossRef]
4. Sun, F.; Hao, Y.L.; Nowak, S.; Gloriant, T.; Laheurte, P.; Prima, F. A thermo-mechanical treatment to improve the superelastic performances of biomedical Ti–26Nb and Ti–20Nb–6Zr (at. %) alloys. *J. Mech. Behav. Biomed. Mater.* **2011**, *4*, 1864–1872. [CrossRef]
5. Ramarolahy, A.; Castany, P.; Prima, F.; Laheurte, P.; Péron, I.; Gloriant, T. Microstructure and mechanical behavior of superelastic Ti–24Nb–0.5 O and Ti–24Nb–0.5 N biomedical alloys. *J. Mech. Behav. Biomed. Mater.* **2012**, *9*, 83–90. [CrossRef] [PubMed]
6. Niinomi, M.; Nakai, M.; Hieda, J. Development of new metallic alloys for biomedical applications. *Acta Biomater.* **2012**, *8*, 3888–3903. [CrossRef] [PubMed]
7. Gepreel, M.A.H.; Niinomi, M. Biocompatibility of Ti-alloys for long-term implantation. *J. Mech. Behav. Biomed. Mater.* **2013**, *20*, 407–415. [CrossRef]

8. Rao, S.; Okazaki, Y.; Tateishi, T.; Ushida, T.; Ito, Y. Cytocompatibility of new Ti alloy without Al and V by evaluating the relative growth ratios of fibroblasts L929 and osteoblasts MC3T3-E1 cells. *Mater. Sci. Eng. C Mater. Biol. Appl.* **1997**, *4*, 311–314. [CrossRef]
9. Kikuchi, M.; Takahashi, M.; Okuno, O. Mechanical properties and grindability of dental cast Ti-Nb alloys. *Dent. Mater. J.* **2003**, *22*, 328–342. [CrossRef]
10. Chang, L.L.; Wang, Y.D.; Ren, Y. In-situ investigation of stress-induced martensitic transformation in Ti–Nb binary alloys with low Young's modulus. *Mater. Sci. Eng. A* **2016**, *651*, 442–448. [CrossRef]
11. Schulze, C.; Weinmann, M.; Schweigel, C.; Keßler, O.; Bader, R. Mechanical properties of a newly additive manufactured implant material based on Ti-42Nb. *Materials* **2018**, *11*, 124. [CrossRef] [PubMed]
12. Zhang, Q.; Chen, J.; Tan, H.; Lin, X.; Huang, W.D. Microstructure evolution and mechanical properties of laser additive manufactured Ti-5Al-2Sn-2Zr-4Mo-4Cr alloy. *Trans. Nonferr. Metal. Soc.* **2016**, *26*, 2058–2066. [CrossRef]
13. Zhang, L.C.; Klemm, D.; Eckert, J.; Hao, Y.L.; Sercombe, T.B. Manufacture by selective laser melting and mechanical behavior of a biomedical Ti–24Nb–4Zr–8Sn alloy. *Scripta Mater.* **2011**, *65*, 21–24. [CrossRef]
14. Wu, X.H.; Liang, J.; Mei, J.F.; Mitchell, C.; Goodwin, P.S.; Voice, W. Microstructures of laser-deposited Ti–6al–4V. *Mater. Des.* **2004**, *25*, 137–144. [CrossRef]
15. Baufeld, B.; Brandl, E.; Biest, O.V.D. Wire based additive layer manufacturing: Comparison of microstructure and mechanical properties of Ti–6Al–4V components fabricated by laser-beam deposition and shaped metal deposition. *J. Mater. Process. Technol.* **2011**, *211*, 1146–1158. [CrossRef]
16. Fischer, M.; Joguet, D.; Robin, G.; Peltier, L.; Laheurte, P. In situ elaboration of a binary Ti-26Nb alloy by selective laser melting of elemental titanium and niobium mixed powders. *Mater. Sci. Eng. C Mater. Biol. Appl.* **2016**, *62*, 852–859. [CrossRef] [PubMed]
17. Wang, Q.; Han, C.; Choma, T.; Wei, Q.; Yan, C.; Song, B.; Shi, Y. Effect of nb content on microstructure, property and in vitro apatite-forming capability of Ti-Nb alloys fabricated via selective laser melting. *Mater. Des.* **2017**, *126*, 268–277. [CrossRef]
18. Sercombe, T.; Jones, N.; Day, R.; Kop, A. Heat treatment of Ti-6Al-7Nb components produced by selective laser melting. *Rapid Prototyp. J.* **2008**, *14*, 300–304. [CrossRef]
19. Vrancken, B.; Thijs, L.; Kruth, J.P.; Humbeeck, J.V. Heat treatment of Ti6Al4V produced by selective laser melting: Microstructure and mechanical properties. *J. Alloy. Compd.* **2012**, *541*, 177–185. [CrossRef]
20. Zhu, Y.Y.; Chen, B.; Tang, H.B. Influence of heat treatments on microstructure and mechanical properties of laser additive manufacturing Ti-5Al-2Sn-2Zr-4Mo-4Cr titanium alloy. *Trans. Nonferr. Metal. Soc.* **2018**, *28*, 36–46. [CrossRef]
21. Wang, T.; Zhu, Y.Y.; Zhang, S.Q.; Tang, H.B.; Wang, H.M. Grain morphology evolution behavior of titanium alloy components during laser melting deposition additive manufacturing. *J. Alloy. Compd.* **2015**, *632*, 505–513. [CrossRef]
22. Li, R.; Shi, Y.; Liu, J. Selective laser melting W–10 wt. % Cu composite powders. *Int. J. Adv. Manuf. Technol.* **2010**, *48*, 597–605. [CrossRef]
23. Kasperovich, G.; Haubrich, J.; Gussone, J.; Requena, G. Correlation between porosity and processing parameters in tial6v4 produced by selective laser melting. *Mater. Des.* **2016**, *105*, 160–170. [CrossRef]
24. Yadroitsev, I.; Krakhmalev, P.; Yadroitsava, I.; Johansson, S.; Smurov, I. Energy input effect on morphology and microstructure of selective laser melting single track from metallic powder. *J. Mater. Process. Technol.* **2013**, *213*, 606–613. [CrossRef]
25. Chen, W.; Chen, C.; Zi, X.; Cheng, X.; Zhang, X.; Lin, Y.C. Controlling the microstructure and mechanical properties of a metastable β titanium alloy by selective laser melting. *Mater. Sci. Eng. A* **2018**, *726*, 240–250. [CrossRef]
26. Liu, Y.J.; Li, S.J.; Wang, H.L.; Hou, W.T.; Hao, Y.L.; Yang, R. Microstructure, defects and mechanical behavior of beta-type titanium porous structures manufactured by electron beam melting and selective laser melting. *Acta Mater.* **2016**, *113*, 56–67. [CrossRef]
27. Leuders, S.; Thöne, M.; Riemer, A.; Niendorf, T.; Tröster, T.; Richard, H.A. On the mechanical behaviour of titanium alloy Ti6Al4V manufactured by selective laser melting: Fatigue resistance and crack growth performance. *Int. J. Fatigue* **2013**, *48*, 300–307. [CrossRef]

materials

MDPI

Article

Assay of Secondary Anisotropy in Additively Manufactured Alloys for Dental Applications

Elena Bassoli * and Lucia Denti

Department of Engineering "Enzo Ferrari", University of Modena and Reggio Emilia, via Pietro Vivarelli 10, 41125 Modena, Italy; lucia.denti@unimore.it

* Correspondence: elena.bassoli@unimore.it; Tel.: +39-059-2056252

Received: 22 August 2018; Accepted: 19 September 2018; Published: 26 September 2018

check for updates

Abstract: Even though additive manufacturing (AM) techniques have been available since the late 1980s, their application in medicine is still striving to gain full acceptance. For the production of dental implants, the use of AM allows to save time and costs, but also to ensure closer dimensional tolerances and higher repeatability, as compared to traditional manual processes. Among the several AM solutions, Laser Powder Bed Fusion (L-PBF) is the most appropriate for the production of metal prostheses. The target of this paper was to investigate the mechanical and microstructural characteristics of Co–Cr–Mo and Ti–6Al–4V alloys processed by L-PBF, with a specific focus on secondary anisotropy that is usually disregarded in the literature. Tensile specimens were built in the EOSINT-M270 machine, along different orientations perpendicular to the growth direction. Density, hardness, and tensile properties were measured and the results combined with microstructural and fractographic examination. For both alloys, the results provided evidence of high strength and hardness, combined with outstanding elongation and full densification. Extremely fine microstructures were observed, sufficient to account for the good mechanical response. Statistical analysis of the mechanical properties allowed to attest the substantial absence of secondary anisotropy. The result was corroborated by the observations of the microstructures and of the failure modes. Overall, the two alloys proved to be high-performing, in very close agreement with the values reported in the datasheets, independently of the build orientation.

Keywords: Powder Bed Fusion; Titanium alloys; Cobalt–Chrome alloys; anisotropy

1. Introduction

Additive Manufacturing (AM) techniques allow the production of objects with complex geometry. Fabrication can be started straightforwardly by using a three-dimensional Computer Aided Design (CAD) model, without tools. The basic idea is to think of every object as consisting of thin layers, usually in the range of 0.03–0.05 mm. The part is built up by progressive addition of material, which enables unprecedented ease of manufacturing of extremely complex shapes, since the three-dimensional manufacturing issues are simplified to two-dimensional problems [1,2]. It becomes, thus, possible to produce parts with cavities and undercuts that, by conventional subtractive methods, would have been unfeasible or would have caused great manufacturing hurdles and costs. AM technologies were introduced in industry in the late 1980s to realize models and prototypes, but nowadays the advances in materials and technology are sufficient to make the production of end products of major interest [3,4]. The great potential and good evolution of techniques led to introduce AM in medicine, where the need for parts that are customized for each patient, with a high degree of personalization, allows full exploitation of the inherent benefits of additive processes [5]. In particular, the Laser-based Powder Bed Fusion (L-PBF) process can be successfully used for the production of prostheses [6], including for example long-span and cantilever metal-ceramic-fixed partial dentures for maxillary and mandibular prosthodontics [7,8]. Apart from restorations, also surgical guides aimed at operation planning can take advantage of the quick production by AM [6,9].

This paper focuses on the characterization of alloys developed on purpose for maxillo-facial surgery, in particular for oral implants. Implantology is a field under continuous innovation, where research efforts are concurrently dedicated to finding new materials, new components, new fabrication processes, with the aim to improve the duration, the aesthetics and the functionality of prostheses and thus ensuring a better quality of life for patients. An easy example is that of gold crowns that have been superseded by metal-ceramic or ceramic crowns [10–12]. An increasing variety of metallic biomaterials is being developed, ranging from commercially pure Titanium [13] and Titanium alloys, through stainless steels, to Cobalt–Chrome alloys. Promising results have recently been attained for innovative β-type Ti alloys with increased wear resistance and lower elastic modulus, so as to better match that of the human bone and prevent the stress shielding effect [14]. Some of these novel alloys exhibit composite microstructures where β–Ti dendrites are surrounded by intermetallic phases so that efficient reinforcing phenomena are established [15]. The promising outlook of these Titanium-based matrix composites for AM has been very recently reviewed [16].

L-PBF process can be used in the construction of metal-ceramic fixed partial dentures (FPDs). Traditionally, the manufacture of the metallic part of FPDs involves a large series of manual operations performed by the dental technician, and the result is often strongly determined by his ability. For metal-prostheses fabrication, the adoption of L-PBF grants a much higher repeatability and predictability with respect to the manual process [17]. Compared to other powder-based methods that require molds, PBF offers outstanding personalization capabilities, in extremely short times and with low costs. On the other hand, the L-PBF process is quite complex and many factors are involved in order to achieve good part quality [18]. Despite the extremely diffused studies on this process, many efforts are still needed to better understand the relation between microstructure, processing, and properties for parts built by L-PBF [19–21].

In L-PBF, at each step, a thin layer of metallic powder is evenly distributed onto the previous layer and a laser selectively scans the regions corresponding to the cross-section of the part. As a consequence, the powder melts and then consolidates into a solid slice. Inherent in the process are two types of possible anisotropy: a primary one, due to the superimposition of layers in the direction that is usually called Z; and a secondary one that may manifest as direction-dependence of properties even within the XY plane, that is to say parallel to the layers [22]. The latter is usually ignored by machine- and material suppliers, and has been disregarded by scientific literature until now. Secondary anisotropy may be caused for example by the action of the recoater blade that spreads the powder in the bed, or by the inert gas flux that blows the melting slags away from the build area [23]. Each of the two phenomena usually acts along either the X or the Y direction, depending on the specific machine architecture. The investigation of secondary anisotropy is markedly important if the intended application is the production of FPDs, because complete prostheses are not straight structures but develop along the maxillary/mandibular arch, hence they involve material properties in several directions of the XY plane. A robust design of the restoration requires a reliable knowledge of any direction-dependent feature.

This research tackles the mechanical properties and the microstructure of two L-PBF fabricated dental alloys, namely Co–Cr–Mo and Ti–6Al–4V, by proposing a statistically-based enquiry of secondary anisotropy.

2. Materials and Methods

Tensile specimens were produced by L-PBF using the two alloys Ti–6Al–4V (EOS GmbH, Krailling, Germany) and Co–Cr–Mo (EOS Cobalt Chrome MP1, EOS GmbH, Krailling, Germany).

The specimens were fabricated on the L-PBF machine EOSINT-M270, by using the following process parameters:

- for Ti–6Al–4V: laser power 340 W, laser spot diameter 0.1 mm, layer thickness 30 μm, scan speed 1250 mm/s, hatch distance 0.12 mm, protective atmosphere (max 0.1% oxygen);

- for Co–Cr–Mo: laser power 200 W, laser spot diameter 0.2 mm, layer thickness 20 μm, scan speed 7000 mm/s, hatch distance 0.3 mm, protective atmosphere (max 1.5% oxygen).

For both alloys, tensile specimens were built in three different orientations relative to the machine distinctive directions, all of the three parallel to the layers and perpendicular to the growth direction. The three groups, each of 6 specimens, are specified as follows:

- "X" group: the axis of the specimens aligns to the direction which the recoater blade spreads the powder in the bed along;
- "Y" group: the axis of the specimens aligns to the direction of the inert gas flux on the powder bed;
- "XY" group: the axis of the specimens is angled 45° with respect that of X and Y groups.

The size and geometry for the tensile test specimens conformed to the prescription specified in standard ASTM E8M [24]. Details are reported in Figure 1.

Of the powders, the nominal physical/mechanical properties and chemical composition are listed in Tables 1 and 2, respectively. The powders were characterized by means of laser granulometry (Malvern Mastersizer 3000, Malvern Panalytical Ltd., Malvern, UK) to assess their size distribution, according to ISO 13320 standard [25].

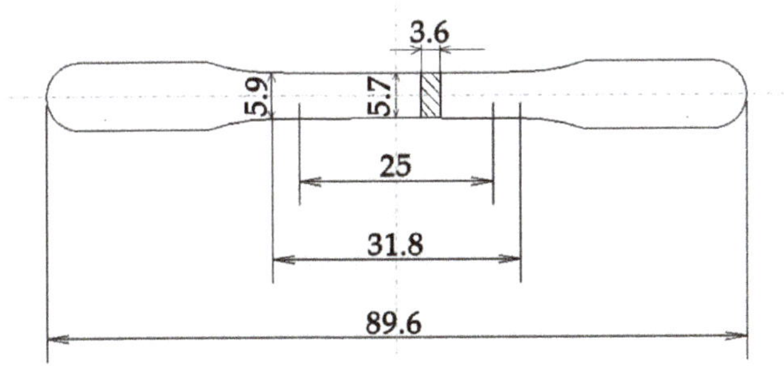

Figure 1. Tensile specimen. Dimensions are expressed in millimeters.

Table 1. Nominal physical and mechanical properties of the two alloys.

Property	Ti-6Al-4V [26]	Co–Cr–Mo [27]
Melting point (°C)	1634–1664	1350–1430
Density (kg/dm^3)	4.43	8.3
Tensile strength (MPa) *	1150	1350
Total extension at fracture (%) *	11	11

* in horizontal direction, as built condition.

Table 2. Nominal chemical composition of the two alloys.

Alloy	Weight Percentage						
Ti-6Al-4V [26]	Ti	Al	V	C	Fe	N	O
-	88–90.2	5.5–6.8	3.5–4–5	<0.08	<0.30	<0.05	<0.2
Co–Cr–Mo [27]	Co	Cr	Mo	Si	Mn	other	
-	59.5	31.5	5.0	2.0	1.0	1.0	

The specimens were tested in the as built condition, without any heat treatment, so as to avoid any smoothing of the secondary anisotropy produced by the L-PBF process.

Before the tensile tests, the Archimedes principle was used to measure the density of all the samples (6 for each group), with an analytical electronic balance having a resolution of 0.1 mg

(Pioneer textsuperscript® Plus PA124C, OHAUS GmbH, Greifensee, Switzerland). The residual porosity was then calculated by using the nominal density of each alloy.

Tensile tests were performed on a SCHENK HYDROPULS PSB testing machine (SCHENCK RoTec GmbH, Darmstadt, Germany) with a capacity of 250 kN, using a crosshead speed of 5 mm/min. Five samples were tested for each alloy and orientation, and one extra specimen of each group was used to measure hardness and to obtain the metallographic sections. The choice of the hardness scale was made according to ISO standard 4498 [28]. Rockwell C was selected and performed following the specifications of standard ISO 6508 [29], by repeating five measurements on each sample. Numerical results for hardness (HRC), tensile strength (UTS) and total extension at fracture (ε_b) were processed through statistical tools (Statistica 8, Statsoft, Hamburg, Germany): the *t*-test with a level of significance of 0.05 was performed to investigate the presence of significant differences between the groups of specimens produced along different orientations.

After tensile tests, rupture surfaces were observed by using a scanning electron microscope, SEM (ESEM, Quanta FEI, Thermo Fisher Scientific, Eindhoven, The Netherlands), in order to investigate the failure mechanisms and the joining phenomena between the particles.

Metallographic sections of the samples were obtained and observed by an optical microscope (OM) (Eclipse LV150N, Nikon, Tokyo, Japan), to get a cross-check of residual porosity and compare the results with those obtained by the Archimedes method. A comparative assessment of the two methods is raising the interests of the scientific community [30], growingly as the two techniques are more and more diffused in industry for the control of AM parts. Preparation of the metallographic sections consisted of micro-cutting, embedding in epoxy resin and polishing till a fine grinding. The final step was carried out with a plan cloth and 1 μm diamond suspension. Several micrographs were acquired through a CCD camera, made binary and analyzed through a software tool for image analysis to determine:

- percentage porosity, calculated as the area fraction of pores out of the overall area;
- the average pore area.

After OM observation, polished sections of Ti–6Al–4V underwent chemical etching with the Dix-Keller reactant (HF 2% vol, HCl 1.5% vol, HNO_3 2.5% vol; water bal.); while metallographic sections of the Co–Cr–Mo alloy were subjected to electrochemical etching (HCl 0.1 M, 2 V, 2 min). Microstructures were observed on the etched samples by means of OM and SEM.

3. Results and Discussion

3.1. Powder Particle Size Distribution

The results of laser granulometry of the two powders are shown in Figure 2. Co–Cr–Mo powder displays a wider distribution, with an average value of the order of 80 μm and a relatively large number of particles in the range 10–30 μm causing a negative skew in the curve. The size distribution for Ti–6Al–4V powder is instead symmetrical, with an average particle dimension of 30 μm.

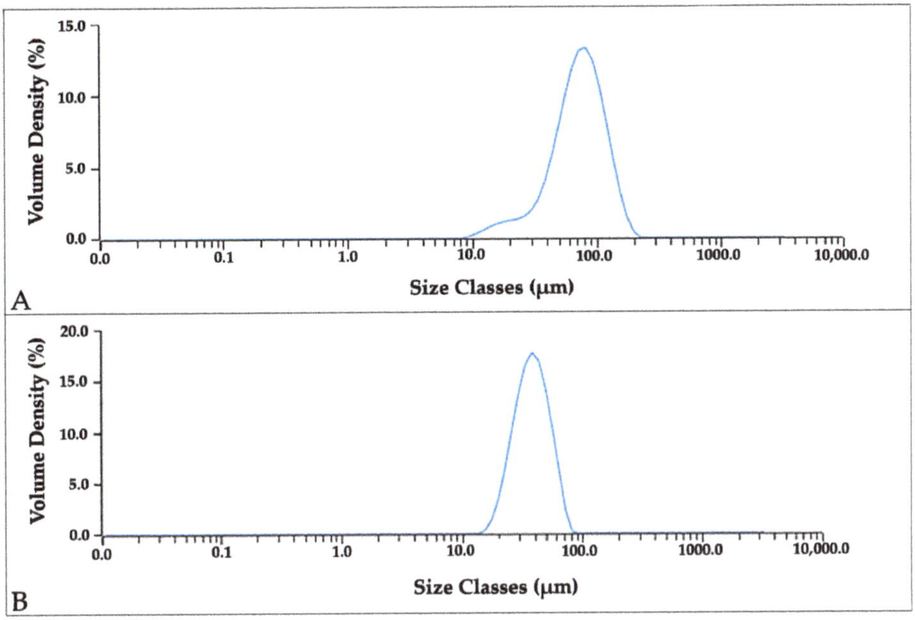

Figure 2. Particle size distribution of the Co–Cr–Mo (**A**) and Ti-6Al-4V (**B**) powders.

3.2. Density and Residual Porosity

Table 3 lists the results for density and for residual porosity measured by the Archimedes principle, as well as by OM observations of metallographic sections. Figures 3 and 4 show, respectively for Ti–6Al–4V and Co–Cr–Mo, examples of the OM images on which residual porosity was calculated by image analysis. The nominal densities of the two alloys are available in Table 1 for comparison. While the data obtained by the Archimedes method are normally distributed, the data by microscopic analysis exhibit the asymmetrical distributions shown in Figure 5. Hence, mean values and standard deviations are listed in Table 4 for the Archimedes figures, whereas median and mean values are reported for the analysis of OM images. Density figures are, for both alloys, very close to the nominal values, with extremely narrow deviations and no evident direction dependence, as the differences between the values calculated for the X, Y, and XY groups are contained within the standard deviations. Residual porosity is in all cases well below 1%, with no distinction for the various orientations. If porosity is calculated by comparing the Archimedes density with the nominal one, the values are slightly higher than those obtained by metallographic observations, with the only exception of Ti–6Al–4V X specimens. Based on these results, the Archimedes method can be reckoned conservative if applied to density control of L-PBF fabricated parts. This remark is in very good agreement with the results attained by Spierings et al. [30], who found the Archimedes measurement highly accurate and repeatable for the control of metal parts produced by PBF. The same study also concludes that, in contrast, microscopic analysis of cross sections can give inconsistent values of density, with variations of up to 4% in the direction of an underestimate of the residual porosity. The two methods are found comparable by Spierings et al. only for low porosities.

Table 3. Density and porosity determined by the Archimedes principle and by OM observations. For Archimedes measurements, standard deviations are given in brackets next to the mean values computed over 6 measurements. For the results obtained by OM, median and mean values are provided.

Density and Porosity	Ti–6Al–4V		Co–Cr–Mo	
Archimedes	density (kg/dm^3) mean (SD)	residual porosity (%) mean	density (kg/dm^3) mean (SD)	residual porosity (%) mean
X	4.43 (0.00)	0	8.24 (0.01)	0.72
XY	4.41 (0.02)	0.45	8.26 (0.01)	0.48
Y	4.40 (0.02)	0.68	8.25 (0.06)	0.60
Analysis of OM images	average pore area (μm^2)	residual porosity (%)	average pore area (μm^2)	residual porosity (%)
	median–mean	median–mean	median–mean	median–mean
X	1.73–4.58	0.00–0.04	5.45–15.87	0.31–0.55
XY	1.82–2.04	0.19–0.21	5.03–10.09	0.17–0.30
Y	13.67–17.60	0.21–0.28	7.76–18.78	0.25–0.43

Table 4. Results of the hardness and tensile tests. Standard deviations are given in brackets next to the mean values.

Hardness and Tensile Tests Results	Ti–6Al–4V			Co–Cr–Mo		
-	HRC	UTS (MPa)	ε_b (%)	HRC	UTS (MPa)	ε_b (%)
X	39.8 (2.58)	1110 (1)	12.1 (1.5))	46.9 (1.13)	1282 (11)	12.8 (0.4)
XY	38.7 (3.85)	1098 (25)	12.7 (0.6)	46.9 (0.93)	1290 (6)	14.1 (0.2)
Y	38.9 (2.53)	1080 (5)	11.4 (0.9)	47.0 (0.76)	1300 (7)	12.9 (0.6)

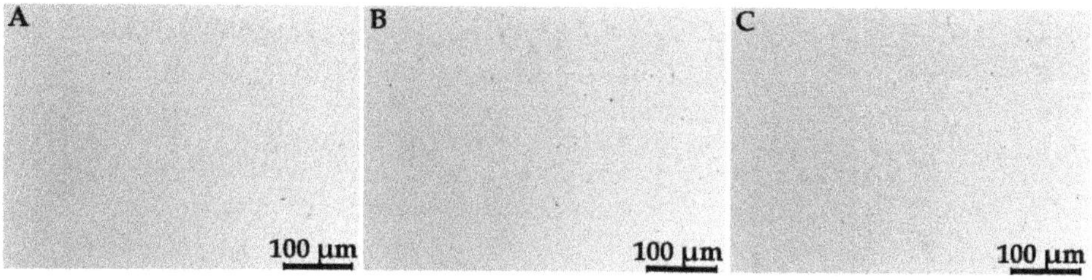

Figure 3. OM images of metallographic sections of Ti–6Al–4V specimens of the X (**A**), Y (**B**), and XY (**C**) groups.

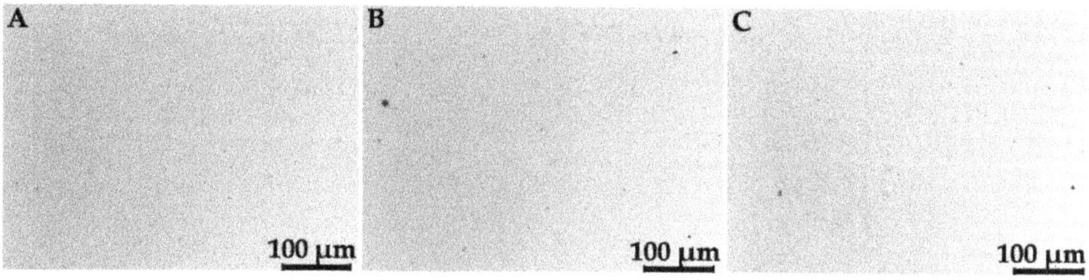

Figure 4. OM images of metallographic sections of Co–Cr–Mo specimens of the X (**A**), Y (**B**), and XY (**C**) groups.

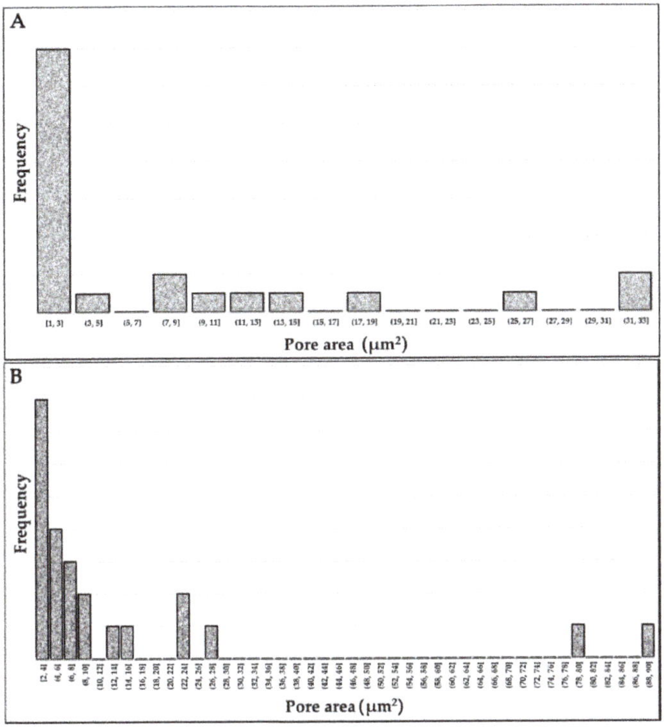

Figure 5. Distribution of the area of pores detected by the analysis of OM images: (**A**) Ti–6Al–4V, (**B**) Co–Cr–Mo.

3.3. Hardness and Tensile Tests

The results of hardness and tensile tests are listed in Table 4. Representative tensile stress–strain curves are shown in Figure 6. Necking is observed for Ti–6Al–4V, whereas the Co–Cr–Mo graphs are bilinear. As a term of comparison, mechanical properties reported in literature for the same alloys are given in Table 5. Normality of data distribution was verified by using the Shapiro-Wilk test, for all the mechanical characteristics recorded in Table 4. Then, the t-test was performed by grouping the mechanical properties according to the variable "orientation". Table 6 registers the results, expressed in terms of probability values (p-values). When lower than 0.05, the p-values can be taken as a decision to reject the null hypothesis of absence of significant differences between the groups, that is to say of absence of anisotropy. In other terms, when the p-value is lower than 0.05 the mechanical response in different orientations can be regarded as non-equivalent.

Table 5. Tensile properties reported in literature for the Ti–6Al–4V and the Co–Cr–Mo alloys.

Ti–6Al–4V			Co–Cr–Mo		
AM Process	**UTS (MPa)**	**ε_b (%)**	**AM Process**	**UTS (MPa)**	**ε_b (%)**
Minimum [31]	896	4	L-PBF [32]	912	10.7
EBM [33]	946	13.2	L-PBF [34]	817	10.5
L-PBF [35]	1250	7	-	-	-
L-PBF + HIP [33]	997	11.4	-	-	-

Table 6. Values resulting from the *t*-test for the variables HRC, UTS, ε_b among the groups built in different orientations. Records below the level of significance of 0.05 are bold.

p-Values	Ti–6Al–4V			Co–Cr–Mo		
-	HRC	UTS (MPa)	ε_b (%)	HRC	UTS (MPa)	ε_b (%)
X vs. XY	0.29	0.31	0.39	0.76	0.18	0.37
X vs. Y	0.26	0.00	0.66	0.91	0.01	0.38
XY vs. Y	0.86	0.16	0.75	0.62	0.05	0.46

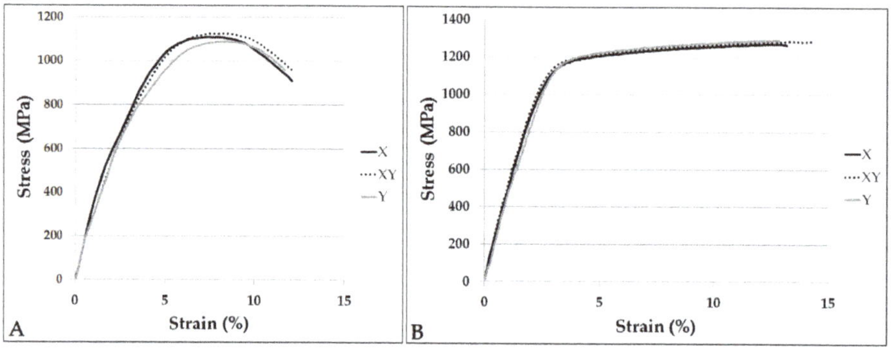

Figure 6. Example stress–strain curves of the Ti–6Al–4V (**A**) and Co–Cr–Mo (**B**) specimens built in the three orientations.

HRC hardness is nearly 39 for Titanium alloy samples and 47 for Co–Cr–Mo specimens, with no statistically-significant differences between samples produced in different orientations. Mean UTS of Titanium specimens resulted in the range 1080–1110 MPa, and extension is about 12.5%. As for the Co–Cr–Mo alloy, mean UTS is between 1280 and 1300 MPa and values of 13–14%were obtained for total extension at fracture. All groups exhibit good test repeatability, with very low standard deviations, except for Ti–6Al–4V XY samples that show a relatively high scattering of the measured UTS. A slight anisotropy can be noticed for Ti–6Al–4V, consisting of a decreasing trend of strength from the X- towards the Y-orientation. Nevertheless, the variation is proportionately low (2.8%) and is pointed out as significant by the *t*-test only when two extreme groups, X and Y, are compared. A similar consideration is valid for Co–Cr–Mo samples, but in this case the opposite trend is observed: strength increases as the build orientation varies from the X- to the Y-direction. Even if the *t*-test is positive in two of the three cases, yet the overall deviation is as little as 1.6%. From a practical standpoint, for any industrial application the mechanical properties in the three directions would be considered undifferentiated, as a deviation of few per cents is by far absorbed by the factor of safety. As to ductility, for all the specimens the values are remarkably high if compared either to the nominal characteristics or to the typical properties of L-PBF fabricated parts. Furthermore, no direction dependence is evidenced for total extension at fracture.

3.4. Fractography

For Ti–6Al–4V specimens, failure occurs by a variety of mechanisms that can be observed comprehensively on macro-views of the rupture surfaces, as in the example in Figure 7A. The rupture shown is a typical failure mode of ductile materials, usually designated as cup and cone rupture. This form of ductile failure begins after necking and develops through sequential steps. At the outset, small micro-voids appear in the innermost zone of the specimen. Then, as plastic deformation proceeds, the micro-voids expand and merge into a crack. Figure 7B,C allow to appreciate, at high magnifications, the failure morphologies of the areas that are marked in Figure 7A as "B" and "C", respectively. As to the first, laterally-fine but raised dimples can be observed, with a large amount of plastic deformation.

This morphology can also be designated as cellular fracture and is often detected on the rupture surfaces of additively manufactured multi-phase materials [36]. On the last-breaking areas, as in Figure 7C, dimples are much flatter, indicating a low energy dissipation. Figure 7D shows a zone of transition between the two described failure modes. Lack of fusion defects are exceptionally spotted on the rupture surfaces (Figure 7E) [24]. Failure modes and rupture morphologies are identical for the specimens produced in the three orientations, as can be reckoned by comparing Figure 7 (X specimen) with Figure 8 (Y and XY specimens).

Co–Cr–Mo specimens exhibit a homogeneous morphology across the rupture surfaces. A representative case for each of the three build orientations is shown in Figure 9, where failure seems to occur mainly by transgranular cleavage, even if the values of extension at fracture would suggest a more ductile mode. In these cases, the term "quasi-cleavage" is usually adopted, to identify a rupture that combines cleavage-like features with evidence of plastic deformation. Also, in this case, as for Ti–6Al–4V, the rupture surfaces of the three groups of samples are totally equal to each other.

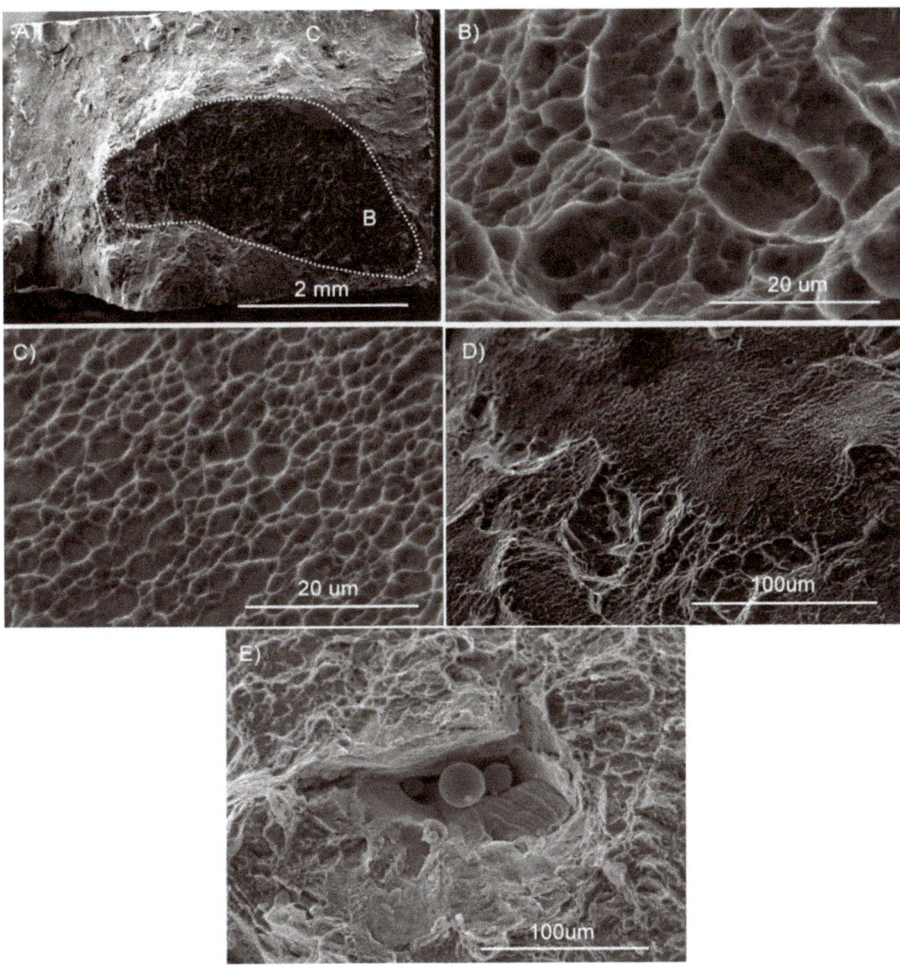

Figure 7. SEM observations of the rupture surface of a Ti–6Al–4V X specimen: (**A**) overall view; (**B**) dimpled rupture morphology; (**C**) quasi-flat rupture morphology; (**D**) mixed morphology; (**E**) lack of fusion defect.

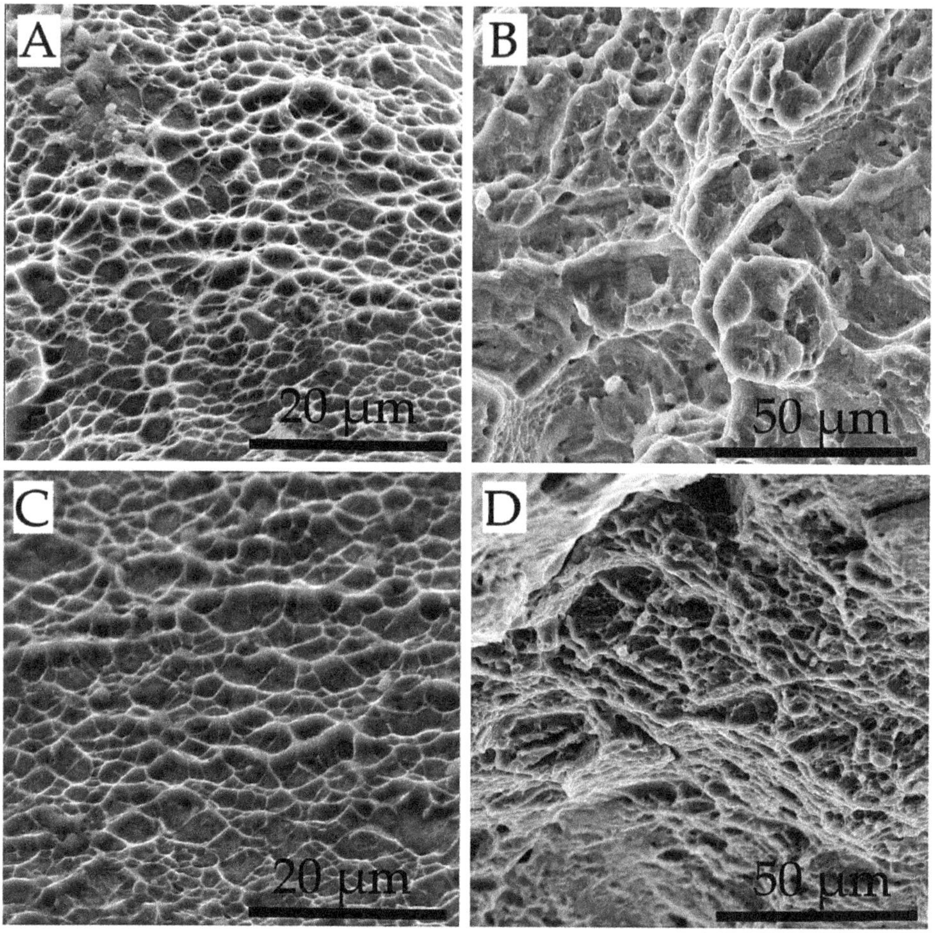

Figure 8. SEM observations of the rupture surfaces of a Ti–6Al–4V Y (**A**,**B**) and XY (**C**,**D**): (**A**,**C**) quasi-flat rupture morphology; (**B**,**D**) dimpled rupture morphology.

For both alloys, if the different building orientations are compared, different rupture morphologies can be perceived in the macroscale, but the micro-mechanisms are in effect equal. This result is consistent with the substantial equivalence of the mechanical properties that has been discussed in Section 3.3. Overall, fractography suggests an extremely fine grain structure, which will be verified in the next section by means of observations of etched sections.

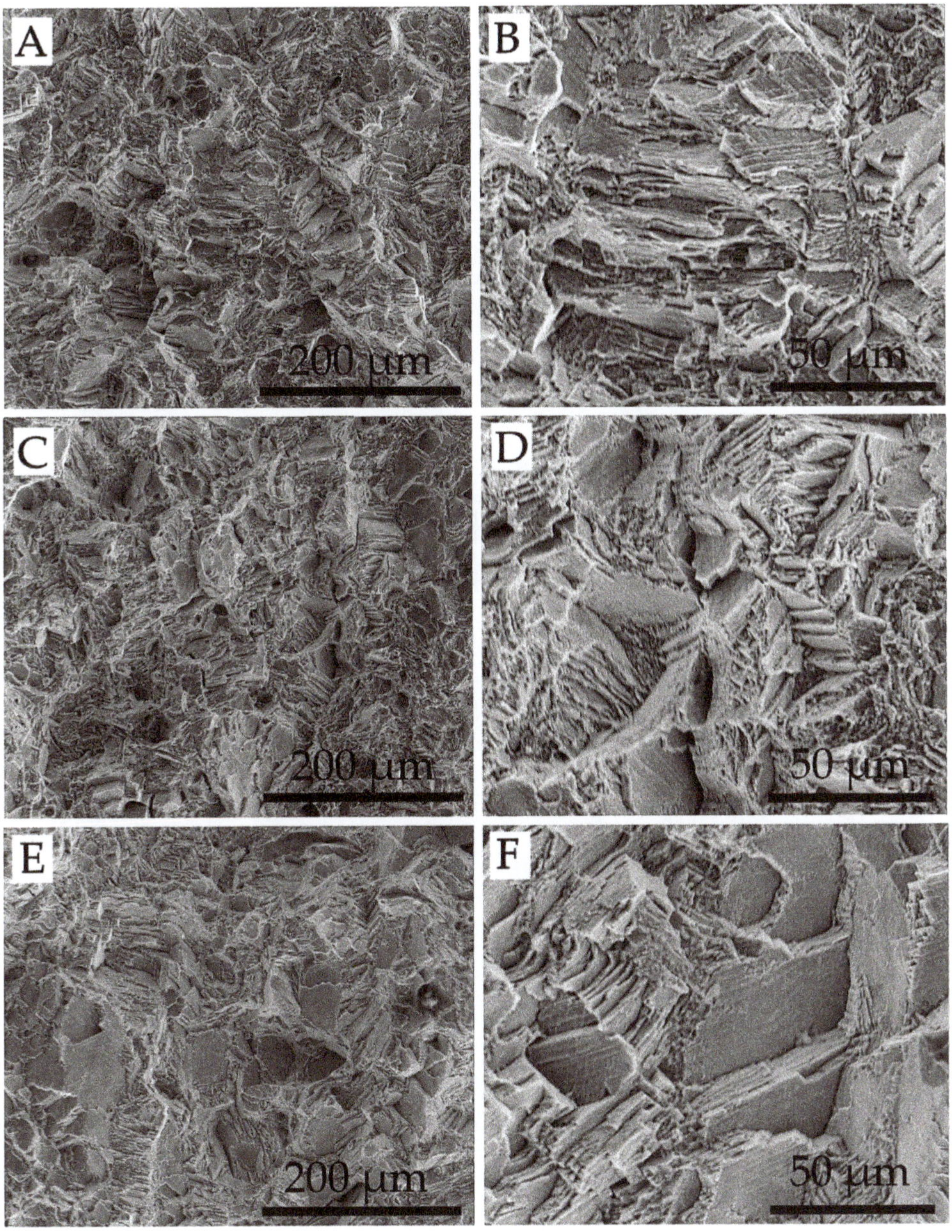

Figure 9. SEM observations of the rupture surface of Co–Cr–Mo specimens: (**A**,**B**) X; (**C**,**D**) Y; (**E**,**F**) XY.

3.5. Microstructure

The microstructure observed for Ti–6Al–4V samples in the different orientations is visible in Figure 10. The sections in Figure 9B,C and Figure 10A are perpendicular to the axes of a Y, X, and XY specimen, respectively. Figure 10D allows to appreciate the microstructure in the XY plane. The Ti–6Al–4V specimens exhibit a uniform acicular α' martensite microstructure [13], which forms under elevated cooling rates. Rapid cooling is frequently reported in literature for

AM processes [19,37,38]. As an example, Criales et al. in a study on the L-PBF process of Inconel 625 measured cooling rates in the order of 150 °C/ms [38]. For Ti–6Al–4V, a rapid quenching is known in literature to cause a martensitic transformation, leading to a very fine needle-like microstructure [39]. No evidence of direction-dependent features is noticed.

Images of the etched sections of Co–Cr–Mo specimens, referred to all the orientations already shown for Ti–6Al–4V samples, are provided in Figure 11A,D. For this alloy, etching reveals the boundaries of the melt pools. In all directions, within each melt pool an ultra-fine columnar microstructure can be appreciated at high magnification (Figure 11E,F), with varying grain orientation. The columnar grains have sub-micron diameter. No evidence of a martensitic structure is visible in Figure 11, however these observations are not conclusive on the point, in the absence of specific investigations, as for example X-ray diffractions.

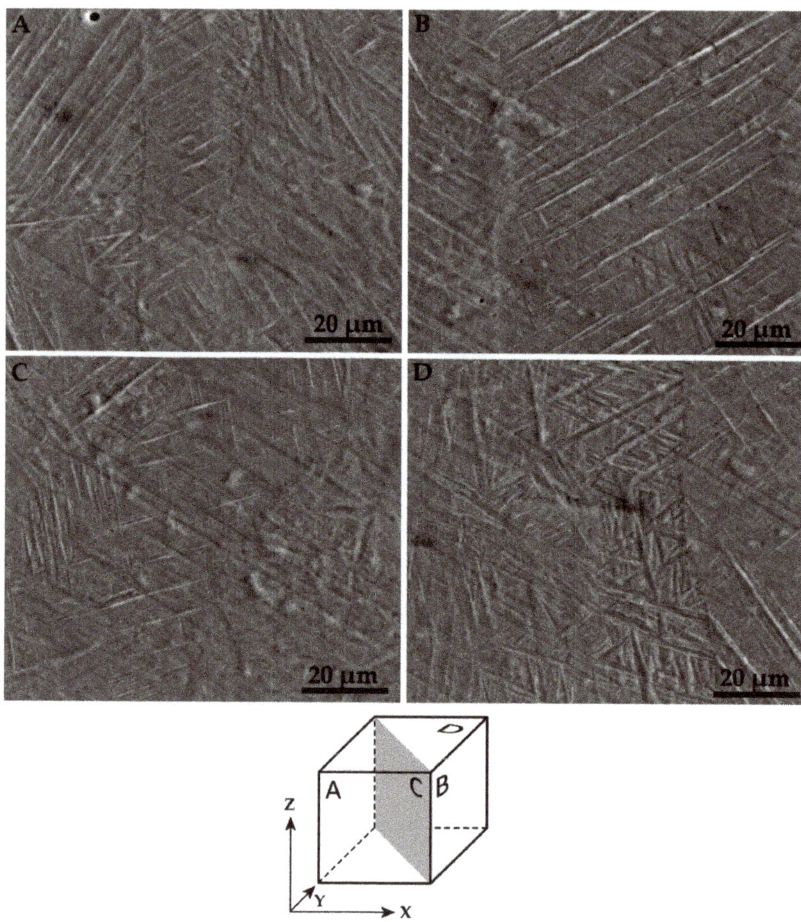

Figure 10. SEM images of etched sections of Ti–6Al–4V specimens. Alignment of the (**A–D**) images is schematized in the cube.

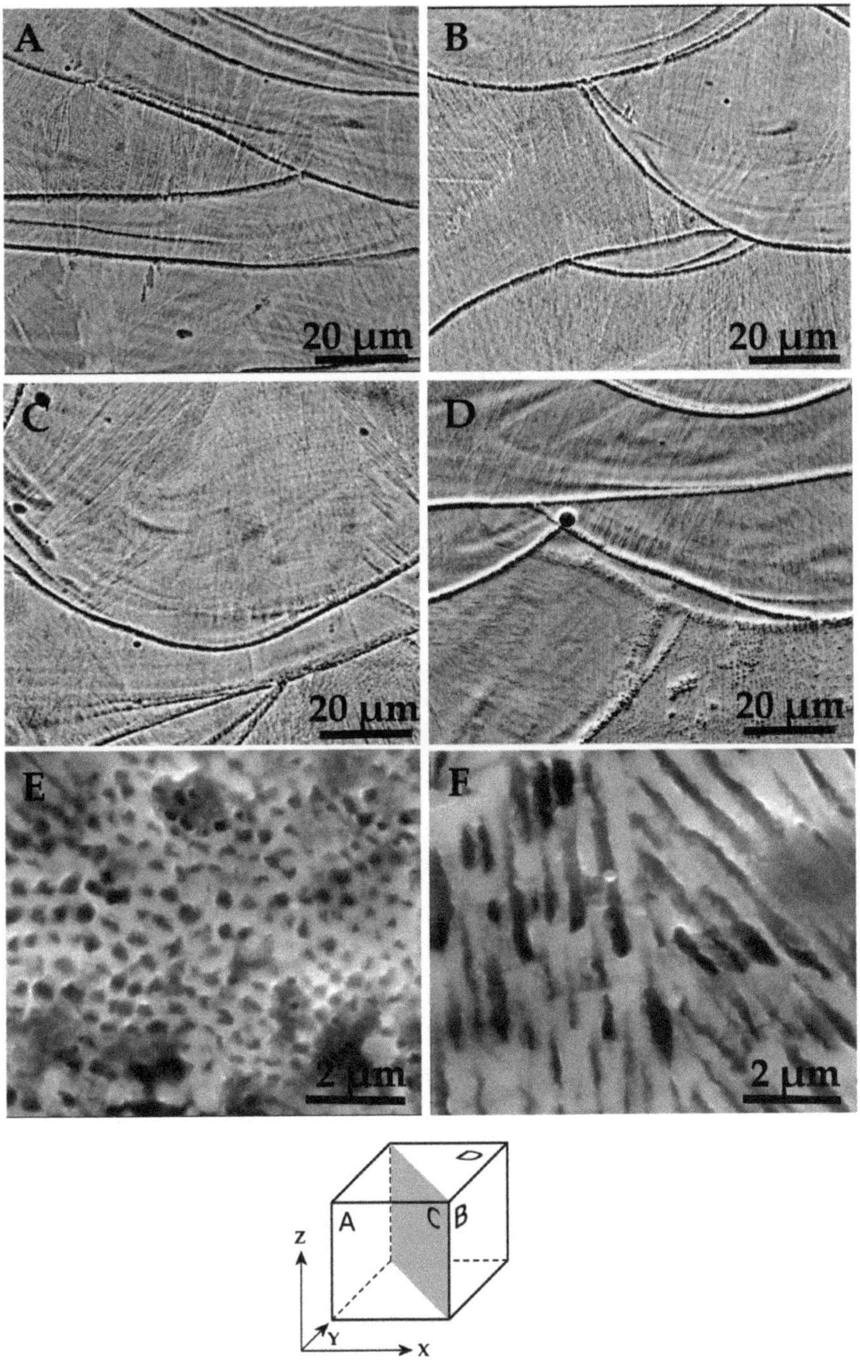

Figure 11. SEM images of etched sections of Co–Cr–Mo specimens. Alignment of the (**A**–**D**) images is schematized in the cube. (**E**,**F**) Detail of the columnar grains.

4. Conclusions

In view of the increasing interest in the use of additively manufactured parts for dental prostheses, the mechanical behavior and the microstructure of Ti–6Al–4V and Co–Cr–Mo parts, built by L-PBF, were investigated, with a specific focus on the evaluation of secondary anisotropy.

For both alloys, the measured hardness and strength were in good agreement with those reported in the datasheets; ductility was remarkably high and nearly full densification was measured. The observed microstructures, typical of the extreme cooling rates experienced by the materials during L-PBF processes, allow to account for the outstanding mechanical properties that were appraised in this study. Statistical analysis of the mechanical properties allowed to attest the substantial absence of secondary anisotropy and the result was confirmed by the observation of identical failure modes of the specimens produced in the different orientations.

On the whole, the results enable the conclusion that the two alloys considered here may achieve exceptionally high properties if manufactured by L-PBF, and that secondary anisotropy is negligible if not totally absent.

Author Contributions: Conceptualization, Methodology, Writing: Review and Editing E.B.; Validation, Investigation, Writing: Original Draft Preparation, L.D.

Funding: This research received no external funding.

Conflicts of Interest: The authors declare no conflict of interest.

References

1. Lancaster, R.; Davies, G.; Illsley, H.; Jeffs, S.; Baxter, G. Structural integrity of an electron beam melted titanium alloy. *Materials* **2016**, *9*, 470. [CrossRef] [PubMed]
2. Bassoli, E.; Sewell, N.; Denti, L.; Gatto, A. Investigation into the failure of inconel exhaust collector produced by laser consolidation. *J. Eng. Fail. Anal.* **2013**, *35*, 397–404. [CrossRef]
3. Weller, C.; Kleer, R.; Piller, F.T. Economic implications of 3D printing: Market structure models in light of additive manufacturing revisited. *Int. J. Prod. Econ.* **2015**, *164*, 43–56. [CrossRef]
4. Atzeni, E.; Iuliano, L.; Marchiandi, G.; Minetola, P.; Salmi, A.; Bassoli, E.; Denti, L.; Gatto, A. Additive manufacturing as a cost-effective way to produce metal parts. In *High Value Manufacturing: Advanced Research in Virtual and Rapid Prototyping, Proceedings of the 6th International Conference on Advanced Research and Rapid Prototyping, Leiria, Portugal, 1–5 October 2013*; CRC Press: Boca Raton, FL, USA, 2013; pp. 3–8.
5. Glantz, P.O.; Ryge, G.; Jendresen, M.D.; Nilner, K. Quality of extensive fixed prosthodontics after five years. *J. Prosthet. Dent.* **1984**, *52*, 475–479. [CrossRef]
6. Walton, T.R. An up to 15-year longitudinal study of 515 metal-ceramic FPDs: Part 1. Outcome. *Int. J. Prosthodont.* **2002**, *15*, 439–445. [PubMed]
7. Laurell, L.; Lundgren, D.; Falk, H.; Hugoson, A. Long-term prognosis of extensive polyunit cantilevered fixed partial dentures. *J. Prosthet. Dent.* **1991**, *66*, 545–552. [CrossRef]
8. Eliasson, A.; Arnelund, C.F.; Johansson, A. A clinical evaluation of cobalt-chromium metal-ceramic fixed partial dentures and crowns: A three- to seven-year retrospective study. *J. Prosthet. Dent.* **2007**, *98*, 6–16. [CrossRef]
9. Pjetursson, B.E.; Tan, K.; Lang, N.P.; Brägger, U.; Egger, M.; Zwahlen, M. A systematic review of the survival and complication rates of fixed partial dentures (FPDs) after an observation period of at least 5 years: I. Implant-supported FPDs. *J. Clin. Oral Implants Res.* **2004**, *15*, 625–642. [CrossRef] [PubMed]
10. Schmalz, G.; Garhammer, P. Biological interactions of dental cast alloys with oral tissues. *Dent. Mater.* **2002**, *18*, 396–406. [CrossRef]
11. Al-Hiyasat, A.S.; Bashabsheh, O.M.; Darmani, H. An investigation of the cytotoxic effects of dental casting alloys. *Int. J. Prosthodont.* **2003**, *16*, 8–12. [PubMed]
12. Odén, A.; Andersson, M.; Krystek-Ondracek, I.; Magnusson, D. Five-year clinical evaluation of Procera AllCeram crowns. *J. Prosthet. Dent.* **1998**, *80*, 450–456. [CrossRef]

13. Attar, H.; Calin, M.; Zhang, L.C.; Scudino, S.; Eckert, J. Manufacture by selective laser melting and mechanical behavior of commercially pure titanium. *Mater. Sci. Eng. A* **2014**, *593*, 170–177. [CrossRef]
14. Ehtemam-Haghighi, S.; Prashanth, K.G.; Attar, H.; Chaubey, A.K.; Cao, G.H.; Zhang, L.C. Evaluation of mechanical and wear properties of tixnb7fe alloys designed for biomedical applications. *Mater. Des.* **2016**, *111*, 592–599. [CrossRef]
15. Okulov, I.V.; Volegov, A.S.; Attar, H.; Bönisch, M.; Ehtemam-Haghighi, S.; Calin, M.; Eckert, J. Composition optimization of low modulus and high-strength tinb-based alloys for biomedical applications. *J. Mech. Behav. Biomed. Mater.* **2017**, *65*, 866–871. [CrossRef] [PubMed]
16. Attar, H.; Ehtemam-Haghighi, S.; Kent, D.; Dargusch, M.S. Recent developments and opportunities in additive manufacturing of titanium-based matrix composites: A review. *Int. J. Mach. Tools Manuf.* **2018**, *133*, 85–102. [CrossRef]
17. Bibb, R.; Eggbeer, D.; Williams, R. Rapid manufacture of removable partial denture frameworks. *Rapid Prot. J.* **2006**, *12*, 95–99. [CrossRef]
18. Van Elsen, M.; Al-Bender, F.; Kruth, J. Application of dimensional analysis to selective laser melting. *Rapid Prot. J.* **2008**, *14*, 15–22. [CrossRef]
19. Frazier, W.E. Metal additive manufacturing: A review. *J. Mater. Eng. Perform.* **2014**, *23*, 1917–1928. [CrossRef]
20. Facchini, L.; Magalini, E.; Robotti, P.; Molinari, A.; Höges, S.; Wissenbach, K. Ductility of a Ti–6Al–4V alloy produced by selective laser melting of prealloyed powders. *Rapid Prot. J.* **2010**, *16*, 450–459. [CrossRef]
21. Kruth, J.-P.; Mercelis, P.; Van Vaerenbergh, J.; Froyen, L.; Rombouts, M. Binding mechanisms in selective laser sintering and selective laser melting. *Rapid Prot. J.* **2005**, *11*, 26–36. [CrossRef]
22. Bassoli, E.; Gatto, A.; Iuliano, L. Joining mechanisms and mechanical properties of pa composites obtained by selective laser sintering. *Rapid Prot. J.* **2012**, *18*, 100–108. [CrossRef]
23. Denti, L. Additive manufactured A357.0 samples using the laser powder bed fusion technique: Shear and tensile performance. *Metals* **2018**, *8*, 670. [CrossRef]
24. ASTM International. *ASTM E8/E8M-09 Standard Test Methods for Tension Testing of Metallic Materials*; ASTM International: West Conshohocken, PA, USA, 2011.
25. ISO International Organization for Standardization. *ISO 13320:2009 Particle Size Analysis—Laser Diffraction Methods*; International Organization for Standardization: Geneva, Switzerland, 2009.
26. EOS GmbH. Material Data Sheet EOS Titanium Ti64. Available online: http://www.harbec.com/wp-content/uploads/2014/08/ti-ti64_m270_material_data_sheet_12-08_en.pdf (accessed on 8 September 2018).
27. EOS GmbH. Material Data Sheet EOS CobaltChrome MP1. Available online: http://ip-saas-eos-cms.s3.amazonaws.com/public/4b839242298b3d77/721463526ca053889c9784ec989f3c88/EOS_CobaltChrome_MP1_en.pdf (accessed on 8 September 2018).
28. International Organization for Standardization. *Sintered Metal Materials, Excluding Hardmetals—Determination of Apparent Hardness and Microhardness*; ISO 4498; International Organization for Standardization: Geneva, Switzerland, 2010.
29. International Organization for Standardization. *Metallic Materials—Rockwell Hardness Test—Part 1: Test. Method*; ISO 6508-1:2015; International Organization for Standardization: Geneva, Switzerland, 2015.
30. Spierings, A.B.; Schneider, M.; Eggenberger, R. Comparison of density measurement techniques for additive manufactured metallic parts. *Rapid Prot. J.* **2011**, *17*, 380–386. [CrossRef]
31. SAE International. *AMS 4999 Titanium Alloy Laser Deposited Products 6Al-4V Annealed*; ASTM International: West Conshohocken, PA, USA, 2002.
32. Koutsoukis, T.; Zinelis, S.; Eliades, G.; Al-Wazzan, K.; Rifaiy, M.A.; Al Jabbari, Y.S. Selective laser melting technique of Co-Cr dental alloys: A review of structure and properties and comparative analysis with other available techniques. *J. Prosthodont.* **2015**, *24*, 303–312. [CrossRef] [PubMed]
33. Rengers, S. Electron beam melting [EBM] vs. Direct metal laser sintering [DMLS]. In Proceedings of the SAMPE Midwest Chapter, Direct Part Manufacturing Workshop, Fairborn, OH, USA, 13–14 November 2012.
34. Averyanova, M.; Bertrand, P.; Verquin, B. Manufacture of Co-Cr dental crowns and bridges by selective laser melting technology. *Virtual Phys. Prot.* **2011**, *6*, 179–185. [CrossRef]
35. Vrancken, B.; Thijs, L.; Kruth, J.-P.; Van Humbeeck, J. Heat treatment of Ti6Al4V produced by selective laser melting: Microstructure and mechanical properties. *J. Alloys Compd.* **2012**, *541*, 177–185. [CrossRef]
36. Bassoli, E.; Denti, L.; Comin, A.; Sola, A.; Tognoli, E. Fatigue behavior of as-built L-PBF A357.0 parts. *Metals* **2018**, *8*, 634. [CrossRef]

37. Mower, T.M.; Long, M.J. Mechanical behavior of additive manufactured, powder-bed laser-fused materials. *Mater. Sci. Eng. A* **2016**, *651*, 198–213. [CrossRef]
38. Criales, L.E.; Arısoy, Y.M.; Lane, B.; Moylan, S.; Donmez, A.; Özel, T. Laser powder bed fusion of nickel alloy 625: Experimental investigations of effects of process parameters on melt pool size and shape with spatter analysis. *Int. J. Mach. Tools Manuf.* **2017**, *121*, 22–36. [CrossRef]
39. Matsumoto, H.; Yoneda, H.; Sato, K.; Kurosu, S.; Maire, E.; Fabregue, D.; Konno, T.J.; Chiba, A. Room-temperature ductility of Ti–6Al–4V alloy with α′ martensite microstructure. *Mater. Sci. Eng. A* **2011**, *528*, 1512–1520. [CrossRef]

materials

MDPI

Article

Effect of Gradient Energy Density on the Microstructure and Mechanical Properties of Ti6Al4V Fabricated by Selective Electron Beam Additive Manufacture

Ta-I Hsu [1], Yu-Ting Jhong [2] and Meng-Hsiu Tsai [3,*]

1 The Genomics Research Center, Academia Sinica, Taipei 115, Taiwan; cardiosea@gmail.com
2 School of Dentistry, College of Dental Medicine, Kaohsiung Medical University, Kaohsiung 807, Taiwan; steelwillpower0130@gmail.com
3 Department of Mold and Die Engineering, National Kaohsiung University of Science and Technology, Kaohsiung 807, Taiwan
* Correspondence: tmh@nkust.edu.tw

Received: 25 February 2020; Accepted: 23 March 2020; Published: 26 March 2020

Abstract: Selective Electron Beam Additive Manufacturing (SEBAM) is a promising powder bed fusion additive manufacturing technique for titanium alloys that select particular area melting in different energy density for producing complexly shaped biomedical devices. For most commercial Ti6Al4V porous medical devices, the gradient energy density is usually applied to manufacture in one component during the SEBAM process which selects different energy density built on particular zones. This paper presents gradient energy density base characterization study on an SEBAM built rectangular specimen with a size of 3 mm × 20 mm × 60 mm. The specimen was divided into three zones were built in gradient energy density from 16 to 26.5 J/mm^3. The microstructure and mechanical properties were investigated by means of scanning electron microscopy, X-ray diffraction, transmission electron microscopy and mechanical test. The α' martensitic and lack of fusion were observed in the low energy density (LED) built zone. However, no α' phase and no irregular pores were observed both in overlap energy density (OED) and high energy density (HED) built zones located at the middle and bottom of the specimen respectively. This implies the top location and lower energy density have positive effects on the cooling rate but negative effects on densification. The subsequence mechanical properties result also supports this point. Moreover, the intermetallic Ti_3Al found in the bottom may be due to the heat transfer from the following melting layer. Furthermore, the microstructure evolution in gradient energy built zones is discussed based on the findings of the microstructure and thermal history correlation analysis.

Keywords: selective electron beam additive manufacture; Ti6Al4V ELI alloy; phase transformation; spatial; gradient energy density; martensitic decomposition; Ti_3Al intermetallic compound; fracture analysis

---

1. Introduction

Selective electron beam additive manufacture (SEBAM) is one of powder bed fusion technology that builds metallic parts with complex geometries by using the electron beam as the heat source to selectively preheat and melt over the metallic powder bed via layer by layer based on 3D CAD model input. SEBAM has attracted much attention over the past few years for its advantages, such as high material utilization, low porosity and fast production rate without post heat treatment in particular cases, and so on. Furthermore, SEBAM is an advanced manufacturing technology that has widely fabricated titanium, nickel base, CoCrMo and difficult forming alloys [1,2].

Ti6Al4V is commonly used in the SEBAM process. Its excellent corrosion resistance, mechanical property and biocompatibility are widely applied to medical applications such as orthopedic [3], dental [4] and spinal implants [5]. Porous medical metallic parts by SEBAM process, which possess low elastic Young's modulus matching to bone tissue and are capable of providing space for in growth of bony tissue to achieve better fixation, have been widely used for medical implants due to reduced stress shielding effect [6] and increased bone osseointegration [7,8]. Therefore, most researchers adjusted the porosity and geometric to improve the benefit biocompatible reticulated meshes and foams with an interconnected porous structure which has porosity open cell and foams (porosity between 55%~90%) [9–11]. As mentioned above, previous studies have fabricated porous complex structure by SEBAM method in low energy density parameters. The reasons for the low energy built could be the post treatment of the porous structure easily, i.e., it is commonly seen that the support structure is connected state plate for heat conduction, lower thermal deformation in normal energy density built [12], but it is difficult to remove the support of the porous structure. Therefore, the low energy density was applied to build the porous structure without the support structure. The effect of microstructure and mechanical property were reported during SEBAM process by in changing various parameters [13–18]. Above the line energy density 0.18 J/m built, the minimum relative density of 99.5% was observed [13]. Optimized overlap distance were 0.25–0.75 mm without fused line defects and the average value of microhardness was 360 HV [14]. However, over this range, it decreased ~340 HV on average. Different built orientations in XY, ZX, ZY, XY 30° and XY 60° inclination to the start plate, the microhardness, nearly 330 HV in the ZX-P specimen, was higher than the value ~275 HV in ZY [15]. With increasing build height and thickness, the tensile strength was decreased 3% in the top [16] and the thicker one had a lower microhardness of 320 HV than 362 HV of the thin sample [17]. The impeller was built with double influence factors to effective the microhardness with more than 10% change was reported [18], the microhardness between ~340 HV to ~360 HV depended on the locations of impeller. The unique microstructure may be due to complex cyclic thermal process in three main stages as rapid cooling from the molten state to the layer temperature, followed by a near isothermal stage at the local temperature until completion of the build, and finishing with a slow cooling to the room temperature [19,20]. In summary of microstructure evolution as mentioned above, micro size β grain transformation and α' formation in the first stage, followed by the α' martensitic being fully or partially transformed into an $\alpha + \beta$ structure. Moreover, the previous melted layer experiences a thermal history for each subsequent layer which will result in different microstructure depends on phase transformation path during the cooling. However, in previous research showed one parameter with one working process or the many samples compared with one specific parameter. In our experiments demonstrated that 3 parameters in one sample to indicative one workpiece made by more reasons to modify the finally commercial products manufacture [13–15,18,21].

It is worth noting that most efforts were made to successfully develop optimized porous structures for medical metallic implants with suitable combination process parameters in order to obtain better bone osseointegration and avoid the stress shielding effect. For example, as shown in Figure 1, the geometry of orthopedic implants is divided into solid and porous parts in gradient energy density built for specified purpose by using SEBAM process. The solid structure built in high density built for dense and strength purpose, but built in low energy built for non-support demands. Transition zone between solid and porous structure, namely of overlap zone with built in mixed high energy and low energy density. Therefore, gradient energy has been widely used for porous medical metallic implant production. To the best of the authors' knowledge, no study has been published to date on the overall characterization of gradient energy built parts, though, there are medical industrial demands for such parts fabricated by SEBAM. The specific research question this study addressed concerns the comprehension process of the effect on the microstructure and mechanical properties in gradient energy density built by SEBAM process. In this study, gradient energy density including three different of energy density was selectively built in different zones (top, middle and top) of one rectangular sample by SEBAM method using Ti6Al4V (ELI) alloy powders on the effect of the microstructure and mechanical properties. It is supposed to be of significance to evaluate the performance of SEBAM-built parts. More importantly, a thorough understanding of gradient energy on microstructure and mechanical properties obtained from this study will aid to further propel the practice application of SEBAM-built metallic parts.

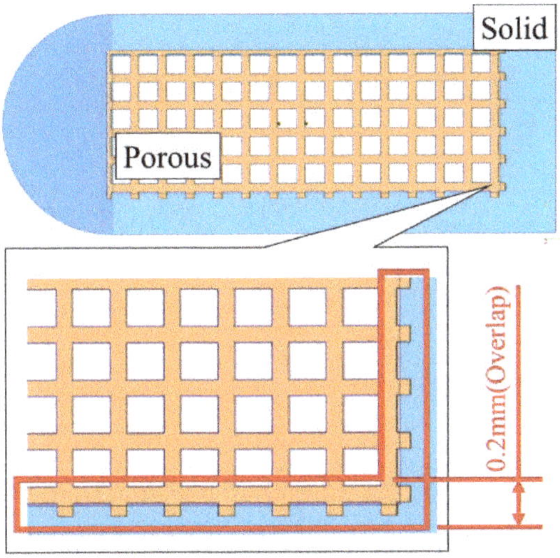

Figure 1. Schematic diagram of commercial porous fusion cage structure was fabricated by selective electron beam additive manufacture (SEBAM) method using gradient energy density on three different zones. Solid and porous parts were made independently by higher and lower energy density. The overlap site with 0.2 mm in width between solid and porous parts was made sequentially by higher and lower energy density.

2. Materials and Methods

2.1. Sample Preparation

In order to investigate the microstructure and micro-harness evolution of one SEBAMed (Kaohsiung, Taiwan) metallic part was built by SEBAM method in gradient energy density. A rectangular sample which 3 mm (thickness) × 20 mm (width) × 60 mm (height) in size was used and selectively in setting three different energy density namely high energy density (HED), low energy density (LED) and overlap energy density (OED) built on bottom ($0 \leq z \leq 20$), middle ($20 \leq z \leq 40$) and top $40 \leq z \leq 60$ of the specimen respectively along the z axis building direction as shown in Figure 2. The energy density parameters in detail were described in the next section.

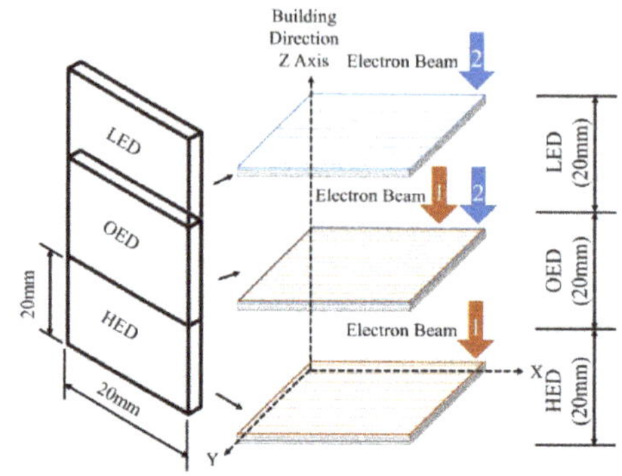

Figure 2. Schematic of SEBAMed samples made by higher, lower and overlap energy density from the bottom to top which labeled as high energy density (HED), overlap energy density (OED) and low energy density (LED) along build direction fabricated by electron beam additive manufacture. HED, LED built zones independently 26.5 and 16 J/mm^3. OED built zone in 26.5 J/mm^3 first and build sequentially 16 J/mm^3.

2.2. Selective Electron Beam Additive Manufacturing

The Extra Low Interstitial Ti-6Al-4V (Grade 23) powder supplied by Arcam AB company was used; the powder size distribution was quoted as 45 to 100 µm. The chemical composition of Ti-6Al-4V powder was supplied as showed in Table 1 followed as ASTM F3001. Recycling of non-melted powder and/or sintered powder (Figure A1 in Appendix A) was achieved via powder recovery system and a mechanical vibrating sieve which mesh size ≤ 150 µm. All SEBAM parts were fabricated on the Arcam Q10 machine (Producer: Arcam EBM GE Additive company, Gothenburg, Sweden). The SEBAM process was implemented in the vacuum chamber with pressure below 5×10^{-3} mbar in the beginning and finished with a pressure of 2×10^{-5} mbar. Each layer was setting preheated 730 °C by fast scanning with the defocused electron beam and layer thickness of 50 µm. Gradient energy density in building parameters was set as follows: the standard parameter of the beam current from 3 mA to 15 mA and scan speed from 1500 mm/s to 4530 mm/s with fixed layer thickness 50 µm, beam diameter of 100 µm and hatch spacing of 150 µm were performed. The energy density can be approximated as:

$$E(J/mm3) = \frac{V(kV) \times C(mA)}{t\ (mm) \times h(mm) \times S(mm/s)} \quad (1)$$

where energy density E, operation voltage V, beam current C, layer thickness t, hatch distance h and scan speed S in Table 2. The high energy density (HED) resulting from the reduced scan speed and increased beam current was therefore 63% higher compared to low energy density (LED) setups. For the transition overlapping 2 mm in width was applied overlap energy density (OED), i.e., HED in the first scan and LED in the after.

Table 1. The chemical composition (in wt %) of Ti-6Al-4V powder.

Element	Ti	Al	V	C	Fe	O	N	H	Y
ASTM F3001	Bal.	5.50–6.50	3.50–4.50	≤0.08	≤0.25	≤0.13	≤0.05	≤0.012	≤0.005
Powder	Bal.	6.49	4.01	0.02	0.18	0.07	0.02	0.001	<0.001

Table 2. Main process parameters with fixed layer thickness 50 μm, beam diameter of 100 μm and hatch spacing of 150 μm used for the SEBAM process.

Items	Scan Speed	Beam Current	Volume Energy Density
HED	(Scan1) 4530 mm/s	15 mA	26.5 J/mm^3
LED	(Scan2) 1500 mm/s	3 mA	16.0 J/mm^3
OED	(Scan1) 4530 mm/s(Scan2) 1500 mm/s	15 mA3 mA	26.5 J/mm^{3}16.0 J/mm^3

2.3. Microstructure Observation

Optical microscopy and scanning electron microscopy (SEM) of SEBAMed samples prepared by grinding and polishing- diamond abrasive disc with water used as coolant in the grinding process and polished with silicon carbide and 0.3 μm Al_2O_3 suspension. Optical and SEM samples were etched in 2.5% HNO_3 + 5% HCl + 92.5% ethanol reagent. Secondary electron image SEM was carried on JEOL JSM-6380 (Tokyo, Japan) at 15 kV. Phase analysis was conducted using X-Ray Diffraction (XRD) (Panalytical B.V., Almelo, The Netherlands) and transmission electron microscopy (TEM) (PHILPLIES CM200 and JEOL 2000 FX, Tokyo, Japan). The characteristic CuKα radiation (λ = 1.5412 Å) in the 2θ range from 20 to 100 degrees having voltage 40 keV and current 40 mA was used. TEM samples were prepared using the following procedures: thin samples sections were manually ground to 0.06 ± 0.02 mm in thickness with silicon carbide paper and then the thin foils for TEM were electro polish using Automatic Twin-Jet Electropolisher Model 110 (Yokohama, Japan) at 25 V in a solution bath consisting of 12% perchloric acid, 15% glacial acetic acid and 75% ethanol reagent.

2.4. Mechanical Properties

Vicker microhardness (500 g, 15 s hold) of 15 individual measurements in each HED, OED and LED were performed on the metallographic samples using an Akashi MVK-H100 (Osaka, Japan) machine. The tensile tests on an MTS-10t (Eden Prairie, MN, USA) Tester using cylinder specimen with gauge length of 30 mm of specimen 3 (Figure A2 in Appendix B) according to ASTM-E8M specifications. The specimens were subjected to impact testing with the use of a Tinius Olsen model IT 504 polymeric impact testing machine (Tinius Olsen, Horsham, PA, USA). All tests were conducted at a pendulum capacity of 15 J, a drop height of 609.6 mm and velocity of 3.46 m/s.

3. Results

3.1. Effect of Gradient Energy Density on the Defects

Figure 3 shows representative specimens of the gradient energy from high energy density to low energy density built zones along the z axis building direction. The irregular pores which size exceeds 100 μm was only observed in the LED built specimen (Figure 3a) which is caused by the lack of fusion. Above line energy density more than 100 J/m is necessary for full densification of Ti6Al4V and below 100 J/m SEBAMed specimens contain more than 1% porosity as earlier reported [13]. This means that irregular pores did not obviously exist both in the HED and OED built zones, implying that highly dense sample can be fabricated by the SEBAM process if the energy density were more than 100 J/m. The irregular pores would greatly reduce yield stress, tensile stress and ductility [13].

In contrast the typical spherical pores several μm in size were found in all HED, OED and LED built samples in Figure 3b–d. It means no relation between the spherical pores and energy density and is mainly caused by the entrapped argon gas inside the Ti6Al4V powder during the production of plasma wire gas atomization [22]. The presence of limited small pores cannot be eliminated by hot isostatic press treatment [23] and did not significantly affect tensile properties. It might be decreased fatigue life.

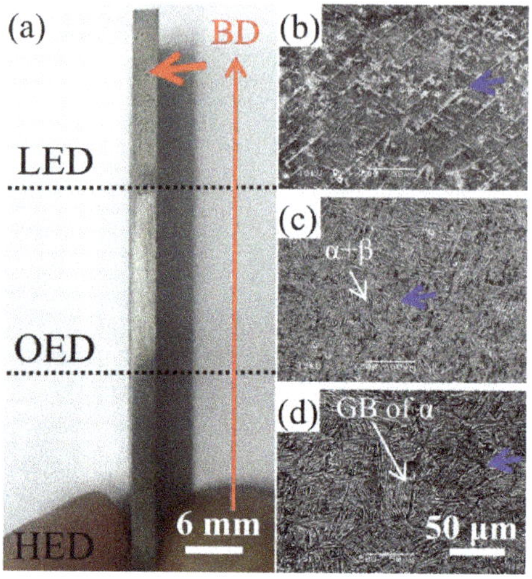

Figure 3. Side view of specimen from bottom to top along the build direction (BD) denoted in red arrow of (**a**) showing the high visible pores which range from several hundred μm to several mm in size denoted by red arrow in LED built zone. SEM secondary electron images shows smaller pores several μm in size denoted by blue arrows in (**b**) LED, (**c**) OED and (**d**) HED built zones. Besides, $\alpha+\beta$ lamella and grain bound (GB) of α structure denoted by white arrows both were observed in (**c**) OED and (**d**) HED built zones.

3.2. Effect of Gradient Energy Density on the Microstructure Evolution

In order to understand the influence of the gradient energy density on three different sites, the microstructure was studied along the build direction. Figure 3b shows partially acicular α' martensitic and α + β were developed by LED built zone where is at the top of the specimen. At the middle of the specimen built in the OED parameter, a lamella mixture of $\alpha+\beta$ structure was observed (Figure 3c). A coarser $\alpha+\beta$ lamella structure in the prior β grain was also observed at the bottom where is built in HED parameter (Figure 3d). No obvious difference in SEM analysis was found both in OED and HED built zones. However, the acicular α' martensitic only was observed in LED built zone, which indicates cooling rate in LED zone was higher than OED and HED built zones.

Figure 4 shows the XRD analysis of the gradient energy density built by SEBAM method using Ti6Al4V powder. The characteristic peaks of dominant α phase with (101), (002) and small fraction of β phase with (110) were observed in all energy density built zones. Minor fraction of β phase in both OED and LED built zones compared to HED built zones. It is difficult to distinguish α and α' phase by XRD analysis. Based on the mentioned above, the acicular α' martensitic phase was only found in the LED built zone where the present characteristic peaks of α phase overlapped with α'. Interestingly, the intermetallic phase Ti_3Al was observed in HED built zone. Further TEM analysis in Figure 5 revealed that the nano-size intermetallic Ti_3Al dispersed near the α grain boundary, and a fine lamella $\alpha+\beta$ structure. Moreover the crystallographic relationship between α, Ti_3Al and β phase as $[001]_{\alpha}$ // $[001]_{Ti3Al}$ and $[011]_{\alpha}$ // $[001]_{\beta}$ are similar results to what Barriobero-Vila reported of selective laser melting with intrinsic heat treatments of Ti6Al4V [24]. The analysis of the formation of nanosize Ti_3Al without the post heat treatment or repetitive melting (intrinsic heat treatment) during the SEBAM process will be discussed in the following section.

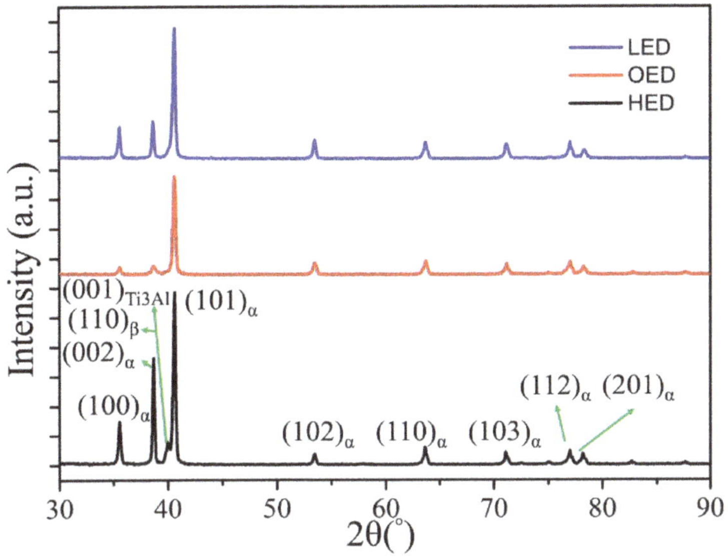

Figure 4. The X-ray diffraction spectrum of HED, LED and OED built samples. The α-Ti6Al4V characteristic (100), (002) and (101) and β-Ti6Al4V characteristic (110) peaks were observed in all samples. The characteristic peaks of α phase are overlapped with α′ martensitic phase only in LED built sample. Minor Ti_3Al phase of its characteristic peak (110) was observed only in HED built sample.

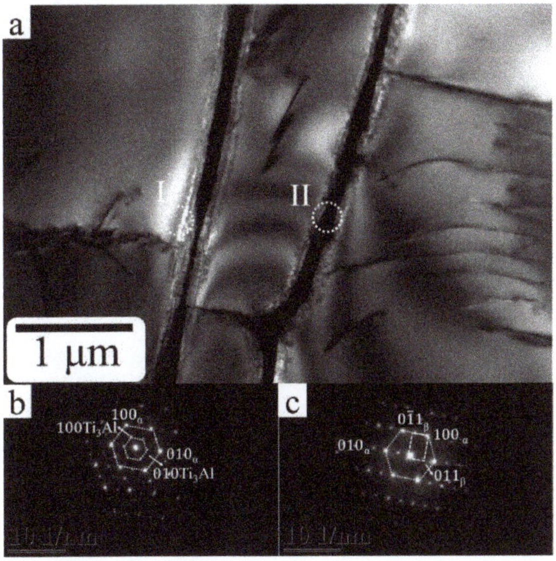

Figure 5. (**a**) Bright field image of ultrafine α+β lamella microstructure found at the center of HED sample. (**b**) Select area diffraction patter from the cycle region (I) in (**a**) showing the present Ti3Al within an α lamella matrix with crystallographic relationship $[001]_{\alpha}//[001]_{Ti3Al}$ (**c**) Select area diffraction patter from the cycle region (II) in (**a**) showing the ultrafine α + β lamella microstructure with crystallographic relationship $[001>]_{\alpha}$ // $[011]_{\beta}$.

3.3. Effects of Gradient Energy Density on the Mechanical Properties

A continuous microhardness on the HED, OED and LED built zones were measured in Figure 6. Apparently, it was divided into two groups via gradient energy density built. The microhardness values in average 320 to 350 HV also revealed no obvious difference between HED and OED built zones. Nevertheless, a relatively high microhardness (from 375 to 420 HV) was observed in the LED built zone. The higher microhardness in the LED built zone was caused by the acicular α' martensitic microstructure (Figure 3b) and it is consistent with those reported for selective laser melting built Ti6Al4V alloy [25,26]. Moreover, the measure average α lath width in the HED and OED built zones were determined to be 0.69 ± 0.04 µm and 0.59 ± 0.14 µm respectively. The result is in good agreement with no obvious difference in mircohardness in the two zones. However, the microhardness in the OED built (double melting) zone showed the finer $\alpha+\beta$ lamella and grain bound of α structure than in the HED built zone. The result was in contrast to the paper showed the coarsening microstructure was caused by double melting in the same build height [14]. This implies that double melting is not only seen to be relevant to grain size. The effect of thermal history from the different building height might have a powerful impact on grain size [18].

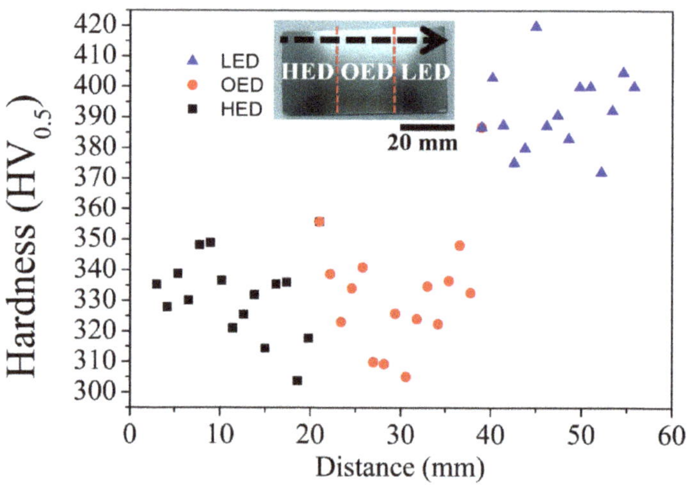

Figure 6. Vickers microhardness profile was measured from bottom to top (black arrow) of the cross section specimen which were built by gradient energy. HED, OED and LED built zones are indicated by black, red and blue symbols.

Stress-strain curves and impact energy of three different energy density built samples were shown in Figure 7. The ultimate tensile strength (UTS) estimated 228 MPa of the LED built sample lower obviously compared to the OED built and HED built samples in Table 2. The presence of the obvious plastic deformation both in OED and HED curves were distinct with LED curve which almost no plastic deformation was observed. Both OED and HED curves had no obvious difference from the UTS and elongation, however in the finally the HED curve was slightly reduced the strain. Higher elastic modulus of OED and HED built samples compared to LED built sample. Interestingly, the highest hardness was in the LED built zone due to rapid cooling leading to acicular α' martensitic but it had the smaller UTS and elongation compared to OED built and HED built sample. To further understand the variation of the mechanical properties, the fracture morphology was observed in Figure 7b–d. The transgranular ductile dimple tearing resulting from the coalescence of microvoids fracture surface was observed in both OED and HED built samples. Crack propagation of fine dimples at the tensile fracture indicated the extent of plastic deformation. Besides, according to the Hall-Petch strength mechanism, smaller grain size provided more grain boundaries, which can impede the movement of the dislocation, demonstrated as OED built sample has slightly higher tensile stress and elongation compared to HED built sample. In contrast, the fracture surface included cleavage, unmelted and partially melted powder were shown in LED built sample. This implied insufficient energy density that was unable to melt the deeper layer had a negative effect on tensile stress and elongation properties. This result is consistent with others reported [13,27].

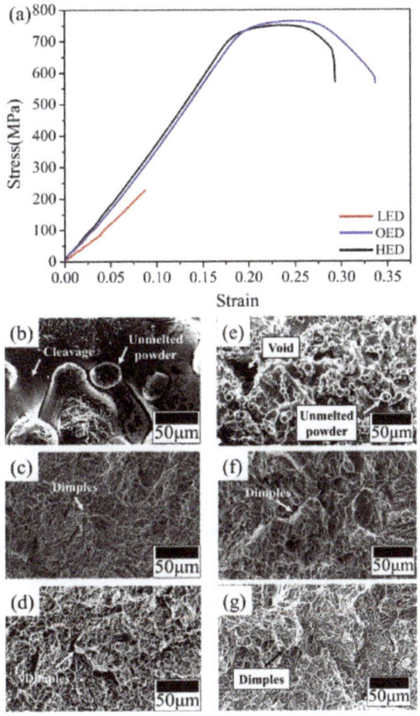

Figure 7. (**a**) Tensile stress-strain curves of gradient energy built samples. LED, OED and HED built samples were respectively denoted by red, blue and black curves. Representative fracture surface of tensile test of different density built in (**b–d**). (**b**) Unmelted powder and brittle cleavage in LED built sample, (**c**,**d**) both tensile fracture morphologies show fine and deeper dimples in HED and OED built samples. Representative fracture surface of impact test of different density built in (**e–g**). (**e**) Unmelted powders and voids were observed in LED samples, (**f**,**g**) both impact fracture morphologies show fine and deeper dimples in HED and OED built samples.

The average fracture toughness properties in different energy density were shown in Table 3. Both OED and HED built samples showed the higher impact energy compared to the LED built sample which had the lower fracture toughness properties (2 to 3.2 J). These are similar results to the tensile stress in the fracture surface. Both fracture surfaces as transgranular ductile dimple fractures from the coalescence of microvoids in OED and HED built samples were observed in Figure 7b–f. However, the lower value of impact energy of fracture surface in LED built sample exhibited pores, unmelted powder and partial melt region (See Figure 7d,g). This means lower energy density may contributed to large area of pores and unmelted powder [13,27].

Table 3. Mechanical properties of SEBAMed Ti6Al4V in varying energy density.

Sample	UTS (MPa)	YS (MPa)	Young's Modulus (GPa)	Elongation (%)	Impact (J)
LED	228 ± 10	NA	74 ± 3	8 ± 1	2.7 ± 0
OED	765 ± 11	725 ± 10	108 ± 1	31 ± 1	12 ± 0
HED	752 ± 19	737 ± 3	116 ± 8	26 ± 1	13 ± 1

4. Discussion

4.1. Formation Gradient Microstructure on the Gradient Energy Density Built Zones

The gradient microstructure was built by gradient energy density from the bottom to top of the specimen. In this work, the microstructure of the particular zones in different energy density built will get the gradient microstructure ($\alpha' + \alpha + \beta$ in LED built zone, slightly finer $\alpha + \beta$ lamella structure in the OED built zone and α+β lamella with minor fraction of nanosize Ti_3Al intermetallic in HED built zone were observed in Figures 3–5). Moreover, all the specimens showed the microhardness values between 304 to 420 HV. The highest microhardness was contributed to the presence of α' martensitic phase (Figure 3b) in LED built zone. However, no α' phase was observed and only fine α+β lamella were observed both in the OED and HED built zones where the average microhardness of 340 HV was shown in Figure 6. The presence of limited small amount Ti_3Al did not significantly affect the microhardness in the HED built zone compared to the microhardness in the OED built zone. In the present study, the thermal events were more complex due to a combination of the building height and gradient energy density.It is showed that the difference of previously report as varying microstructure only in changing building direction [15] process parameters [28,29], building high [18,30], overlap distance [14], heat treatment [19], intrinsic heat treatment [24]. Therefore, the analysis of formation of gradient microstructure/phase in gradient built zones will be discussed in the following section.

4.1.1. Formation α' Martensitic in LED Built Zone

The thermal gradient, G (K/m), and solidification velocity, R (m/s) can be used to predict solidification microstructure where G/R controls the mode of solidification (morphology) and G × R (cooling rate) in SEBAM process [31]. For an increasing energy density, the cooling rate became slower as was shown in solidification map for SEBAM Ti6Al4V melted [31]. Moreover, the cooling rate of the top was higher than the bottom also reported in SEBAMed Ti6Al4V parts [18]. In the present study, the SEM investigations in Figure 3 revealed that formation of acicular α' martensitic in LED-built zone where at the top of the specimen along the build direction. Both the low energy built and the top of the building height caused a high cooling rate. Moreover, during the SEBAM process, as the electron beam scanned over the Ti-6Al-4V powder, a melt pool formed and then was rapidly solidified into β grains. The prior β grains underwent a rapid cooling (a critical cooling rate of >410 °C/s) transforming into martensitic α' phase. Moreover, it its known that the build temperature of 600~650 °C in the chamber was below martensitic start temperature (Ms) 800 °C. Thus, α' martensitic phase should be formed after melting.

4.1.2. Formation α+β in OED Built and HED Built Zones

The observation of the α+β lamella structure inside OED built (Figure 3c) and HED built (Figure 3d) samples also proves the aforementioned process. Following the above mentioned formation of martensitic α', β phase was transformed into an α' ($\beta \rightarrow \alpha'$) reaction on cooling. In addition, α' martensitic will decompose into α and β phases, i.e., $\alpha' \rightarrow \alpha + \beta$, when it is subjected to isothermal annealing in the $\alpha + \beta$ two phase field. It was reported that α' in the thicker sample would be able to decompose into α/β phases due to higher thermal mass causing the slower cooling. During the EBM process, an electron beam was constantly scanning over the Ti6Al4V powder bed and the heat was mainly transferred from top to bottom through the built parts, the temperature of the newly deposited layers must be higher than 650 °C in the middle and bottom of the specimen that was sufficient to enable α' decomposition.

4.1.3. Formation Ti_3Al in HED Built Zones

The formation of the intermetallic Ti3Al phase along the building direction of the bottom of the specimen was revealed using X-ray diffraction by the presence of (001) characteristic peak as shown in Figure 4 as well as by select area diffraction carried out during TEM investigations (Figure 5). According to previous investigations of SLMed Ti6Al4V and SLMed Ti6Al4V with intrinsic heat treatment, the driving force for Ti_3Al formation due to the fast solidification during the SLM process, concentration of aluminum and oxygen [24]. Moreover, Al depletion at the interface during EBM process was reported [29]. In this work, diffraction pattern evidence of Ti_3Al was only observed in HED built zone (the bottom of the specimen) mostly affected by heat transfer from the following layer. Since precipitation of this phase can occur during aging at 500–600 °C for several hours [31], it is reasonable to conclude that this range of temperatures was reached by the scanning of following layers, lead to the precipitation of an intermetallic Ti_3Al.

4.2. Cooling Path in Gradient Energy Density and Subsequence Heat Transfer

From the microstructure point of view as mentioned above in three different zones. The α' + α + β in the LED built zone where it is the top of the specimen, α + β lamella structure in OED built zone where it is the middle of the specimen and α + β lamella with small amount of Ti3Al in HED built zone where it is the bottom of the specimen, were observed due to different thermal history. It implied the cooling rate of the top (LED built zone) was faster than the middle (OED built zone) and the bottom (HED built zone) of the specimen. Further the evidence of slight smaller width of α lath in the middle in Figure 3 implied slight higher cooling rate in the middle compared to the bottom. Based on the literature reported the complex thermal events in the EBM process can be simplified in three main stages [19]. The first step is a rapid cooling from the molten state to the layer temperature, followed by a quasi-isothermal stage at the local cooling until completion of the build, and finishing with a slow cooling to room temperature. Gradient energy density and different locations causing different thermal events were predicted in Figure 8 base on the microstructure analysis as mentioned above. Moreover, the heat coming from the following layer melting will transfer to the previous layer causing the similar effect as aging heat treatment was observed only at the bottom. Those observed results proved the effect combined gradient energy density and different locations on microstructure and phase transformation due to different thermal history.

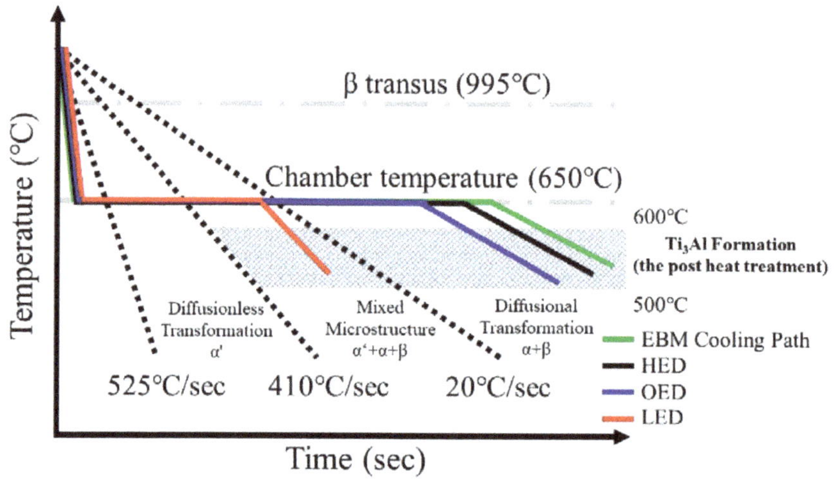

Figure 8. Prediction cooling path and phase transformation based on the literature [19] of LED, OED and LED parameters built in top, middle and bottom of the specimen respectively. The difference in cooling rate result from different energy density and spatial location thus contribute to different microstructure. i.e., α′ + α + β in LED built zone, α + β lamella structure in the OED built zone and α + β + Ti_3Al intermetallic phase in the HED built zone were observed.

5. Conclusions

In this work, gradient energy density was carried out on a specimen from the bottom to the top by SEBAM technology. The effect of energy density on the microstructure and mechanical properties of SEBAMed Ti6Al4V was investigated by SEM, X-ray, TEM, microhardness, tensile and impact tests. Based on the experimental results, the following conclusions can be drawn:

- Gradient energy density (16 to 26.5 J/mm^3) built sample with dimension of 20 mm × 60 mm × 3 mm was fabricated by SEBAM. Small spherical pores coming from powder were found to be independent of the energy density and locations in the built part. Large irregular pores were observed in LED built zone due to lack of fusion but no internal crack and large pores were observed in OED or HED built zones.
- The α′ + α + β, α + β and α + β + Ti_3Al graded microstructure was observed from the top to the bottom. The graded microstructure was caused by the complex heat transfer due to gradient energy density and locations during building. Correspondingly, gradually increased microhardness was achieved from the bottom to the top and there was no obvious difference from the bottom to the middle.
- Lack of fusion defects resulted in a decrease in plasticity in the LED built zone. The unmelted powder and voids in following fracture morphology were observed by SEM.
- A small amount intermetallic Ti3Al was observed only in the bottom (HED built zone). The formation of Ti3Al is attributed to the heat transfer from the following layer melting.

Author Contributions: T.-I.H. and M.-H.T. wrote the main manuscript text and prepared all figures. Y.-T.J. carried out the SEM experiments and prepared all the sample. T.-I.H. performed the TEM analysis. M.-H.T. designed all the experimental and edited the manuscript. All authors have read and agreed to the published version of the manuscript.

Funding: This research was funded by Ministry of Science and Technology, grant number 108-2218-E-992-320-MY2.

Acknowledgments: The authors gratefully acknowledge the sponsorship from Ministry of Science and Technology of Taiwan, ROC, under the project No. MOST 108-2218-E-992-320-MY2. Meanwhile, the authors are thankful for the EBAM lab equipment and financial support from Metal Industry Research & Development Centre (MIRDC).

Conflicts of Interest: The authors declare no conflict of interest.

Appendix A Ti6Al4V Powder Morphology

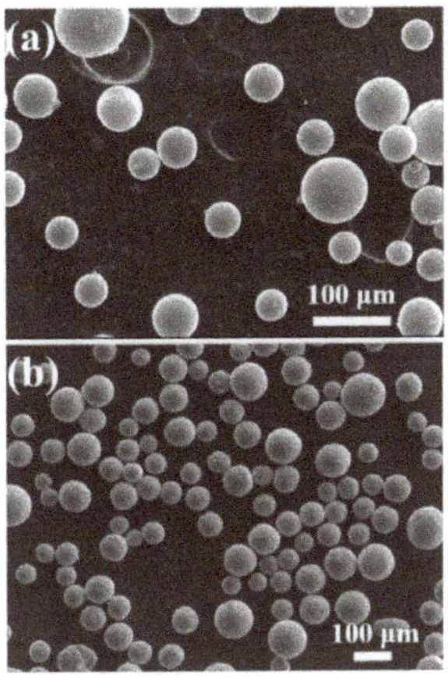

Figure A1. The secondary electron image of Ti6Al4V powder in (**a**) fresh powder and (**b**) recycle powder. Both types of powder were spherically shaped.

Appendix B The Test Sample of Tensile Test

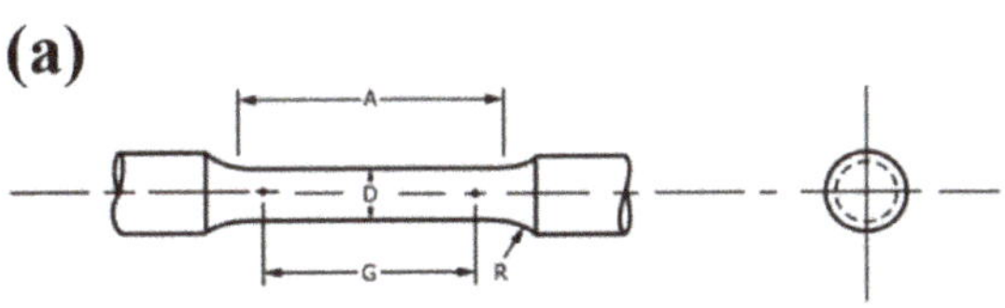

Unit: mm	L	d	A	D	G	R
Standard	None-regulated	None-regulated	36	6	30	6
Printed	100	10	36	6	30	6
Machining	97.89	9.46	36	5.78	30	6

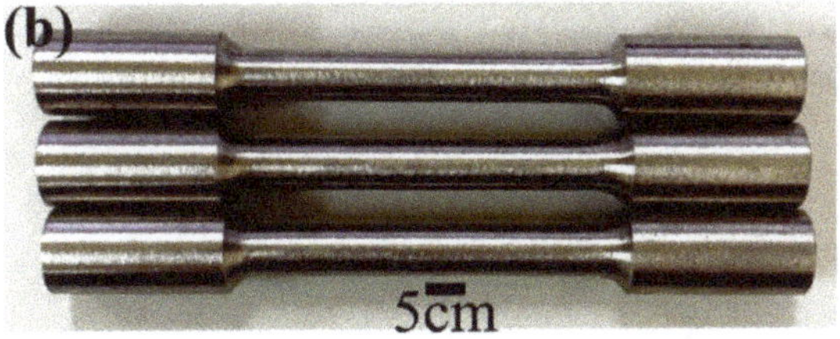

Figure A2. The cylinder test sample according to ASTM E8 standards with standard size, printed and machining size in (**a**). (**b**) The machined built tensile test samples which were built in LED parameter.

References

1. Hao, Y.-L.; Li, S.-J.; Yang, R. Biomedical titanium alloys and their additive manufacturing. *Rare Met.* **2016**, *35*, 661–671. [CrossRef]
2. Murr, L.E.; Martinez, E.; Gaytan, S.M.; Ramirez, D.A.; Machado, B.I.; Shindo, P.W.; Martinez, J.L.; Medina, F.; Wooten, J.; Ciscel, D.; et al. Microstructural Architecture, Microstructures, and Mechanical Properties for a

Nickel-Base Superalloy Fabricated by Electron Beam Melting. *Metall. Mater. Trans. A* **2011**, *42*, 3491–3508. [CrossRef]

3. Marin, E.; Fusi, S.; Pressacco, M.; Paussa, L.; Fedrizzi, L. Characterization of cellular solids in Ti6Al4V for orthopaedic implant applications: Trabecular titanium. *J. Mech. Behav. Biomed. Mater.* **2010**, *3*, 373–381. [CrossRef] [PubMed]
4. Syam, W.; Al-Ahmari, A.; Mannan, M.; Al-Shehri, H.; Al-Wazzan, K. Metallurgical, accuracy and cost analysis of Ti6Al4V dental coping fabricated by electron beam melting process. In Proceedings of the 5th International Conference on Advanced Research in Virtual and Rapid Prototyping, Leiria, Portugal, 28 September–1 October 2011; pp. 375–383. [CrossRef]
5. Wu, S.H.; Li, Y.; Zhang, Y.Q.; Li, X.K.; Yuan, C.F.; Hao, Y.L.; Zhang, Z.Y.; Guo, Z. Porous titanium-6 aluminum-4 vanadium cage has better osseointegration and less micromotion than a poly-ether-ether-ketone cage in sheep vertebral fusion. *Artif. Organs* **2013**, *37*, E191–E201. [CrossRef] [PubMed]
6. Niinomi, M.; Nakai, M. Titanium-Based Biomaterials for Preventing Stress Shielding between Implant Devices and Bone. *Int. J. Biomater.* **2011**, *2011*, 836587. [CrossRef]
7. Li, G.; Wang, L.; Pan, W.; Yang, F.; Jiang, W.; Wu, X.; Kong, X.; Dai, K.; Hao, Y. In vitro and in vivo study of additive manufactured porous Ti6Al4V scaffolds for repairing bone defects. *Sci. Rep.* **2016**, *6*, 34072. [CrossRef]
8. Zhang, Z.; Li, H.; Fogel, G.R.; Liao, Z.; Li, Y.; Liu, W. Biomechanical Analysis of Porous Additive Manufactured Cages for Lateral Lumbar Interbody Fusion: A Finite Element Analysis. *World Neurosurg.* **2018**, *111*, e581–e591. [CrossRef]
9. Parthasarathy, J.; Starly, B.; Raman, S.; Christensen, A. Mechanical evaluation of porous titanium (Ti6Al4V) structures with electron beam melting (EBM). *J. Mech. Behav. Biomed. Mater.* **2010**, *3*, 249–259. [CrossRef]
10. Cansizoglu, O.; Harrysson, O.; Cormier, D.; West, H.; Mahale, T. Properties of Ti–6Al–4V non-stochastic lattice structures fabricated via electron beam melting. *Mater. Sci. Eng. A* **2008**, *492*, 468–474. [CrossRef]
11. Wu, Y.C.; Kuo, C.N.; Shie, M.Y.; Su, Y.L.; Wei, L.J.; Chen, S.Y.; Huang, J.C. Structural design and mechanical response of gradient porous Ti-6Al-4V fabricated by electron beam additive manufacturing. *Mater. Des.* **2018**, *158*, 256–265. [CrossRef]
12. Hussein, A.; Hao, L.; Yan, C.; Everson, R.; Young, P. Advanced lattice support structures for metal additive manufacturing. *J. Mater. Process. Technol.* **2013**, *213*, 1019–1026. [CrossRef]
13. Kirchner, A.; Klöden, B.; Luft, J.; Weißgärber, T.; Kieback, B. Process window for electron beam melting of Ti-6Al-4V. *Powder Metall.* **2015**, *58*, 246–249. [CrossRef]
14. Wang, P.; Nai, M.L.S.; Sin, W.J.; Lu, S.; Zhang, B.; Bai, J.; Song, J.; Wei, J. Effect of overlap distance on the microstructure and mechanical properties of in situ welded parts built by electron beam melting process. *J. Alloys Compd.* **2019**, *772*, 247–255. [CrossRef]
15. Bruno, J.; Rochman, A.; Cassar, G. Effect of Build Orientation of Electron Beam Melting on Microstructure and Mechanical Properties of Ti-6Al-4V. *J. Mater. Eng. Perform.* **2017**, *26*, 692–703. [CrossRef]
16. Zhou, J.; Wang, P.; Nai, M.L.S.; Sin, W.J.; Wei, J.; Adiguzel, O. Effect of Building Height on Microstructure and Mechanical Properties of Big-Sized Ti-6Al-4V Plate Fabricated by Electron Beam Melting. *MATEC Web Conf.* **2015**, *30*, 02001. [CrossRef]
17. Tan, X.; Kok, Y.; Tan, Y.J.; Vastola, G.; Pei, Q.X.; Zhang, G.; Zhang, Y.-W.; Tor, S.B.; Leong, K.F.; Chua, C.K. An experimental and simulation study on build thickness dependent microstructure for electron beam melted Ti–6Al–4V. *J. Alloys Compd.* **2015**, *646*, 303–309. [CrossRef]
18. Wang, P.; Tan, X.; Nai, M.L.S.; Tor, S.B.; Wei, J. Spatial and geometrical-based characterization of microstructure and microhardness for an electron beam melted Ti–6Al–4V component. *Mater. Des.* **2016**, *95*, 287–295. [CrossRef]
19. Galarraga, H.; Warren, R.J.; Lados, D.A.; Dehoff, R.R.; Kirka, M.M.; Nandwana, P. Effects of heat treatments on microstructure and properties of Ti-6Al-4V ELI alloy fabricated by electron beam melting (EBM). *Mater. Sci. Eng. A* **2017**, *685*, 417–428. [CrossRef]
20. Safdar, A.; Wei, L.Y.; Snis, A.; Lai, Z. Evaluation of microstructural development in electron beam melted Ti-6Al-4V. *Mater. Charact.* **2012**, *65*, 8–15. [CrossRef]
21. Tammas-Williams, S.; Zhao, H.; Léonard, F.; Derguti, F.; Todd, I.; Prangnell, P.B. XCT analysis of the influence of melt strategies on defect population in Ti–6Al–4V components manufactured by Selective Electron Beam Melting. *Mater. Charact.* **2015**, *102*, 47–61. [CrossRef]

22. Cunningham, R.; Nicolas, A.; Madsen, J.; Fodran, E.; Anagnostou, E.; Sangid, M.D.; Rollett, A.D. Analyzing the effects of powder and post-processing on porosity and properties of electron beam melted Ti-6Al-4V. *Mater. Res. Lett.* **2017**, *5*, 516–525. [CrossRef]
23. Barriobero-Vila, P.; Gussone, J.; Haubrich, J.; Sandlobes, S.; Da Silva, J.C.; Cloetens, P.; Schell, N.; Requena, G. Inducing Stable alpha + beta Microstructures during Selective Laser Melting of Ti-6Al-4V Using Intensified Intrinsic Heat Treatments. *Materials* **2017**, *10*, 268. [CrossRef] [PubMed]
24. Khorasani, A.; Gibson, I.; Awan, U.S.; Ghaderi, A. The effect of SLM process parameters on density, hardness, tensile strength and surface quality of Ti-6Al-4V. *Addit. Manuf.* **2019**, *25*, 176–186. [CrossRef]
25. Vrancken, B.; Thijs, L.; Kruth, J.-P.; Van Humbeeck, J. Heat treatment of Ti6Al4V produced by Selective Laser Melting: Microstructure and mechanical properties. *J. Alloys Compd.* **2012**, *541*, 177–185. [CrossRef]
26. Tan, X.; Kok, Y.; Tan, Y.J.; Descoins, M.; Mangelinck, D.; Tor, S.B.; Leong, K.F.; Chua, C.K. Graded microstructure and mechanical properties of additive manufactured Ti–6Al–4V via electron beam melting. *Acta Mater.* **2015**, *97*. [CrossRef]
27. Hrabe, N.; Quinn, T. Effects of processing on microstructure and mechanical properties of a titanium alloy (Ti–6Al–4V) fabricated using electron beam melting (EBM), Part 2: Energy input, orientation, and location. *Mater. Sci. Eng. A* **2013**, *573*, 271–277. [CrossRef]
28. Rafi, H.K.; Karthik, N.V.; Gong, H.; Starr, T.L.; Stucker, B.E. Microstructures and Mechanical Properties of Ti6Al4V Parts Fabricated by Selective Laser Melting and Electron Beam Melting. *J. Mater. Eng. Perform.* **2013**, *22*, 3872–3883. [CrossRef]
29. Tan, X.; Kok, Y.; Toh, W.Q.; Tan, Y.J.; Descoins, M.; Mangelinck, D.; Tor, S.B.; Leong, K.F.; Chua, C.K. Revealing martensitic transformation and alpha/beta interface evolution in electron beam melting three-dimensional-printed Ti-6Al-4V. *Sci. Rep.* **2016**, *6*, 26039. [CrossRef]
30. Al-Bermani, S.S.; Blackmore, M.L.; Zhang, W.; Todd, I. The Origin of Microstructural Diversity, Texture, and Mechanical Properties in Electron Beam Melted Ti-6Al-4V. *Metall. Mater. Trans. A* **2010**, *41*, 3422–3434. [CrossRef]
31. Boyer, R.; Welsch, G.; Collings, E.W. *Materials Properties Handbook: Titanium Alloys*; ASM International: Materials Park, OH, USA, 1994.

materials

MDPI

Article

Comprehensive Evaluation of the Properties of Ultrafine to Nanocrystalline Grade 2 Titanium Wires

Jan Palán [1,*], Radek Procházka [1], Jan Džugan [1], Jan Nacházel [1], Michal Duchek [1], Gergely Németh [2,3], Kristián Máthis [2], Peter Minárik [2] and Klaudia Horváth [2,3]

1 COMTES FHT a.s., Průmyslová 995, 334 41 Dobřany, Czech Republic; radek.prochazka@comtesfht.cz (R.P.); jan.dzugan@comtesfht.cz (J.D.); jan.nachazel@comtesfht.cz (J.N.); michal.duchek@comtesfht.cz (M.D.)

2 Charles University, Faculty of Mathematics and Physics, Ke Karlovu 5, 121 16 Prague, Czech Republic; gergely1227@gmail.com (G.N.); mathis@met.mff.cuni.cz (K.M.); peter.minarik@mff.cuni.cz (P.M.); horviklaudi@gmail.com (K.H.)

3 Nuclear Physics Institute of the CAS, Husinec—Řež 130, 250 68 Řež, Czech Republic

* Correspondence: jan.palan@comtesfht.cz; Tel.: +420-377-197-366

Received: 21 November 2018; Accepted: 10 December 2018; Published: 11 December 2018

check for updates

Abstract: This paper describes the mechanical properties and microstructure of commercially pure titanium (Grade 2) processed with Conform severe plastic deformation (SPD) and rotary swaging techniques. This technology enables ultrafine-grained to nanocrystalline wires to be produced in a continuous process. A comprehensive description is given of those properties which should enable straightforward implementation of the material in medical applications. Conform SPD processing has led to a dramatic refinement of the initial microstructure, producing equiaxed grains already in the first pass. The mean grain size in the transverse direction was 320 nm. Further passes did not lead to any additional appreciable grain refinement. The subsequent rotary swaging caused fine grains to become elongated. A single Conform SPD pass and subsequent rotary swaging resulted in an ultimate strength of 1060 MPa and elongation of 12%. The achieved fatigue limit was 396 MPa. This paper describes the production possibilities of ultrafine to nanocrystalline wires made of pure titanium and points out the possibility of serial production, particularly in medical implants.

Keywords: titanium; ECAP; Conform; continuous extrusion; wire; medical implants

1. Introduction

Much work has recently been done in the field of severe plastic deformation (SPD) processing of pure titanium [1]. These processes have shown great potential as they have delivered a more than two-fold improvement in mechanical properties [2–5]. With improved mechanical properties, pure titanium has become suitable for applications that require high strength levels (dental implants, bone screws, etc.). Besides that, implants with a smaller cross-section can be designed. For instance, implants with smaller load-bearing cross-sections are suitable for children [6]. In terms of biocompatibility, pure titanium is associated with increased osteoblast proliferation and better osseointegration [6,7]. Kim and Park [7] established a model explaining enhanced surface cell attachment on nanocrystallized commercially pure titanium (CP–Ti). The reason for this is that the number of nodules at the triple-point junctions of the grain microstructure increases as the grain size is refined through equal-channel angular pressing (ECAP). These properties facilitate a connection between the implant and the human tissue [6–10].

Commercialisation of ultrafine-grained materials prepared by SPD techniques is hindered by the fact that these techniques are rarely suitable for production outside the laboratory [1,11,12]. These techniques were generally intended for microstructural evolution observations and assessment of the properties of the investigated materials. Hence, the next step is to transfer laboratory procedures to industrial practice. Some attempts to develop processes for the continuous production of ultrafine-grained materials have taken place recently. The attempts included wire production methods such as equal-channel angular swaging (ECAS) [13] and Conform ECAP [14–16]. Their commercialisation is under way, particularly in the field of pure titanium wire processing. This development is associated with a research group headed by R. Z. Valiev [17], which is responsible for the development of Conform ECAP. Conform ECAP processing is a relatively new modification of the conventional ECAP technique. In this process, the principle used to generate the frictional force to push a work-piece through an ECAP die is similar to the Conform process, while a modified ECAP die design is used so that the work-piece can be repetitively processed to produce ultrafine-grained (UFG) structures. The angle of channel intersection is 120°. The input stock for this method is a square bar of 11×11 mm^2 cross-sectional area [14]. To obtain the desired properties, six passes at 200–400 °C are typically carried out. The resultant semi-finished product is transferred to a wire-drawing machine, where it is worked at an elevated temperature [17,18]. This process is known as the thermomechanical treatment (TMT) of pure titanium. Wire drawing leads to intensive strengthening. Products of the Conform ECAP method have equiaxed grains with high-angle boundaries. During the subsequent drawing operation, these fine grains become elongated, their dislocation density steeply increases and the material gradually loses its plastic deformation capability. For instance, processing of Grade 4 commercially pure (CP) titanium through six passes in Conform ECAP at 200 °C and subsequent drawing at 200 °C leads to a strength of 1267 MPa [17].

An alternative for the continuous production of ultrafine-grained wires is the Conform SPD technique (Figure 1) [19]. This technique is being explored by the company COMTES FHT. Its principle is similar to that of Conform ECAP, as the feedstock is continuously forced through a chamber along an angled channel. Unlike Conform ECAP, however, Conform SPD uses round stock and the angle of its channel intersection is 90° [2–4,19]. The principle of forming titanium using Conform SPD technology to obtain ultrafine-grained products is shown in Figure 1. The feedstock is guided by the coining roll to the gap between the driving wheel and the shoe. High friction forces cause the feedstock to move along the groove in the driving wheel all the way to the abutment. Once the material hits the abutment, it changes direction and exits the Conform SPD machine through the chamber die [20–24].

This paper gives a description of the processing of CP Grade 2 titanium by a combination of Conform SPD and cold rotary swaging, and includes an overview of the final properties. It summarises information which is expected to facilitate the implementation of this material in the serial production of medical implants.

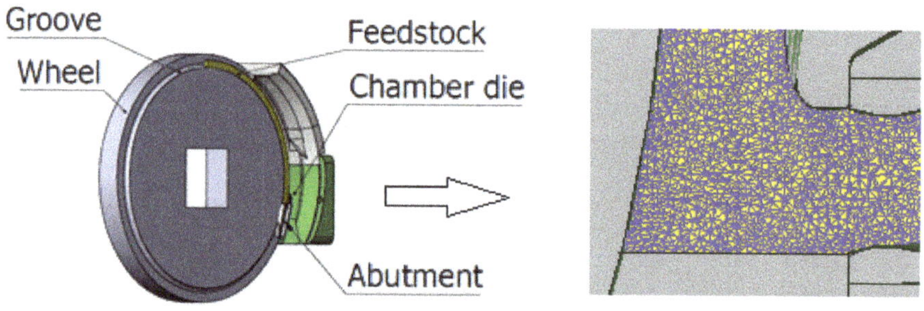

Figure 1. Schematic representation of the Conform severe plastic deformation (SPD) technique.

2. Materials and Methods

This study concerns commercially pure Grade 2 titanium (ASTM B348 Gr2). Its chemical composition is given in Table 1. The composition was determined by means of the Bruker Q4 Tasman optical emission spectrometer. The initial diameter of the wire was 10 mm. The initial material was delivered with respect to the ASTM B348 standard.

Table 1. Chemical composition of feedstock, wt. %.

Element	Fe	O	C	H	N	Ti
Content	0.046	0.12	0.023	0.0026	0.0076	balance

The wire was processed by the Conform SPD technology. Figure 2 shows the distribution of the strain rate, material velocity and the temperature in the die chamber, respectively. They were calculated with the DEFORM-3D software. Detailed conditions for these calculations are described in the publication [25]. The calculations described in this publication are for information only. The feedstock had a round cross-section of 10 mm diameter. At the start of the processing, it was at room temperature and no external heating was provided. Up to three passes were completed (1 pass, 2 passes, 3 passes). The driving wheel ran at 0.5 rpm. The A route was used, i.e., the feedstock orientation was identical in all passes.

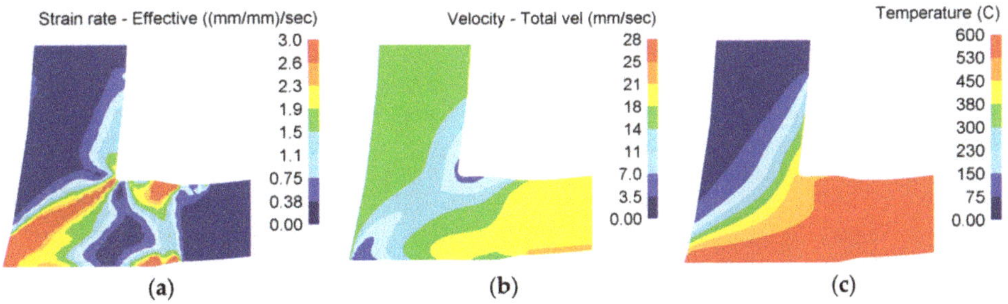

Figure 2. Numerical modelling of the Conform SPD process: (**a**) Strain rate distribution (s^{-1}); (**b**) Velocity distribution (mm/s); (**c**) Temperature distribution (°C).

In order to further increase the mechanical properties of the material, cold rotary swaging was applied in the next step. Rotary swaging was performed on the as-received material and on the material after a single pass through Conform SPD. The purpose was to compare the final properties after rotary swaging combined with Conform SPD, and after rotary swaging alone. The diameter of the initial feedstock was the same as for Conform SPD—10 mm. One operation involved the reduction of the cross-sectional area by 20%. The total reduction in cross-sectional area was up to 80%.

Metallographic specimens were prepared using a standard procedure involving grinding and subsequent polishing. The microstructure was revealed by etching with Kroll reagent. A Carl Zeiss—Observer, the Z1m optical microscope, and bright-field illumination were used for microstructure observation.

Fractographic examination was performed using the JEOL JSM 6380 scanning electron microscope. Fracture surfaces of fatigue test specimens were observed using secondary electron imaging.

Thin foils were prepared for transmission electron microscope (TEM) observations. The final electrolytic thinning was performed with the use of a Tenupol 5 device, using a solution of 300 mL CH3OH + 175 mL 2-butanol + 30 mL HClO4 at −10 °C and a voltage of 40 V. The TEM analysis was performed in a JEOL 200CX instrument, with an acceleration voltage of 200 kV. Selective electron diffraction was used for determining phases. The grain size was measured using the linear intercept method.

The electron-backscatter diffraction (EBSD) observations were performed on the cross-section of the samples using an FEI Quanta 200 FX scanning electron microscope (SEM, Thermo Fisher Scientific, Brno, Czech Republic). The EBSD measurements were conducted at a working distance of 13 mm using a step size of 50 nm at 10 kV acceleration voltage. The exact sizes of the scanned areas are indicated in the captions of particular figures. Prior to the EBSD measurement, the surfaces of the samples were ground on SiC papers (from 320 to 1200 grit) and subsequently polished for 24 h in a three-step vibratory polishing procedure. The surfaces of the specimens were finally ion-beam-polished on a Leica EM RES102 system (Leica Mikrosysteme, Wetzlar, Germany) before observation. In order to observe the microstructure of the rotary swaged sample transmission, the Kikuchi diffraction (TKD) method was used. The TKD method is also referred to as "transmission EBSD". For ultra-fine grained materials with grain size from a few tens to hundreds of nanometers, the resolution of conventional EBSD is not sufficient since the interaction volume is comparable to the grain size. Moreover, in heavily deformed materials, EBSD analysis is obscured by high dislocation density, residual strains and lattice rotations. Transmission EBSD was measured in a Zeiss Auriga Compact FIB-SEM (Jena, Germany) using a step of 10 nm at 30 kV acceleration voltage on a standard TEM foil.

Tensile testing was carried out in an electromechanical testing machine under quasi-static loading conditions, at a constant strain rate of 0.0002/s at room temperature. Round bar tensile specimens with a diameter of 3 mm and 5 mm, respectively, were employed. Samples of 5 mm in diameter were used for wires after Conform processing. Samples of 3 mm in diameter were used for wires after Conform processing and rotary swaging. A mechanical extensometer was used for strain measurement for both testing specimen geometries.

The conventional fatigue testing procedure consists of cyclic load application on sub-sized round-bar specimens of an hourglass shape of 1.5 mm in diameter, with sinusoidal waveform at constant amplitude in the tension-compression mode. Specimens were intended to be tested under cyclic loading, with stress ratio $R = -1$ at room temperature. Due to the nature of bulk ultrafine-grained (UFG) materials, it was possible to use a test frequency up to 50 Hz without signs of self-heating in the high cyclic fatigue (HCF) regime verified by thermography measurements. Tests were conducted on an INOVA servo-hydraulic testing machine with a loading capacity of 15 kN. Tests were run until failure (N_f) or up to 10 million cycles (red circles in graphs). The test results are plotted on semi-log scale in Figure 8. Due to the data scattering, a statistical approach using a 90% two-sided confidence interval was added to the data plot. The fatigue limit of 396 MPa was estimated based on two specimens surviving 10 million cycles without failure.

3. Results and Discussion

3.1. Microstructure Evolution in Pure Titanium after Conform SPD and Rotary Swaging

The microstructure of the feedstock consisted of equiaxed recrystallized grains with annealing twins (Figure 3). These annealing twins can be seen in Figure 3a. They are a characteristic feature in materials with a hexagonal crystallographic structure due to their limited potential for deformation slip at low deformations [26]. The twins are broad and lenticular.

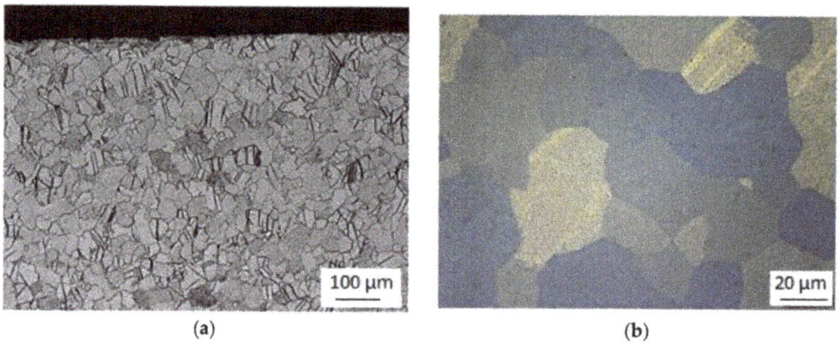

(a) (b)

Figure 3. (**a**) Micrograph of the as-received structure in the transverse direction; (**b**) detailed micrograph of the as-received structure.

Figure 4 shows transmission electron micrographs of the material processed by Conform SPD with various numbers of passes. Table 2 lists average grain sizes after various numbers of passes through the Conform SPD machine. Figure 4a,b show a general view of the substructure in the transverse and longitudinal direction after the first pass. There are polyhedral grains with non-uniform dislocation density (Figure 4a) and regions with elongated grains with low-angle boundaries (Figure 4b). Variations in the grain morphology were found over the volume and the microstructure is not homogenous after the first pass. The variations in grain morphology are an indication of the non-uniform thermomechanical conditions (temperature distribution, stress state, and strain rate) within the material—the non-homogeneous character of the process is clearly visible in Figure 2.

The second and third pass produced equiaxed grains (Figure 4, from c to f). The third pass even resulted in a larger mean grain size of 420 nm in the longitudinal direction (Table 2). This increase can be attributed, in part, to non-uniform deformation and to the high surface activity of the UFG structure. High surface activity and dislocation density (strain magnitude) lower the temperatures of softening processes [4]. Certain grain growth can be expected to occur during formation due to deformation heat and the heat retained in the die chamber. This effect proves that the deformation heat (Figure 2c) is very likely to be a contributing factor in the recrystallization (post dynamic recrystallization) of the refined microstructure. The generation and effects of deformation heat on softening processes are often ignored in SPD applications. Furthermore, the decrease in grain size after the two passes is smaller than after the first pass. The refinement effect of Conform SPD seems to be limited when the initial microstructure is in the sub-micron scale [27]. A recent study [28] has shown that small grains may cause low stability when the metals are heavily deformed, and additional straining may occur as Conform SPD does not further reduce the grain size because of the intrinsic instability of nano-sized (below 100 nm) and submicrometre-sized (between 100 and 1000 nm) grains [11]. The dynamic balance of grain refinement between structure refinement and recovery at ambient temperature occurred, which was already proven in the stated study [27]. Thus, the grain refinement phenomena of Conform SPD exhibit a limit, and with additional passes, the grain size and dislocation density will reach their ultimate values. Conform SPD processing leads to substantial grain refinement and to mostly equiaxed grains in the work-piece, even in a single pass. These grains are so small that they only rotate in the subsequent passes, after which the microstructure becomes more homogeneous [29]. The mechanism by which the subgrains rotate is not so well understood. Wu et al. [11] describe a process in which dislocation motion becomes restricted due to the small subgrain size and grain rotation becomes more energetically favourable [30]. Mishra et al. [29] proposed a slightly different explanation, in which the rotation is aided by diffusion along the grain boundaries (which is much faster than through the grain interior) [29]. Conform SPD therefore reaches its grain refinement limit in just one pass. Further passes mostly lead to nothing more than homogenization of the microstructure. However, according to some studies, the maximum (limit) grain refinement is only achieved after multiple passes [11,17,18].

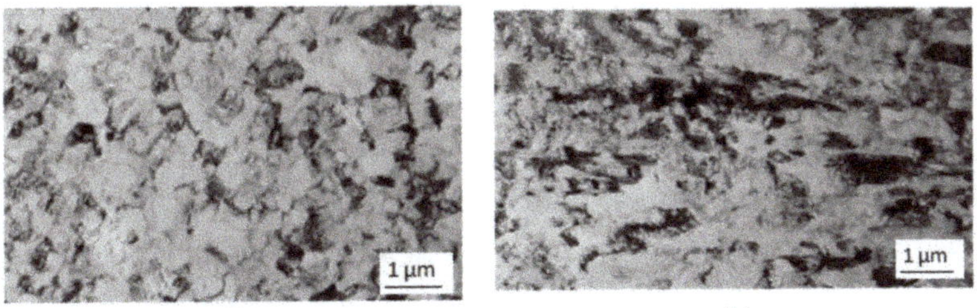

(a) (b)

Figure 4. *Cont.*

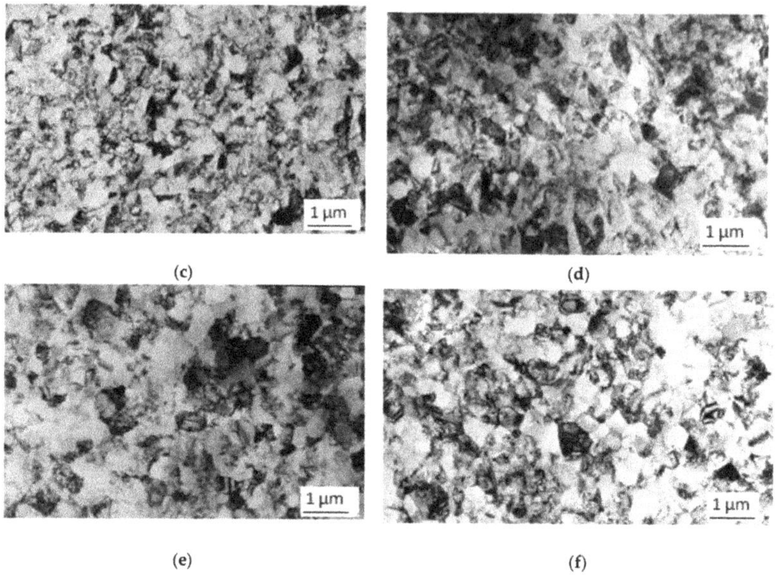

Figure 4. (**a**) Substructure in the transverse direction after the first pass; (**b**) substructure in the longitudinal direction after the first pass; (**c**) substructure in the transverse direction after the second pass; (**d**) substructure in the longitudinal direction after the second pass; (**e**) substructure in the transverse direction after the third pass; (**f**) substructure in the longitudinal direction after the third pass.

Table 2. Mean grain size for different processing steps.

	Transverse Direction	**Longitudinal Direction**
As received	28.95 µm	
1 pass	320 ± 35 nm	310 ± 30 nm
2 passes	250 ± 25 nm	310 ± 30 nm
3 passes	330 ± 30 nm	420 ± 30 nm

In order to further decrease the grain size, cold rotary swaging was employed on the material which had already been processed using the Conform device. The product of a single pass through the Conform SPD machine was rotary-swaged with 80% total reduction in the cross-sectional area. Small grains thus extended in the longitudinal direction and their dislocation density steeply increased (darker areas). This is illustrated in Figure 5.

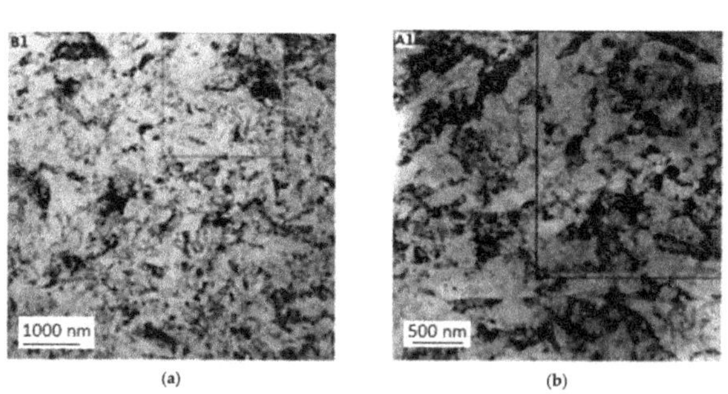

Figure 5. (**a**) Substructure in the longitudinal direction after one pass through the Conform SPD machine and rotary swaging; (**b**) substructure in the longitudinal direction after one pass and rotary swaging.

3.2. Texture Evaluation

The results of the EBSD measurements conducted in the centres of samples after one, two and three Conform SPD passes are shown in Figure 6. With increasing numbers of passes, significant grain refinement occurs (please note the different scales of the images). The texture evolution is also shown in the inverse pole figure (IPF) maps in Figure 6. After the first pass, an intensive maximum can be observed close to the (0001) pole. With further passes, this maximum weakens and a new component in the inclined area between the horizontal and vertical axes is observed. This new component is dominant in the sample after three Conform SPD passes. A similar texture evolution was observed in ECAPed Mg alloys containing rare earth elements, where the intensity of this maximum was inclined by $\approx 55°$ from the processing direction [31]. The reason for the appearance of this component is most probably given by the activation of non-basal slip systems.

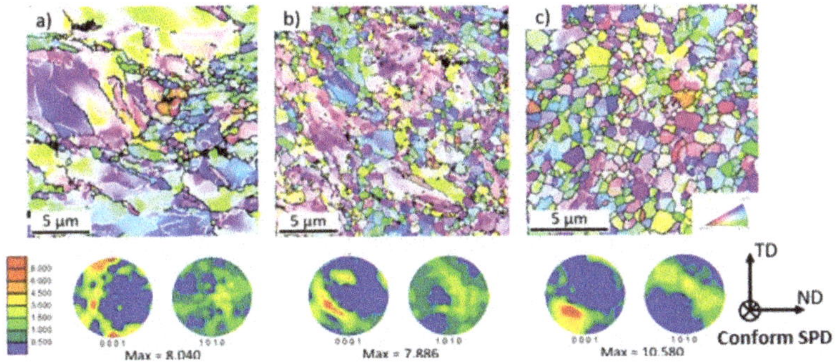

(a) (b) (c)

Figure 6. Cross-section orientation image maps (OIM) and inverse pole figure (IPF) maps for samples after (**a**) first pass (50×50 µm), (**b**) second pass (20×20 µm), and (**c**) third pass (15×15 µm) through Conform SPD. In the figure on the left, the scale of texture intensities is shown as multiples of the random density (m.r.d.) from 0–8.000. The maximum value of each texture is listed below the IPF maps. The orientation triangle for the electron-backscatter diffraction (EBSD) maps is shown in the right-hand corner of the figure.

Figure 7 presents the transmission EBSD map of the sample after one SPD pass followed by rotary swaging. The microstructure, similarly to that after one SPD pass, is not homogeneous. It contains some coarse grains with some low-angle grain boundaries near the recrystallized area. The sample has a very intensive fibre texture with basal planes oriented parallel to the SPD direction. This texture is not only observed in the coarse grains but also in the recrystallized ones.

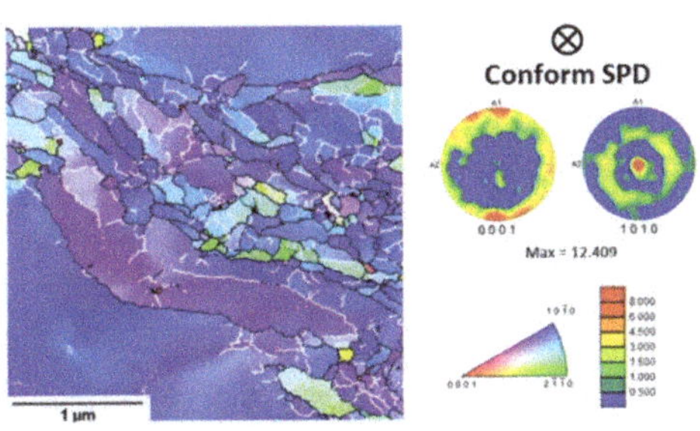

Figure 7. Transmission EBSD map of the sample after one Conform SPD pass and rotary swaging (3×3 µm). High-angle grain boundaries (>15°) are marked by black lines, while the low-angle grain boundaries (between 4° and 15°) are marked by white lines. The texture intensities are carried out as the multiples of random density (m.r.d.) from 0–8.000, with maximum at 12.409 m.r.d.

3.3. The Effect of Conform SPD and Rotary Swaging on Tensile Properties

Table 3 lists the tensile tests results. Materials in several differing conditions were tested after various numbers of passes through the Conform SPD machine and after rotary swaging with a reduction in cross-sectional area. The first pass through the Conform SPD machine led to an increase in Ultimate Tensile Strength (UTS) from 480 MPa for the as-received material to 580 MPa, whereas only a minimum decrease in A_5 elongation occurred. Further passes did not bring any significant strengthening. After the second pass, the strength was 600 MPa and after the third pass it reached 623 MPa. These increases in strength were not accompanied by decreases in A_5 elongation. These results are in agreement with the trends in grain size, since further passes caused no additional refinement and therefore there was no additional increase in strength. The cross-sectional area of a wire processed with one pass through the Conform SPD machine was further reduced by rotary swaging. It led to a notable increase in ultimate strength and yield strength. The ultimate strength was 1060 MPa, which is twice as high as in the as-received material. On the other hand, elongation was a mere 12%, in contrast to the 25% achieved with just the Conform SPD processing. This can be attributed to a much higher dislocation density and the elongation of the initially equiaxed grains [4]. The reduction in area was greater than in the initial material. This finding is in agreement with the study reported in [4]. This improvement in the reduction of area was observed for all post-Conform SPD conditions. The rotary-swaged specimen identified as "rotary swaging 80% area reduction" had a strength of 964 MPa. This is approximately 100 MPa less than in the specimen after one pass through the Conform SPD machine and rotary swaging to the same reduction level of the cross-sectional area. The elongation was 3% lower and the reduction in area was almost 24% lower in the first specimen than in the latter one. These values suggest that ductility increases for Conform SPD-processed specimens.

Table 3. Mechanical properties after Conform SPD processing and after Conform SPD + rotary swaging. Ultimate tensile strength (UTS); offset yield (OYS); reduction in area (RA); elongation (A_5).

Condition	0.2 OYS [MPa]	UTS [MPa]	A_5 [%]	RA [%]
As received	370 ± 9.4	480 ± 7.7	25 ± 1.3	52 ± 1.9
Conform SPD—1 pass	540 ± 5.8	580 ± 6.1	23 ± 1.2	62 ± 2.3
Conform SPD—2 passes	560 ± 1.6	600 ± 5.6	23 ± 1.3	62 ± 2.2
Conform SPD—3 passes	570 ± 1.8	623 ± 4.8	23 ± 1.4	62 ± 2.1
Conform SPD 1 pass + Rotary Swaging (80% area reduction)	975 ± 2.3	1060 ± 4.6	12 ± 1.4	58 ± 2.3
Rotary Swaging (80% area reduction)	890 ± 2.1	964 ± 4.4	9 ± 1.3	34.2 ± 2.4

3.4. Evaluation of Fatigue Properties

Figure 8 shows fatigue data for the specimen that underwent one pass in Conform SPD and was then rotary-swaged to an 80% reduction in cross-sectional area. This type of processing gives high values of mechanical properties while maintaining the process productivity. Its fatigue strength was 396 MPa. The fatigue strength of Grade 2 titanium in the ASTM B348 condition is approximately 240 MPa [32]. Comparing those values indicates a demonstrable increase in fatigue strength and the impact of the grain size on fatigue. Similar results were reported by other authors [32,33]. The results show that the fatigue strength of ultra-fine to nano-grained titanium at 10^7 cycles is 60 MPa higher than conventional CP titanium Grade 4 but does not exceed that of the Ti-6Al-4V alloy, which has 530 MPa. It is clear that increasing the fatigue strength of CP titanium depends on tensile strength, this relationship being characteristic of titanium, as opposed to wavy-slip fcc materials [5]. The explanation is that the cross-slip of dislocations is more difficult in the hcp lattice. Therefore, the fatigue strength of Ti depends on the parameters of the size and shape of the grains and the type of boundaries. The twinning mechanism does not play a key role in the cyclic deformation of UFG titanium, and the fatigue mechanisms are likely related to the grain boundaries. Figure 9 shows images of the fracture surface of a ruptured fatigue test sample. It is obvious that the fatigue area covers about 60% of the fracture surface. The arrow points in the crack propagation direction. The fracture surface has been strongly smoothed out [5,27].

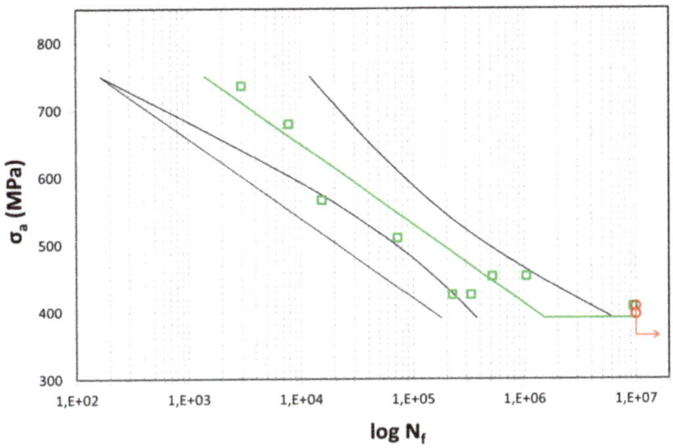

Figure 8. S–N curve for the sample after Conform SPD one pass + rotary swaging (80% area reduction).

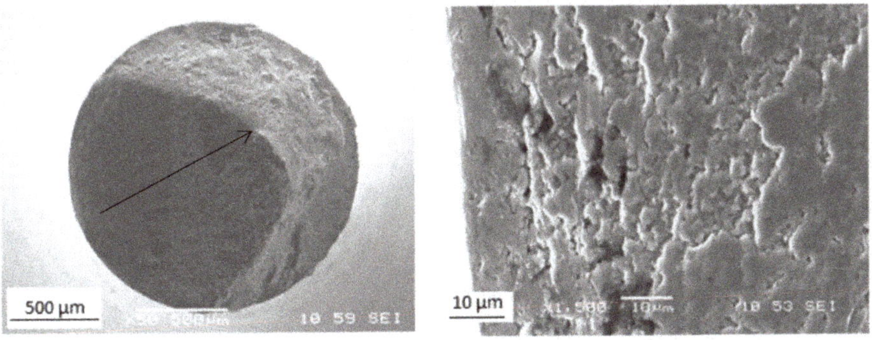

(**a**) (**b**)

Figure 9. Fracture surface of the sample (Conform SPD one pass + rotary swaging (80% area reduction)) after (**a**) fatigue testing and (**b**) fracture initiation area.

In summary, ultra-fine to nano-grained Ti (Grade 2) can replace conventional Ti Grade 4 with the assumption of increased component life, as the fatigue strength is roughly 60 MPa higher. The feasibility of replacing Ti-6Al-4V is still in question because its fatigue strength is about 130 MPa higher. On the other hand, the ultimate strength is about 200 MPa higher. Another advantage of commercially pure Ti is its enhanced biocompatibility as compared to the Ti-6Al-4V alloy. Recent studies, in particular, point out the toxic effects of Al and V after release of these elements into the human body. An assessment of fatigue strength of a dental implant made from ultra-fine to nano-grained titanium Grade 2 is already being evaluated by COMTES FHT and will be the subject of further articles and studies.

4. Conclusions

This paper presents the successful processing of commercially pure Grade 2 titanium with Conform SPD and rotary swaging techniques. The proposed route enables high-strength wires with an ultrafine to nanocrystalline microstructure to be produced in a continuous process. The article primarily describes the characteristics of the high-strength wire. The following conclusions can be stated:

- Processing with Conform SPD already leads to dramatic grain refinement in the first pass. The average grain size was 320 nm. Subsequent rotary swaging further reduced the average grain size. Grains were preferentially elongated in the longitudinal direction and the sample has a very intensive fibre texture, with basal planes oriented parallel to the longitudinal direction.

- Combined processing with Conform SPD and rotary swaging leads to a significant increase in mechanical properties. The ultimate strength was 1060 MPa. The increase is mainly given by the refinement of the initial grain structure and by an increased dislocation density. The fatigue limit achieved for room temperature was 396 MPa.
- The proposed technological route gives the possibility to produce high-strength wires with an ultrafine to nanocrystaline structure. The presented paper gives an idea of how to produce such a material in a continuous way and describes the properties of the final product.

Author Contributions: J.P. and M.D. conceived and designed the technological experiments and conducted the final evaluation of the results. R.P. and J.D. conceived and designed the fatigue tests. J.N. carried out the metallographic evaluation. G.N., P.M., K.H. and K.M. conceived and designed the texture evaluation.

Funding: This research was developed under the project entitled Development of West-Bohemian Centre of Materials and Metallurgy No.: LO1412, which is financed by the Ministry of Education of the Czech Republic.

Acknowledgments: This paper was developed under the project entitled Development of West-Bohemian Centre of Materials and Metallurgy No.: LO1412, which is financed by the Ministry of Education of the Czech Republic. K.M. is grateful for support from the Czech Science Foundation (GB 14-36566G). G.N. and K.H. acknowledge the support of the Operational Programme Research, Development and Education, The Ministry of Education, Youth and Sports (OP RDE, MEYS) [CZ.02.1.01/0.0/0.0/16_013/0001794].

Conflicts of Interest: The authors declare no conflict of interest.

References

1. Mishnaevsky, L.; Levashov, E.; Valiev, R.Z.; Segurado, J.; Sabirov, I.; Enikeev, N.; Prokoshkin, S.; Solov'yov, A.V.; Korotitskiy, A.; Gutmanas, E.; et al. Nanostructured titanium-based materials for medical implants: Modeling and development. *Mater. Sci. Eng. R Rep.* **2014**, *81*, 1–19. [CrossRef]
2. Palán, J.; Kubina, T.; Motyčka, P. The effect of annealing on mechanical and structural properties of UFG titanium grade 2. *IOP Conf. Ser. Mater. Sci. Eng.* **2017**, *179*, 012055. [CrossRef]
3. Zemko, M.; Kubina, T.; Dlouhý, J.; Kover, M.; Hodek, J. Technological aspects of preparation of nanostructured titanium wire using a CONFORM machine. *IOP Conf. Ser. Mater. Sci. Eng.* **2014**, *63*, 012049. [CrossRef]
4. Palán, J.; Šutta, P.; Kubina, T.; Dománková, M. Effect of severe plastic and heavy cold deformation on the structural and mechanical properties of commercially pure titanium. *Mater. Tehnol.* **2017**, *51*, 849–853. [CrossRef]
5. Semenova, I.P.; Valiev, R.Z.; Yakushina, E.B.; Salimgareeva, G.H.; Lowe, T.C. Strength and fatigue properties enhancement in ultrafine-grained Ti produced by severe plastic deformation. *J. Mater. Sci.* **2008**, *43*, 7354–7359. [CrossRef]
6. Ostrovska, L.; Vistejnova, L.; Dzugan, J.; Slama, P.; Kubina, T.; Ukraintsev, E.; Kubies, D.; Kralickova, M.; Hubalek Kalbacova, M. Biological evaluation of ultra-fine titanium with improved mechanical strength for dental implant engineering. *J. Mater. Sci.* **2016**, *51*, 3097–3110. [CrossRef]
7. Zhou, Q.; Wang, L.; Zou, C.-H. Enhanced Surface Precipitates on Ultrafine-Grained Titanium in Physiological Solution. *Metals* **2017**, *7*, 245. [CrossRef]
8. An, B.; Li, Z.; Diao, X.; Xin, H.; Zhang, Q.; Jia, X.; Wu, Y.; Li, K.; Guo, Y. In vitro and in vivo studies of ultrafine-grain Ti as dental implant material processed by ECAP. *Mater. Sci. Eng. C* **2016**, *67*, 34–41. [CrossRef]
9. Estrin, Y.; Ivanova, E.P.; Michalska, A.; Truong, V.K.; Lapovok, R.; Boyd, R. Accelerated stem cell attachment to ultrafine grained titanium. *Acta Biomater.* **2011**, *7*, 900–906. [CrossRef]
10. Kim, T.N.; Balakrishnan, A.; Lee, B.C.; Kim, W.S.; Dvorankova, B.; Smetana, K.; Park, J.K.; Panigrahi, B.B. In vitro fibroblast response to ultra-fine grained titanium produced by a severe plastic deformation process. *J. Mater. Sci. Mater. Med.* **2008**, *19*, 553–557. [CrossRef]
11. Wu, H.; Jiang, J.; Liu, H.; Sun, J.; Gu, Y.; Tang, R.; Zhao, X.; Ma, A. Fabrication of Ultra-Fine Grained Pure Titanium with High Strength and Good Ductility via ECAP plus Cold Rolling. *Metals* **2017**, *7*, 563. [CrossRef]

12. Valiev, R.Z.; Langdon, T.G. Principles of equal-channel angular pressing as a processing tool for grain refinement. *Prog. Mater. Sci.* **2006**, *51*, 881–981. [CrossRef]
13. Görtan, M.O.; Groche, P. Tool design guidelines for the equal channel angular swaging (ECAS) process. *J. Mater. Process. Technol.* **2014**, *214*, 2220–2232. [CrossRef]
14. Raab, G.I.; Valiev, R.Z.; Gunderov, D.V.; Lowe, T.C.; Misra, A.; Zhu, Y.T. Long-Length Ultrafine-Grained Titanium Rods Produced by ECAP-Conform. *Mater. Sci. Forum* **2008**, *584–586*, 80–85. [CrossRef]
15. Xu, C.; Schroeder, S.; Berbon, P.B.; Langdon, T.G. Principles of ECAP–Conform as a continuous process for achieving grain refinement: Application to an aluminum alloy. *Acta Mater.* **2010**, *58*, 1379–1386. [CrossRef]
16. Palán, J.; Maleček, L.; Hodek, J.; Zemko, M.; Dzugan, J. Possibilities of biocompatible material production using Conform SPD technology. *Arch. Mater. Sci. Eng.* **2017**, *1*, 5–11. [CrossRef]
17. Gunderov, D.V.; Polyakov, A.V.; Semenova, I.P.; Raab, G.I.; Churakova, A.A.; Gimaltdinova, E.I.; Sabirov, I.; Segurado, J.; Sitdikov, V.D.; Alexandrov, I.V.; et al. Evolution of microstructure, macrotexture and mechanical properties of commercially pure Ti during ECAP-conform processing and drawing. *Mater. Sci. Eng. A* **2013**, *562*, 128–136. [CrossRef]
18. Polyakov, A.V.; Dluhoš, L.; Dyakonov, G.S.; Raab, G.I.; Valiev, R.Z. Recent Advances in Processing and Application of Nanostructured Titanium for Dental Implants: Recent Advances in Processing and Application. *Adv. Eng. Mater.* **2015**, *17*, 1869–1875. [CrossRef]
19. Kubina, T.; Dlouhý, J.; Kövér, M.; Hodek, J. Study of Thermal Stability of Ultra Fine-Grained Commercially Pure Titanium Wire Prepared in Conform Equipment. *Mater. Sci. Forum* **2014**, *782*, 415–420. [CrossRef]
20. Palán, J.; Procházka, R.; Zemko, M. The microstructure and mechanical properties evaluation of UFG Titanium Grade 4 in relation to the technological aspects of the CONFORM SPD process. *Procedia Eng.* **2017**, *207*, 1439–1444. [CrossRef]
21. Thomas, B.M.; Derguti, F.; Jackson, M. Continuous extrusion of a commercially pure titanium powder via the Conform process. *Mater. Sci. Technol.* **2017**, *33*, 899–903. [CrossRef]
22. ALkhazraji, H.; El-Danaf, E.; Wollmann, M.; Wagner, L. Enhanced Fatigue Strength of Commercially Pure Ti Processed by Rotary Swaging. *Adv. Mater. Sci. Eng.* **2015**, *2015*, 1–12. [CrossRef]
23. Wang, H.-F.; Han, J.-T.; Hao, Q.-L. Influence of Mandrel on the Performance of Titanium Tube with Cold Rotary Swaging. *Mater. Manuf. Process.* **2015**, *30*, 1251–1255. [CrossRef]
24. Knauer, E.; Freudenberger, J.; Marr, T.; Kauffmann, A.; Schultz, L. Grain Refinement and Deformation Mechanisms in Room Temperature Severe Plastic Deformed Mg-AZ31. *Metals* **2013**, *3*, 283–297. [CrossRef]
25. Hodek, J.; Kubina, T.; Dlouhy, J. *FEM Model of Continuous Extrusion of Titanium in DEFORM Software*; Tanger Ltd.: Plzeň, Czech Republic, 2013; pp. 347–351.
26. Humphreys, F.J.; Hatherly, M. *Recrystallization and Related Annealing Phenomena*, 2nd ed.; Elsevier: Amsterdam, The Netherlands; Boston, MA, USA, 2004; ISBN 978-0-08-044164-1.
27. Stolyarov, V.V.; Zhu, Y.T.; Lowe, T.C.; Valiev, R.Z. Microstructure and properties of pure Ti processed by ECAP and cold extrusion. *Mater. Sci. Eng. A* **2001**, *303*, 82–89. [CrossRef]
28. Lu, K. Stabilizing nanostructures in metals using grain and twin boundary architectures. *Nat. Rev. Mater.* **2016**, *1*, 16019. [CrossRef]
29. Mishra, A.; Kad, B.K.; Gregori, F.; Meyers, M.A. Microstructural evolution in copper subjected to severe plastic deformation: Experiments and analysis. *Acta Mater.* **2007**, *55*, 2563–2564. [CrossRef]
30. An, X.H.; Lin, Q.Y.; Wu, S.D.; Zhang, Z.F. Microstructural evolution and shear fracture of Cu–16at.% Al alloy induced by equal channel angular pressing. *Mater. Sci. Eng. A* **2010**, *527*, 4510–4514. [CrossRef]
31. Minárik, P.; Král, R.; Čížek, J.; Chmelík, F. Effect of Different c/a Ratio on the Microstructure and Mechanical Properties in Magnesium Alloys Processed by ECAP. *Acta Mater.* **2016**, *107*, 83–95. [CrossRef]
32. Estrin, Y.; Vinogradov, A. Fatigue Behaviour of Light Alloys with Ultrafine Grain Structure Produced by Severe Plastic Deformation: An Overview. *Int. J. Fatigue* **2010**, *32*, 898–907. [CrossRef]
33. Semenova, I.P.; Polyakov, A.V.; Raab, G.I.; Lowe, T.C.; Valiev, R.Z. Enhanced Fatigue Properties of Ultrafine-Grained Ti Rods Processed by ECAP-Conform. *J. Mater. Sci.* **2012**, *47*, 7777–7781. [CrossRef]

materials

MDPI

Article

Bioactive Sphene-Based Ceramic Coatings on cpTi Substrates for Dental Implants: An In Vitro Study

Hamada Elsayed [1,2], Giulia Brunello [3,4], Chiara Gardin [5,6], Letizia Ferroni [5,6], Denis Badocco [7], Paolo Pastore [7], Stefano Sivolella [4], Barbara Zavan [5,6] and Lisa Biasetto [3,*]

1 Department of Industrial Engineering, University of Padova, Via F. Marzolo 9, 35131 Padova, Italy; hamada.elsayed@unipd.it
2 Ceramics Department, National Research Centre, El-Bohous Street, 12622 Cairo, Egypt
3 Department of Management and Engineering, University of Padova, Stradella San Nicola 3, 36100 Vicenza, Italy; giulia-bru@libero.it
4 Department of Neurosciences, Section of Dentistry University of Padova, Via Giustiniani 2, 35128 Padova, Italy; stefano.sivolella@unipd.it
5 Department of Biomedical Sciences, University of Padova, Via Ugo Bassi 58/B, 35131 Padova, Italy; chiara.gardin@unipd.it (C.G.); letizia.ferroni@unipd.it (L.F.); barbara.zavan@unipd.it (B.Z.)
6 Maria Cecilia Hospital, GVM Care & Research, 48033 Ravenna, Italy
7 Department of Chemical Sciences, University of Padova, Via F. Marzolo 1, 35131 Padova, Italy; denis.badocco@unipd.it (D.B.); paolo.pastore@unipd.it (P.P.)
* Correspondence: lisa.biasetto@unipd.it; Tel.: +39-04-4499-8747

Received: 27 September 2018; Accepted: 5 November 2018; Published: 9 November 2018

Abstract: Titanium implant surface modifications have been widely investigated to favor the process of osseointegration. The present work aimed to evaluate the effect of sphene ($CaTiSiO_5$) biocoating, on titanium substrates, on the in vitro osteogenic differentiation of Human Adipose-Derived Stem Cells (hADSCs). Sphene bioceramic coatings were prepared using preceramic polymers and nano-sized active fillers and deposited by spray coating. Scanning Electron Microscopy (SEM) analysis, surface roughness measurements and X-ray diffraction analysis were performed. The chemical stability of the coatings in Tris-HCl solution was investigated. In vitro studies were performed by means of proliferation test of hADSCs seeded on coated and uncoated samples after 21 days. Methyl Thiazolyl-Tetrazolium (MTT) test and immunofluorescent staining with phalloidin confirmed the in vitro biocompatibility of both substrates. In vitro osteogenic differentiation of the cells was evaluated using Alizarin Red S staining and quantification assay and real-time PCR (Polymerase Chain Reaction). When hADSCs were cultured in the presence of Osteogenic Differentiation Medium, a significantly higher accumulation of calcium deposits onto the sphene-coated surfaces than on uncoated controls was detected. Osteogenic differentiation on both samples was confirmed by PCR. The proposed coating seems to be promising for dental and orthopedic implants, in terms of composition and deposition technology.

Keywords: implant; titanium; osseointegration; biocompatibility; bioactive ceramic coatings; sphene

1. Introduction

Commercially pure titanium (cpTi) and its alloys represent the materials of choice for manufacturing orthopedic and dental implants because of their high mechanical properties, good corrosion resistance and excellent biocompatibility [1–4].

The successful outcome of the dental implants mainly depends on the osseointegration at bone-implant level. In recent years, several surface treatments have been applied to Ti implants with the aim of increasing the speed and success rate of osseointegration [5,6]. Surface modifications included acid etching, isothermal oxidation, hydrothermal synthesis method, and combination of sandblasting and acid etching (SLA) [5,6].

Bioactive coatings appear to be promising materials favoring a faster and more enhanced osseointegration [1]. Several different coatings were reported in the literature, including hydroxyapatite, bioglass, proteins, polysaccharides, drugs, calcium phosphate, and calcium silicate [7–11].

Hydroxyapatite (HA), an osseoconductive material, can induce direct formation of bone tissue around implants, thanks to its unique chemical composition similar to that of the inorganic mineral part of bone tissue [9,12]. Despite its excellent biological properties, there are concerns for using HA-coated Ti-6Al-4V implants. However, the mismatch of their thermal expansion coefficients may cause delamination at coating-substrate interface, compromising the long-term success of osseointegrated implants [13,14].

In addition, CaO–SiO_2-based bioglasses have been investigated as coating materials. However, they have demonstrated poor bonding strength, due to the higher thermal expansion coefficient compared to that of Ti-6Al-4V [15,16], and high degradation rate [17–21].

Cp-Ti possess an average CTE (Coefficient of Thermal Expansion) of 8.9 (10^{-6} C^{-1}), Ti alloys an average CTE of 9.2 (10^{-6} C^{-1}). Hydroxyapatite shows CTE in the range between 10 and 14 (10^{-6} C^{-1}), while sphene has a CTE of 6 (10^{-6} C^{-1}). SiO_2-CaO bioactive glasses show CTE ranging between 8.5 and 19 (10^{-6} C^{-1}).

However, the clinical applications of bioglasses are limited by their low bending strength, high brittleness, low fracture toughness and workability. To overcome the major drawbacks related to the use of bioglass, several silicate-based ceramics were produced and tested as bioactive materials for bone tissue regeneration. Moreover, when compared to $CaSiO_3$ ceramics and tricalcium phosphate TCP (TriCalcium Phosphate), sphene ($CaTiSiO_5$) ceramics drove higher bone-derived cell attachment, spreading, proliferation and differentiation. $CaTiSiO_5$ ceramics have shown significantly higher chemical stability. Also, they have shown better chemical and biological properties compared to HA ceramics.

Sphene ($CaTiSiO_5$) bioactive coatings on titanium substrates were produced using various methods, such as plasma-spraying [14], sol-gel method [22], a hybrid technique of microarc oxidation coupled with heat treatment [23], and airbrush spray coating [24]. These studies have proved their chemical stability, excellent bonding strength (improved compared to HA coatings), bioactivity and cellular biocompatibility. Taken together, these results suggest that sphene ceramics may be potential candidates for bioactive coatings for orthopedic and dental implants. This reduced difference between CTE of cpTi and $CaTiSiO_5$ compared to that of cpTi and HA, is known to be the main property responsible for their improved adhesion [14,22–26].

Recently, our research group deposited sphene coatings from a preceramic polymer precursor "Silicone" onto Ti plates by the airbrush spray technique. Coatings exhibited excellent adhesion to the substrates and homogeneous deposition mode [24,27].

The novelty of this procedure compared to those reported in the literature consists both of the development of sphene synthesis via preceramic polymers route and nano-sized precursors, as well as in the deposition technique. The main advantages of the proposed technology can be summarized as follow:

- The use of preceramic polymer together with nano-sized precursors allowed for the improvement of reaction efficiency to produce sphene as well as an optimized suspension useful for spray coating via airbrush.
- The use of spray coating is a low-cost technique capable of producing coatings possessing a controlled morphology (roughness and thickness) and thereby improved adhesion to the metallic substrate.
- Compared to high temperature coating techniques, such as plasma coating, the spray coating used in this work is performed at room temperature and further heating of the samples at relatively low temperature.

- The optimized composition of the bioceramic coating together with its improved adhesion to the substrate is the result of the synergic effect of synthesis of sphene via preceramic polymer route and the deposition technique we used.

The primary objective of the current research consists of evaluating the effect of polymer-derived sphene-coated Ti substrates on the osteogenic differentiation of Adipose-Derived Stem Cells (ADSCs) in vitro. The secondary aim consists of assessing sphene coating composition, its in vitro chemical stability, and its micrometer-scale profile- and areal-topography.

2. Materials and Methods

2.1. Samples Preparation and Coating Deposition

Sphene ($CaTiSiO_5$) bioactive ceramic coating on Titanium plates (cpTi) was developed by using a preceramic polymer "silicone" approach as previously reported [24,27]. Preceramic silicone "polymethylsiloxane" (commercially available and known as "SILRES® MK", Wacker-Chemie GmbH, München, Germany), isopropanol, nano-sized $CaCO_3$ powder (PlasmaChem, Berlin, Germany, 90 nm) and nano-sized TiO_2 powders (Evonik Degussa GmbH, Essen, Germany, 21 nm) were used as starting materials.

Briefly, to prepare the coating suspension, SILRES® MK silicone resin powder was used as SiO_2 precursor. The MK silicone was placed in isopropyl alcohol. Active fillers of nanoparticles of $CaCO_3$, and TiO_2 were mixed with the MK silicone under magnetic stirring. The total solid content load in the suspension was ~48 vol%. The addition of active fillers was in stoichiometric molar ratio that allows developing the sphene bioceramic as a final ceramic product after sintering at relative lower temperature (i.e., 950 °C).

The suspension was homogenized by sonication for 15 min and then transferred into an automatic airbrush (Prona-RA-C2, PronaTools, Toronto, Canada M3J 3A1) for spray coating deposition. During the coating process, the silicone-fillers mixture was kept homogenous by magnetic stirring.

Rectangular plates of $13 \times 13 \times 3$ mm^3 size of commercially pure Ti grade 2 (Torresin Titanio s.r.l, Padova, Italy) were used as substrates. The chemical composition of the cpTi plates, as given by the producer (wt.%), was: Fe 0.060, O 0.140, N 0.004, H 0.003, C 0.016, Ti 99.78. Before coating deposition, the cpTi plates were ultrasonically cleaned with acetone, alcohol, and deionized water for ten minutes for each passage, and finally the titanium substrates were dried using compressed air. The use of cpTi instead of Ti alloys as substrate was driven by the choice of restricting the field of study to dental implants.

After many preliminary tests, the processing setup was optimized. The distance between the substrate and nozzle distance was fixed to be 350 mm. The diameter of nozzle of the airbrush was of 1 mm. The air inlet was set at three bars and a deposition time of 1 s was chosen. After deposition process, the plates were left to dry in air at room temperature. Then, the coated samples were heat-treated in static air at 950 °C for 1 h (using 5 °C/min as heating rate).

2.2. Surface and Coating Characterization

The morphology of the surface of both uncoated (cpTi) and sphene-coated (Sphene) cpTi plates was investigated by Field Emission Gun Scanning Electron Microscopy (FEG-SEM, Quanta 250 Fei, Eindhoven, The Netherlands).

X-ray diffraction (XRD) patterns of the coatings, after thermal treatment, were collected with an X-ray diffractometer (XRD Bruker D8 Advance, Milano, Italy) operated with Cu-Kα radiation 40 mA and 40 mV. The Rietveld refinement was subsequently applied to quantitatively analyze the XRD data using MAUD (Materials Analysis Using Diffraction) software and COD (Crystallography Open Database).

Surface topography was investigated using a stylus profilometer (Form Talysurf i-Series, Taylor Hobson Ltd., Leicester, UK), a high-resolution system for both profile- and areal-topography measurements. For 2D profile measurements (R_a and R_z), two plates of each group (cpTi and Sphene) were analyzed and a total of 3 scans (evaluation length equal to 5 mm) were extracted from the surface of each sample. For 3D areal measurements, at least 4 squared areas of 0.5 mm × 0.5 mm, extracted from the same surface, in at least two samples per group (cpTi and Sphene) were examined by 3D scanning. Areal parameter S_a and S_z were selected to describe the surface topography. Profile and areal data were filtered to eliminate waviness as well as form components of surface topography by applying a Gaussian filter with a sampling length equal to 0.8 mm, in accordance with ISO 4288 and ISO 25178. Data evaluation was performed with Talymap analysis software (Taylor Hobson Ltd., Leicester, UK).

2.3. Chemical Stability

The chemical stability of the sphene-coated cpTi substrates was evaluated in Tris-HCl buffer solution. The coated samples were immersed in the buffer solution of tris-(hydroxymethyl)-aminomethane (Tris, $(CH_2OH)_3CNH_2$) and hydrochloric acid (HCl) with a pH value of 7.4 and were examined at 1, 3 and 7 days. The ratio of plate surface area to Tris-HCl solution volume was 0.1 cm^2/mL.

Before and after soakings, dried coated samples were analyzed by means of a stylus profilometer (Form Talysurf i-Series) to evaluate changes in surface roughness. The roughness values of 3 different tracks (5 mm evaluation length) and 4 square areas (0.5 mm × 0.5 mm) per sample were recorded. The following profile roughness parameters, R_a and R_z, and areal parameters, S_a and S_z, were selected.

The variation of Ca, Si and Ti concentrations in the Tris-HCl solution was measured by Inductively Coupled Plasma Optical Emission Spectrometry (ICP-OES; iCAP 7400 ICP-OES, Thermo Fisher Scientific, Cambridge, UK) and confirmed by Inductively Coupled Plasma Mass Spectrometry (ICP-MS, Agilent Technologies 7700 × ICP-MS system, Agilent Technologies International Japan, Ltd., Tokyo, Japan). The ICP-MS was tuned daily using a tuning solution containing 10 μg L^{-1} ^{140}Ce, ^{7}Li, ^{205}Tl, and ^{89}Y (Agilent Technologies, UK). A 100 μg L^{-1} solution of ^{45}Sc and ^{115}In (Aristar, BDH, UK) prepared in HNO_3 1.4% was used as internal standard through addition to the sample solution via a T-junction.

All calibrations of Ca, Si and Ti were obtained in Tris-HCl solution using the multi-standard CCS-5 (Inorganic-Ventures) for Si and Ti (100 mg L^{-1}), and multi-standard IMS-120 (Ultra Scientific Multistandard) for Ca (1000 mg L^{-1}). All regressions were linear with a determination coefficient (R^2) larger than 0.999. Five samples for each time point were taken and measured five times. Data were given as mean ± standard deviation. The dissolution kinetics of Ca, Si and Ti were obtained from the released concentrations.

2.4. Human ADSCs Isolation and Cell Culture

Human ADSCs (hADSCs) were isolated from the adipose tissue of healthy patients undergoing cosmetic surgery procedures. Tissue collection protocol received a favorable ethical opinion by the Local Bioethical Committee, Padova University and all participants provided written consent. In addition, all experiments performed with human-derived materials were conducted in accordance to the relevant guidelines and regulations. The tissues were digested, and the cells isolated and expanded as described in Gardin et al. [28].

2.5. Seeding of hADSCs

hADSCs were seeded on both uncoated and sphene-coated samples, in a 12-well plate, with a density of 2 × 10^4 cells/sample. The cells were cultured in DMEM (Dulbecco's Modified Eagle's Medium) High Glucose or Osteogenic Differentiation Medium, at 37 °C and 5% CO_2, for 21 days. Osteogenic Differentiation Medium was made of DMEM High Glucose supplemented with 10 nM dexamethasone, 10 mM b-glycerophosphate, and 10 ng mL^{-1} of basic fibroblast growth factor. Both media were completed with 10% fetal bovine serum and 1% penicillin/streptomycin. After 21 days of culture, Methyl Thiazolyl-Tetrazolium (MTT) test, Scanning Electron Microscopy (SEM) analysis, immunofluorescent staining with phalloidin, Alizarin Red S staining and quantification, and real-time PCR were carried out.

2.6. MTT Assay

To assess the proliferation rate of hADSCs grown on both uncoated and coated samples, a colorimetric mitochondrial viability assay was performed as described by Denizot and Lang [29] with minor modifications. After removing the culture medium, the samples were incubated in 1 mL of 0.5 mg/mL MTT solution in phosphate buffered saline (PBS) for 3 h at 37 °C. The MTT solution was then removed, and each sample was extracted with 0.5 mL of 10% dimethyl sulfoxide in isopropanol for 30 min at 37 °C. For each sample, Optical Density (O.D.) values, at 570 nm, were recorded in duplicate on 200 μL aliquots using a multilabel plate reader (Victor 3, Perkin Elmer, Milan, Italy).

2.7. SEM Analysis

For SEM imaging, hADSCs grown on both uncoated and sphene-coated samples for 21 days were fixed in 2.5% glutaraldehyde in 0.1 M cacodylate buffer for 1 h, then progressively dehydrated in ethanol. Cell spreading and morphology were evaluated using FEG-SEM (Quanta 250 Fei, Eindhoven, The Netherlands).

2.8. Immunofluorescence

Cell adhesion to the scaffolds surface was evaluated by immunofluorescent staining with phalloidin. Briefly, cells were fixed in 4% paraformaldehyde in PBS for 10 min, then permeabilized with 0.1% triton X-100 in PBS for 30 min at room temperature. 5 mg/mL phalloidin were then used for fluorescent staining of actin filaments, whereas nuclear staining was performed with 2 μg mL^{-1} Hoechst H33342 solution for 5 min. Image acquisition was obtained with an inverted optical microscope DMI4000 B (Leica Microsystems, Wetzlar, Germany).

2.9. Alizarin Red S Staining and Quantification

The formation of extracellular mineral deposits onto uncoated and sphene-coated samples was detected by Alizarin Red S staining and quantification. Cells were fixed in 4% paraformaldehyde in PBS for 10 min at room temperature. Then, cells were stained adding 0.5 mL of 40 mM freshly prepared Alizarin Red S solution (pH 4.2) for 20 min at room temperature. After washing with double-distilled water, the Alizarin Red S stained areas were extracted with 0.5 mL of 10% cetylpyridinium chloride solution for 1 h at room temperature under gentle agitation. For each sample, O.D. values at 570 nm were recorded in duplicate on 200 μL aliquots using a Victor 3 plate reader.

2.10. Real-Time PCR

Total RNA was extracted from hADSCs, cultured on both uncoated and sphene-coated samples for 21 days, with a Total RNA Purification Plus Kit (Norgen Biotek Corporation, Thorold, ON, Canada). For the first-strand cDNA synthesis, 1 μg of total RNA of each sample was reverse transcribed with a SensiFAST™ cDNA Synthesis Kit (Bioline, London, UK), following the manufacturer's protocol. Human primers were selected for each target gene using a Primer 3 software. Real-time PCRs were run using the chosen primers at a concentration of 400 nM and SensiFAST™ SYBR No-ROX Kit (Bioline) on a Rotor-Gene 3000 (Corbett Research, Sydney, Australia). The thermal cycling conditions were as follow: 2 min denaturation at 95 °C; 40 cycles of 5 s denaturation at 95 °C; annealing for 10 s at 60 °C; and 20 s elongation at 72 °C. Differences in gene expression were calculated by normalizing to the expression of the Transferrin Receptor (TFRC) housekeeping gene.

2.11. Statistical Analysis

One-way analysis of variance (ANOVA) was used for data analysis. *t*-tests were used to determine data significance ($p < 0.05$). All tests were performed using SPSS 16.0 software (SPSS Inc., Chicago, IL, USA) (licensed to the University of Padua, Padova, Italy).

3. Results

3.1. Surface Characterization

FEG-SEM images of uncoated cpTi and sphene-coated surfaces are shown in Figure 1a,b, respectively. Sphene coating is characterized by a homogenous grey substrate, composed of tetragonal crystals <1 μm, and "white islands" growing on it. These white agglomerates appear to be composed of vertically aggregated spherical particles. As reported in previous work [27], EDS (Energy Dispersive X-ray Spectrometry) analysis identified Ti, O, Si and Ca in areas corresponding to these white spherical structures. Instead, in the grey phase Ti and O were mainly present.

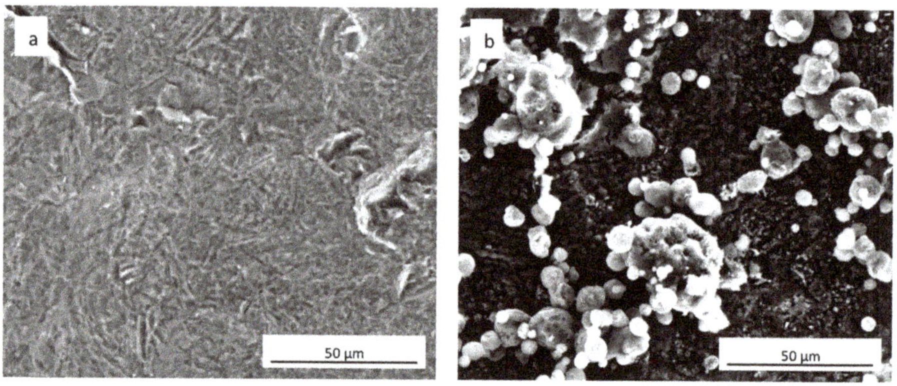

Figure 1. Surface morphology (SEM-FEG) of (**a**) as-received uncoated cpTi and (**b**) sphene-coated cpTi.

The coating composition was further investigated by XRD analysis and Rietveld refinement (Figure 2) that showed the presence of rutile (TiO_2, 60 wt.%) followed by sphene ($CaTiSiO_5$, 31 wt.%) and perovskite ($CaTiO_3$, 9 wt.%). The as-received TiO_2 powders, used as a filler to develop the sphene ceramic, are characterized by a weight percentage ratio of 80 to 20 between anatase and rutile, as claimed in the manufacturer's datasheet. Anatase and rutile are the two main polymorphs of titanium oxide. They both exhibit tetragonal crystalline structure but obey different space groups. Anatase is a metastable phase, and the transformation to the stable rutile structure occurs as irreversible phase transformation in the range between 600 °C and 700 °C [30]. The treatment at 950 °C, performed in air, may have induced the nucleation and growth of rutile crystals before sphene synthesis, to yield a final weight percentage ratio of 60 wt. % rutile. The higher stability of rutile compared to anatase may have inhibited the reaction to produce sphene at this temperature. It has indeed been reported in the literature that the formation of sphene by the sol-gel method is complete at 1300 °C [31]. As previously reported [24], the choice of keeping the temperature at 950 °C was driven by the need for preserving the structure of the cpTi plate without affecting the bonding of the coating to the substrate. Further studies are ongoing aimed at increasing the amount of sphene produced by the preceramic polymer precursor route.

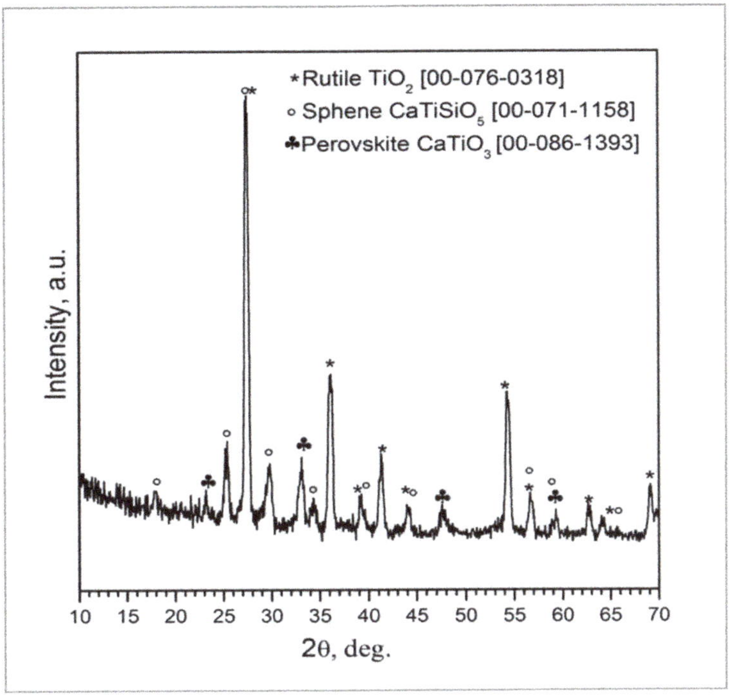

Figure 2. XRD pattern of sphene-coated cpTi after heat treatment at 950 °C in air for 1 h.

Surface roughness was investigated by means of a profilometer. 2D profile measurements (R_a and R_z) and 3D areal measurements (S_a and S_z) of uncoated (cpTi) and sphene-coated (Sphene) cpTi substrates are reported in Table 1 and Figure 3.

A clear difference in the values of R_a and R_z between the cpTi substrate and the sphene-coated substrate can be observed in the 2D measurements. However, this difference is smoothed in the 3D areal measurements. Indeed, the surface of the cpTi substrate is characterized by a flatter morphology (Figure 3a) than the sphene-coated sample (Figure 3b). The sphene coating presents a high number of peaks homogeneously distributed, while the cpTi substrate presents a limited number of peaks but characterized by a wider area. The yellow and red peaks in Figure 3b well correspond to the edges of the white agglomerates detected by SEM observations (Figure 1b).

3.2. Chemical Stability

The 2D roughness data after Tris-HCl immersion test show a slight increase after 1 day of soaking time. However, this increase in roughness values shows an almost constant trend after 3 and 7 days. In accordance to these data, 3D maps of sphene after immersion in Tris-HCl solution (Figure 3c,d) demonstrated a reduction in the numbers of peaks after the dissolution test, with a similar pattern of peak and valley distributions at 1, 3 and 7 days after soaking. This is confirmed by S_a and S_z values that were reported in Table 1.

Table 1. Linear and surface roughness of as-received cpTi, sphene-coated cpTi, and sphene-coated cpTi after 1 and 7 days of chemical etching in Tris-HCl. Values are expressed as mean (standard deviation).

Samples	Ra (μm)	Rz (μm)	Sa (μm)	Sz (μm)
cpTi	2.82 (0.30)	14.30 (1.38)	3.51 (0.65)	25.15 (5.28)
Sphene	3.94 (0.75)	23.70 (3.74)	3.64 (0.47)	28.09 (3.58)
Sphene Tris-HCl, 1 day	4.66 (0.41)	27.80 (2.05)	4.39 (1.07)	37.28 (6.49)
Sphene Tris-HCl, 3 days	3.66 (0.52)	22.29 (2.85)	4.69 (0.37)	37.70 (6.57)
Sphene Tris-HCl, 7 days	4.48 (0.37)	27.74 (2.81)	3.99 (0.45)	34.58 (4.96)

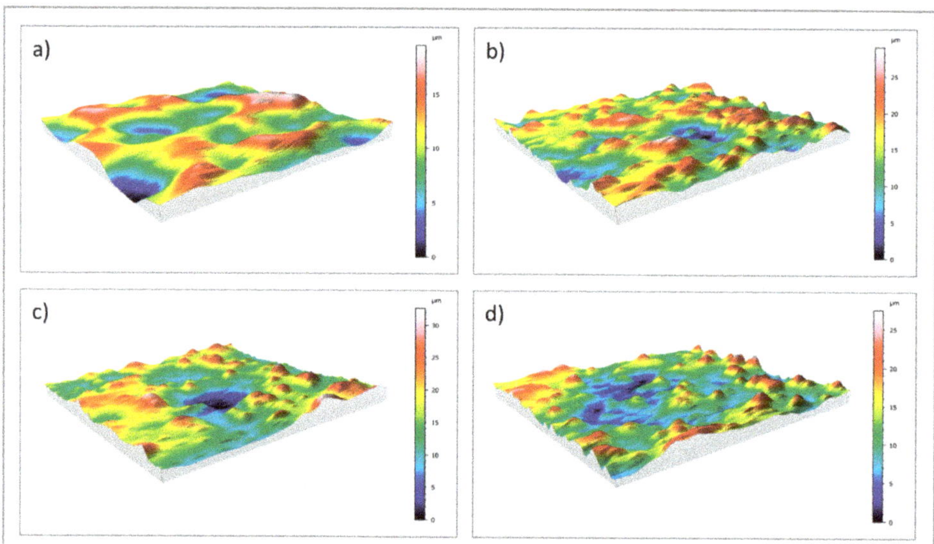

Figure 3. 3D maps of: (**a**) uncoated cpTi; (**b**) sphene-coated cpTi; (**c**) sphene-coated cpTi after 1 day in Tris-HCl buffer solution; (**d**) sphene-coated cpTi after 7 days in Tris-HCl buffer solution.

Figure 4 shows ICP analysis of Ca, Si and Ti ion release of the sphene coating in Tris-HCl solution at different time points. The average concentrations of Ca and Si ions, after 1 day of soaking, are 0.069 mM and 0.045 mM, respectively. Similar concentration levels were found after 3 and 7 days of immersion in Tris-HCl solution. Only trace amount of Ti ions released from the coating was detected (0.001–0.002 mM).

These findings are consistent with the above mentioned areal-topography measurements and suggest that the dissolution occurs only in the first hours after soaking.

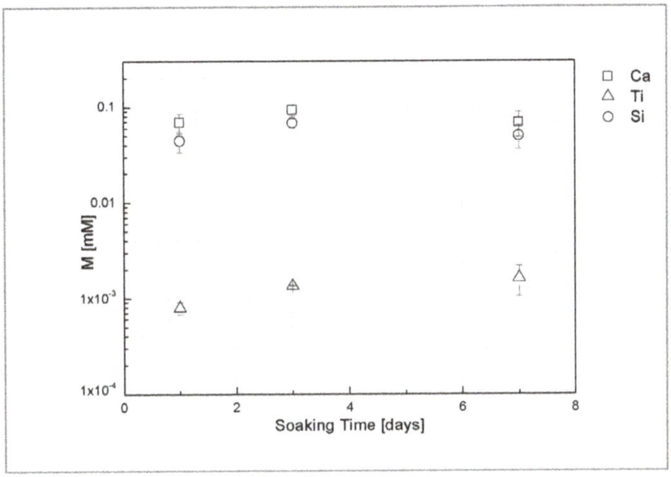

Figure 4. Ca, Ti and Si ion release of sphene-coated samples, after 1, 3, and 7 days of soaking time.

3.3. Cell Proliferation

The MTT assay proved that hADSCs were able to proliferate on both sphene-coated and uncoated samples (Figure 5). The O.D. values recorded for cells loaded on coated samples were found to be significantly ($p < 0.05$) higher to those observed for the controls in both the experimental conditions. Moreover, from MTT results, it was shown that hADSCs' proliferation was higher when hADSCs were cultured in Osteogenic Differentiation Medium compared to in DMEM High Glucose, in particular for the sphene-coated samples.

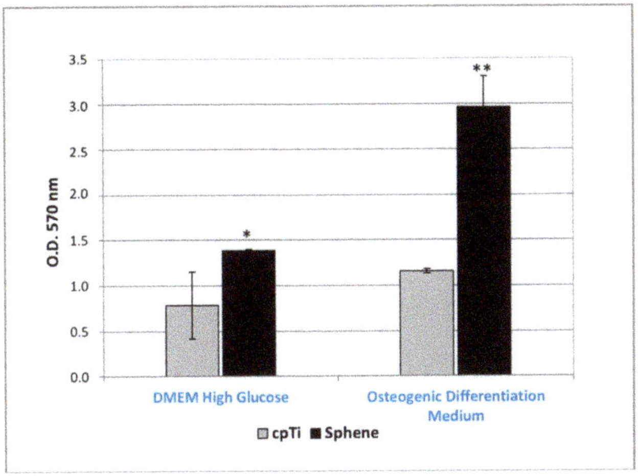

Figure 5. MTT assay of hADSCs cultured for 21 days on uncoated cpTi samples (*cpTi*) and on sphene-coated (*Sphene*) samples in DMEM High Glucose or Osteogenic Differentiation Medium. * $p < 0.05$, ** $p < 0.01$, *** $p < 0.001$.

3.4. Cell Adhesion and Morphology

The SEM analyses showed how hADSCs anchored to the surface of the specimens (Figure 6a–d). The cells were extremely flat with the typical star morphology associated with the osteoblastic-like phenotype and their distribution was similar, independent of the culture medium used (Figure 6a,c). After 21 days of culture onto the sphene-coated surfaces, cells showed short and thin filopodia when grown in DMEM High Glucose (Figure 6b); whereas they appeared flat and overlapped when cultured in the presence of the Osteogenic Differentiation Medium (Figure 6d). In both cases, the sphene coating remained intact during the culturing time, thus demonstrating its chemical stability.

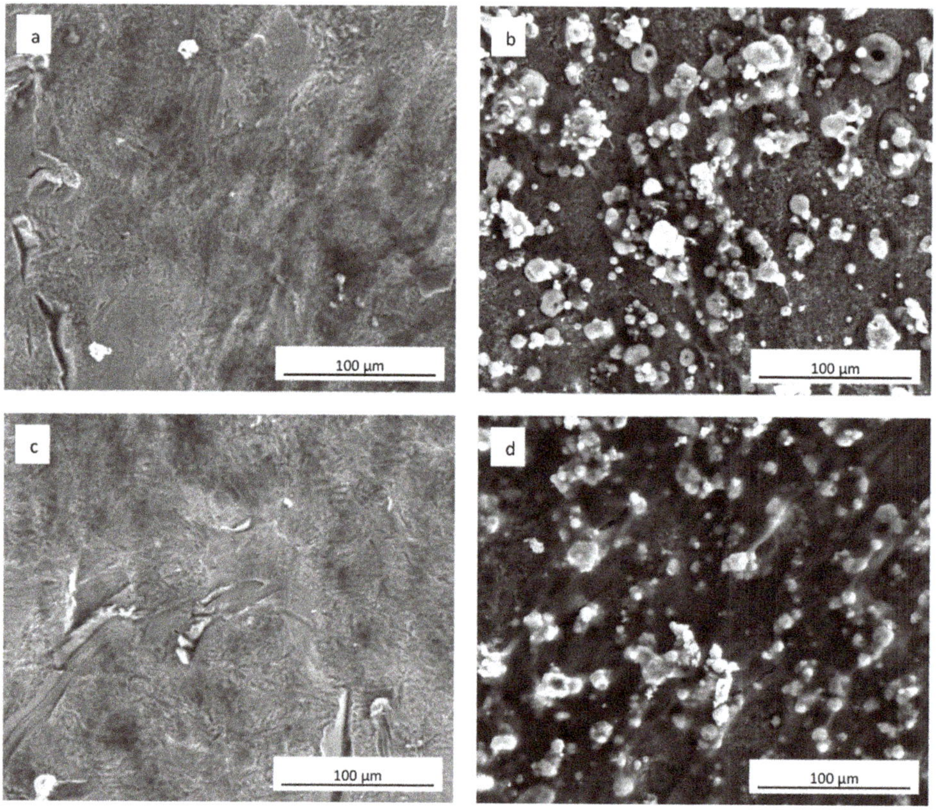

Figure 6. SEM images (1000× magnification) of hADSCs grown for 21 days on (**a**) uncoated cpTi in DMEM High Glucose; (**b**) sphene-coated cpTi in DMEM High Glucose; (**c**) uncoated cpTi in Osteogenic Differentiation Medium; (**d**) sphene-coated cpTi in Osteogenic Differentiation Medium.

3.5. Cytoskeletal Organization

Staining with fluorescent phalloidin showed that cells were able to attach and spread on both the cpTi (Figure 7a,c) and sphene-coated samples (Figure 7b,d). After 21 days of culture, hADSCs had completely colonized the surfaces and formed a continuous layer with no significant differences between the experimental conditions. In vivo, such as in periosteum, osteoblasts are aligned with the collagen they produce and also on system cells are aligned to the collagen substrate. Cytoskeletal stresses and tension increase with increasing ECM (ExtraCellular Matrix) stiffness. The cytoskeleton provides a structural frame for the cell through Focal adhesions. Focal adhesion forms when adapter proteins link the cytoskeleton to integrins, which permits cells to adhere to the substrate. A variety of signaling proteins are also associated with focal adhesions, including focal adhesion kinase (FAK), an important mediator of signaling at these centers. Forces are also transmitted to the substrate at these sites. Osteogenic differentiation can also be affected by the substrate properties where cells are seeded. On our samples, we can confirm that the shape of a cell follows the steps that occur in vivo.

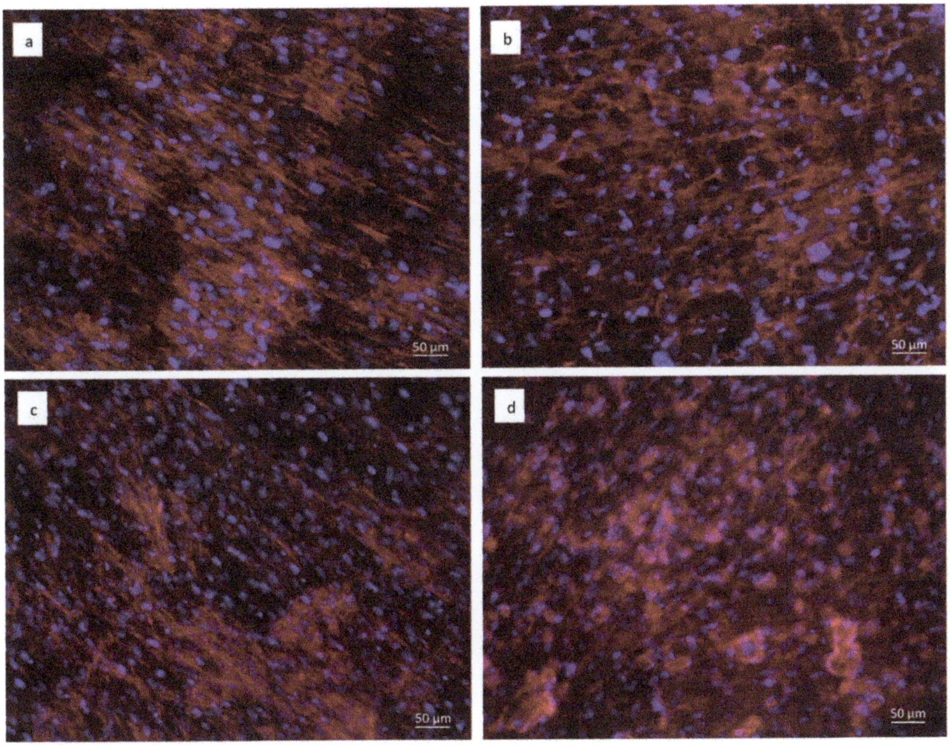

Figure 7. Immunofluorescent staining of the actin filaments with phalloidin (in red). Cell nuclei are counterstained with Hoechst (in blue). hADSCs seeded for 21 days on: (**a**) uncoated cpTi in DMEM High Glucose; (**b**) sphene-coated cpTi in DMEM High Glucose; (**c**) uncoated cpTi in Osteogenic Differentiation Medium; (**d**) sphene-coated cpTi in Osteogenic Differentiation Medium.

3.6. In Vitro hADSC Osteogenic Differentiation

The in vitro osteogenic differentiation of hADSCs was first evaluated by marking of extracellular calcium deposition with Alizarin Red S staining and quantification at 21 days of culture (Figure 8). A difference in the presence of calcium deposits on the sphene-coated surfaces and the cpTi controls was evidenced for hADSCs cultured in the Osteogenic Differentiation Medium. Calcium deposits were significantly ($p < 0.05$) higher in the first experimental condition. The accumulation of calcium deposits by hADSCs was not observed when cells were seeded on the sphene-coated or uncoated samples in DMEM High Glucose.

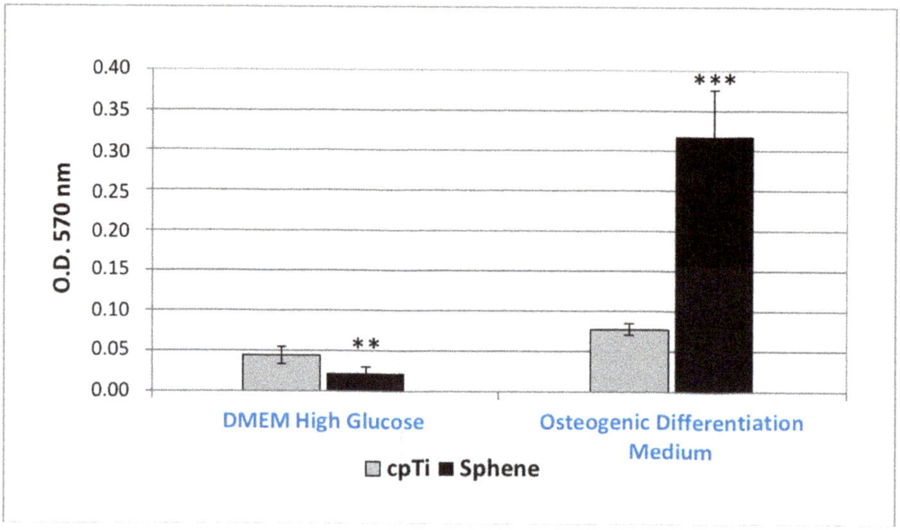

Figure 8. Alizarin Red S quantification of calcium deposits produced by hADSCs seeded onto *cpTi* or *Sphene* surfaces in DMEM High Glucose or Osteogenic Differentiation Medium for 21 days. * $p < 0.05$, ** $p < 0.01$, *** $p < 0.001$.

3.7. Real-Time PCR

Real-time PCR was used to quantify the relative expression level of several osteogenic differentiation markers. In particular, expression of runt-related transcription factor 2 (RUNX2), osterix (OSX), collagen type I (COL1A1), osteocalcin (OC), osteonectin (ON), osteopontin (OPN), alkaline phosphatase (ALPL), and receptor activator of nuclear factor kappa-B ligand (RANKL) has been investigated. Runx2 and Osterix are the main transcription factors involved on osteoblastic commitment. Collagen is the principal component of the bone matrix. Alkaline Phosphatase is involved on the calcification of the Extracellular matrix.

Osteonectin is a protein which role is related to the connection between extracellular matrix component and osteoblast. It is mostly produced during the first phases of bone regeneration or during bone remodeling [32–34].

As evident in Figure 9a,c, the relative expression of RUNX2, OSX, OC, OPN, and RANKL did not change significantly between the uncoated and sphene-coated surfaces, both when cells were grown in DMEM High Glucose or in Osteogenic Differentiation Medium. Similarly, there was no noticeable difference in ALPL (Figure 9a), COL1A1 and ON (Figure 9b) mRNA expression between cpTi and sphene-coated samples in DMEM High Glucose; instead, a slight down-regulation of ALPL (Figure 9c) and COL1A1 (Figure 9d) expression was identified in cells seeded onto the sphene surfaces in the presence of Osteogenic Differentiation Medium.

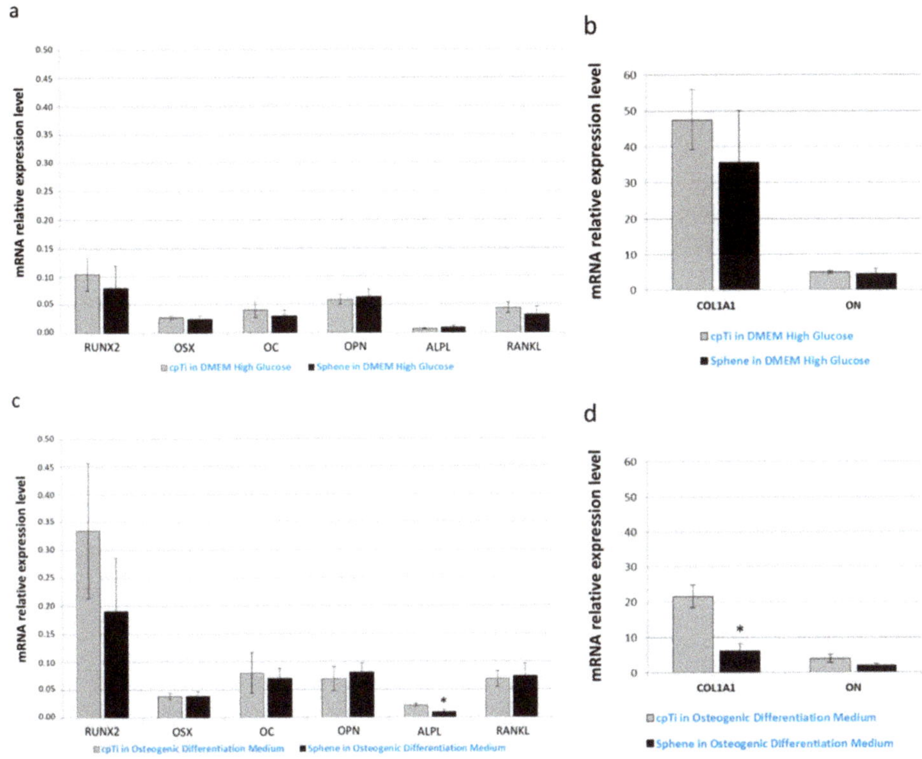

Figure 9. Real-time PCR analysis of the osteogenic differentiation markers runt-related transcription factor 2 (RUNX2), osterix (OSX), collagen type I (COL1A1), osteocalcin (OC), osteonectin (ON), osteopontin (OPN), alkaline phosphatase (ALPL), and receptor activator of nuclear factor kappa-B ligand (RANKL). Data are normalized to the expression of the transferrin receptor (TFRC) internal reference. Gray bars indicate the relative expression level of the selected genes in the hADSCs seeded onto *cpTi* samples in DMEM High Glucose (**a**,**b**) or Osteogenic Differentiation Medium (**c**,**d**) for 21 days. Black bars represent the gene expression level of the same markers in the hADSCs seeded onto *Sphene* samples in DMEM High Glucose (**a**,**b**) or Osteogenic Differentiation Medium (**c**,**d**) for 21 days.

4. Discussion

In the present study, the sphene-based coatings on cpTi substrates showed a rough morphology, with roughness Ra in the range of 3.25 to 4.75 μm. The morphology was characterized by small crystals made of TiO_2 and $CaTiO_3$ (valley of Figure 4) and islands of unregular shape that were detected by EDS analysis to be composed of sphene [27]. The quantitative XRD analysis performed on the coated substrate showed the reaction efficiency and the amount of sphene that was about 31 wt. %.

Micro-roughened surfaces were demonstrated to have a positive effect on early blood cell/implant interactions [35] and osteoblast proliferation [36]. In addition, surface topography was found to influence osteoconduction in both animal studies [37,38] and human retrieval study [39], with good results shown by rough implant surfaces. Sphene coatings examined in the present study possessed a moderately rough surface. The surface roughness quantified by 3D areal measurements showed a good correspondence with 2D measurements in both cpTi and coated cpTi.

Even though the release of Ca and Si ions has been demonstrated to favor osteoblastic proliferation and differentiation [18,19,40], one of the major drawbacks of biosilicate ceramic coatings, such as $CaSiO_3$ coatings, consists of their poor chemical stability and high degradation rate that might compromise their long-term stability [41,42]. Therefore, a limited release of these ions would enhance coating bioactivity and, on the other hand, would not interfere with the integrity of the coating itself [43].

To assess the chemical stability of the coatings, Tris-HCl was chosen due to the absence of Ca, Si or Ti ions in its composition. The results of dissolution tests in Tris-HCl showed that the dissolution happened only during the first hours of soaking and then remained constant after 1, 3 and 7 days. This behavior agrees with the results reported by Wu et al. [22], thus proving that the developed coating is comparable in terms of chemical properties to standard HA. At the same experimental conditions of the present study, a lower released Ca concentration (2.74 ppm) after 7 days of soaking in Tris-HCl was found in the sphene-coated samples here investigated as compared to the values of around 20 ppm and >110 ppm of plasma-sprayed sphene ceramic coatings and HA coatings, respectively [14]. Moreover, as found by Wang et al. [43], a higher concentration of Ca ions released to the solution was measured compared with the concentration of Si ions released after 7 days of immersion. In addition, only an irrelevant amount of Ti ions was detected by Wang et al. [43] in the Tris-HCl after one week of soaking.

The roughness of samples, after dissolution tests slightly increased compared to the starting samples. This is mainly observable in the areal measurements where a sharp increase of Sz was measured from unetched sphene to 1-day soaking time in Tris-HCl. In agreement with ICP analysis, no significant trends on roughness at increasing soaking time can be observed. Thus, this proves that little dissolution of Ca, Ti and Si, happened during the first 24 h.

The chemical composition of implant surfaces profoundly influences cell adhesion, spreading, proliferation, and differentiation [44,45]. Sphene-based substrates better supported hADSCs attachment and proliferation, compared with uncoated cpTi samples, as revealed by the MTT assay. By SEM analysis, a uniform layer of well-spread cells could be observed when cells were cultured onto the sphene-coated surfaces in Osteogenic Differentiation Medium. The results of Alizarin Red S quantification suggested that sphene coating favorably affects the accumulation of mineralized calcium phosphate when cells are cultured in the presence of osteogenic factors. Moreover, cells seeded onto sphene-coated surfaces expressed similar levels of osteoblasts-related genes to those expressed by cells cultured on uncoated cpTi in both culture conditions. We can hypothesize that the combination of the coating and the Osteogenic Differentiation Medium promoted a higher calcium deposition on sphene-coated surfaces than on uncoated cpTi substrates. These findings agree with Wang et al. [43], who demonstrated that sphene glass-ceramic coatings were able to support human osteoblast-like cells (HOBs) attachment, proliferation and differentiation and enhance the expression level of bone-related genes. Furthermore, it was demonstrated that sphene ceramic coatings significantly enhanced HOBs proliferation and ALP activity compared with plasma-sprayed HA coating and uncoated Ti-6Al-4V substrates [14]. Pure dense sphene ceramic disks were found to promote human bone-derived cells attachment and to significantly improve their proliferation and differentiation, as compared with α-$CaSiO_3$ ceramic disks [25]. Pure sphene ceramic disks were not only found to modulate HOBs, but also osteoclasts and endothelial cells [26]. As compared with the study of Ramaswamy et al. [26], in the current work, mesenchymal stem cells (MSCs) were preferred to human osteoblast-like cells, human osteoclasts and human microvascular endothelial to in vitro investigate the osteoconductive properties of the biomaterial, since stem cells are the main actors enrolled during the early stages of tissue regeneration. MSCs are stem cells involved on tissue regeneration, which can migrate to the damage site thanks to their receptor and to coordinate all the phase of tissue regeneration [46]. Hypoxia, which usually occurs at the injured sites due to a disruption in blood supply, has been reported to largely contribute to MSC mobilization and homing. The hypoxia-inducible factors (HIFs) are key regulators of the cellular response to decrease in the local oxygen level, also regulating the expression of genes involved in MSC recruitment and migration to the damaged tissues [32–34,46–48].

In a good agreement with previously published work [14,25,26,43], our results suggest that the ionic environment, determined by the dissolution of Ca and Si ions from the sphene biomaterial, may contribute to stimulate cell proliferation and differentiation. In addition, the in vivo assessment revealed enhanced osseointegration of both sphene-coated and HA-coated Ti-6Al-4V implants, compared to uncoated implants, confirming the in vitro observations and the potential use of sphene coatings in clinical application [26].

5. Conclusions

In conclusion, sphene-based crack-free coatings were successfully produced starting from a preceramic polymer and nano-sized fillers and deposited using an automatic airbrush. The coatings presented good chemical stability, as a minimal dissolution occurred only during the first hours of soaking in Tris-HCl solution and then remained stable over the period considered. The coatings also supported hADSCs attachment, proliferation, and differentiation in vitro. Taken together, our findings indicate a potential for use of sphene ceramics as coating materials for orthopedic and dental implants. In vivo studies should be performed to confirm in vitro observations. Further studies are ongoing to increase the reaction efficiency.

Author Contributions: Conceptualization, H.E., B.Z. and L.B.; Data curation, H.E., G.B., C.G., L.F., D.B. and L.B.; Formal analysis, H.E., G.B., C.G. and L.F.; Supervision, S.S., B.Z. and L.B.; Validation, L.B.; Writing—original draft, H.E. and G.B.; Writing—review and editing, H.E., G.B., C.G., L.F., D.B., P.P., S.S., B.Z. and L.B.

Funding: This research received no external funding.

Acknowledgments: The authors would like to thank Prof. Roberto Meneghello of Department of Management and Engineering for his contribution in surface characterization.

Conflicts of Interest: The authors declare no competing interests.

References

1. Kirmanidou, Y.; Sidira, M.; Drosou, M.E.; Bennani, V.; Bakopoulou, A.; Tsouknidas, A.; Michailidis, N.; Michalakis, K. New Ti-Alloys and Surface Modifications to Improve the Mechanical Properties and the Biological Response to Orthopedic and Dental Implants: A Review. *Biomed. Res. Int.* **2016**, *2016*, 2908570. [CrossRef] [PubMed]
2. Quirynen, M.; Al-Nawas, B.; Meijer, H.J.; Razavi, A.; Reichert, T.E.; Schimmel, M.; Storelli, S.; Romeo, E. ; Roxolid Study Group. Small-diameter titanium Grade IV and titanium-zirconium implants in edentulous mandibles: Three-year results from a double-blind, randomized controlled trial. *Clin. Oral Implants Res.* **2015**, *26*, 831–840. [CrossRef] [PubMed]
3. Elias, C.N.; Fernandes, D.J.; Resende, C.R.; Roestel, J. Mechanical properties, surface morphology and stability of a modified commercially pure high strength titanium alloy for dental implants. *Dent. Mater.* **2015**, *31*, e1–e13. [CrossRef] [PubMed]
4. Niinomi, M. Mechanical biocompatibilities of titanium alloys for biomedical applications. *J. Mech. Behav. Biomed. Mater.* **2008**, *1*, 30–42. [CrossRef] [PubMed]
5. Zreiqat, H.; Valenzuela, S.M.; Nissan, B.B.; Roest, R.; Knabe, C.; Radlanski, R.J.; Renz, H.; Evans, P.J. The effect of surface chemistry modification of titanium alloy on signalling pathways in human osteoblasts. *Biomaterials* **2005**, *26*, 7579–7586. [CrossRef] [PubMed]
6. Liu, X.; Chu, P.K.; Ding, C. Surface modification of titanium, titanium alloys, and related materials for biomedical applications. *Mater. Sci. Eng. R Rep.* **2004**, *47*, 49–121. [CrossRef]
7. Wennerberg, A.; Bougas, K.; Jimbo, R.; Albrektsson, T. Implant coatings: New modalities for increased osseointegration. *Am. J. Dent.* **2013**, *26*, 105–112. [PubMed]
8. Narayanan, R.; Seshadri, S.K.; Kwon, T.Y.; Kim, K.H. Calcium phosphate-based coatings on titanium and its alloys. *J. Biomed. Mater. Res. B Appl. Biomater.* **2008**, *85*, 279–299. [CrossRef] [PubMed]
9. Lin, D.Y.; Wang, X.X. A novel method to synthesize hydroxyapatite coating with hierarchical structure. *Colloids Surf. B Biointerfaces* **2011**, *82*, 637–640. [CrossRef] [PubMed]

10. Liu, X.; Tao, S.; Ding, C. Bioactivity of plasma sprayed dicalcium silicate coatings. *Biomaterials* **2002**, *3*, 963–968. [CrossRef]
11. Mohammadi, H.; Sepantafar, M. Ion-Doped Silicate Bioceramic Coating of Ti-Based Implant. *Iran. Biomed. J.* **2016**, *20*, 189–200. [CrossRef] [PubMed]
12. Xuereb, M.; Camilleri, J.; Attard, N.J. Systematic review of current dental implant coating materials and novel coating techniques. *Int. J. Prosthodont.* **2015**, *28*, 51–59. [CrossRef] [PubMed]
13. Xue, W.; Liu, X.; Zheng, X.; Ding, C. Plasma-sprayed diopside coatings for biomedical applications. *Surf. Coat. Technol.* **2004**, *185*, 340–345. [CrossRef]
14. Wu, C.; Ramaswamy, Y.; Liu, X.; Wang, G.; Zreiqat, H. Plasma-sprayed $CaTiSiO_5$ ceramic coating on Ti-6Al-4V with excellent bonding strength, stability and cellular bioactivity. *J. R. Soc. Interface* **2009**, *6*, 159–168. [CrossRef] [PubMed]
15. Gomez-Vega, J.M.; Saiz, E.; Tomsia, A.P.; Marshall, G.W.; Marshall, S.J. Bioactive glass coatings with hydroxyapatite and Bioglass particles on Ti-based implants. 1. Processing. *Biomaterials* **2000**, *21*, 105–111. [CrossRef]
16. Bloyer, D.R.; Gomez-Vega, J.M.; Saiz, E.; McNaney, J.M.; Cannon, R.M.; Tomsia, A.P. Fabrication and characterization of a bioactive glass coating on titanium implant alloys. *Acta Mater.* **1999**, *47*, 4221–4224. [CrossRef]
17. Silver, I.A.; Deas, J.; Erecinska, M. Interactions of bioactive glasses with osteoblasts in vitro: Effects of 45S5 Bioglassw, and 58S and 77S bioactive glasses on metabolism, intracellular ion concentrations and cell viability. *Biomaterials* **2001**, *22*, 175–185. [CrossRef]
18. Sun, J.; Wei, L.; Liu, X.; Li, J.; Li, B.; Wang, G.; Meng, F. Influences of ionic dissolution products of dicalcium silicate coating on osteoblastic proliferation, differentiation and gene expression. *Acta Biomater.* **2009**, *5*, 1284–1293. [CrossRef] [PubMed]
19. Valerio, P.; Pereira, M.M.; Goes, A.M.; Leite, M.F. The effect of ionic products from bioactive glass dissolution on osteoblast proliferation and collagen production. *Biomaterials* **2004**, *25*, 2941–2948. [CrossRef] [PubMed]
20. Xynos, I.D.; Edgar, A.J.; Buttery, L.D.K.; Hench, L.L.; Polak, J.M. Gene-expression profiling of human osteoblasts following treatment with the ionic products of Bioglassw 45S5 dissolution. *J. Biomed. Mater. Res.* **2001**, *55*, 151–157. [CrossRef]
21. Wu, C.; Chen, Z.; Yi, D.; Chang, J.; Xiao, Y. Multidirectional effects of Sr, Mg and Si-containing bioceramic coatings with high bonding strength on inflammation, osteoclastogenesis and osteogenesis. *ACS Appl. Mater. Interfaces* **2014**, *6*, 4264–4276. [CrossRef] [PubMed]
22. Wu, C.; Ramaswamy, Y.; Gale, D.; Yang, W.; Xiao, K.; Zhang, L.; Yin, Y.; Zreiqat, H. Novel sphene coatings on Ti-6Al-4V for orthopedic implants using sol-gel method. *Acta Biomater.* **2008**, *4*, 569–576. [CrossRef] [PubMed]
23. Cheng, S.; Wei, D.; Zhou, Y. Formation and structure of sphene/titania composite coatings on titanium formed by a hybrid technique of microarc oxidation and heat-treatment. *Appl. Surf. Sci.* **2011**, *257*, 3404–3411. [CrossRef]
24. Biasetto, L.; Elsayed, H.; Bonollo, F.; Colombo, P. Polymer-derived sphene biocoating on cpTi substrates for orthopedic and dental implants. *Surf. Coat. Technol.* **2016**, *301*, 140–147. [CrossRef]
25. Wu, C.; Ramaswamy, Y.; Soeparto, A.; Zreiqat, H. Incorporation of titanium into calcium silicate improved their chemical stability and biological properties. *J. Biomed. Mater. Res. A* **2008**, *86*, 402–410. [CrossRef] [PubMed]
26. Ramaswamy, Y.; Wu, C.; Dunstan, C.R.; Hewson, B.; Eindorf, T.; Anderson, G.I.; Zreiqat, H. Sphene ceramics for orthopedic coating applications: An in vitro and in vivo study. *Acta Biomater.* **2009**, *5*, 3192–3204. [CrossRef] [PubMed]
27. Biasetto, L.; Elsayed, H. Sphene silicate ceramic coatings on cpTi substrates: Process upgrade. *Surf. Coat. Technol.* **2017**, *321*, 416–424. [CrossRef]
28. Gardin, C.; Bressan, E.; Ferroni, L.; Nalesso, E.; Vindigni, V.; Stellini, E.; Pinton, P.; Sivolella, S.; Zavan, B. In vitro concurrent endothelial and osteogenic commitment of adipose-derived stem cells and their genomical analyses through comparative genomic hybridization array: Novel strategies to increase the successful engraftment of tissue-engineered bone grafts. *Stem Cells Dev.* **2012**, *21*, 767–777. [CrossRef] [PubMed]

29. Denizot, F.; Lang, R. Rapid colorimetric assay for cell growth and survival. Modifications to the tetrazolium dye procedure giving improved sensitivity and reliability. *J. Immunol. Methods* **1986**, *89*, 271–277. [CrossRef]
30. Hanaor, D.A.H.; Sorrell, C.C. Review of the anatase to rutile phase transformation. *J. Mater. Sci.* **2011**, *46*, 855–874. [CrossRef]
31. Muthuraman, M.; Patil, K.C. Synthesis, properties, sintering and microstructure of sphene, $CaTiSiO_5$: A comparative study of coprecipitation, sol–gel and combustion processes. *Mater. Res. Bull.* **1998**, *33*, 655–661. [CrossRef]
32. Liu, H.; Li, D.; Zhang, Y.; Li, M. Inflammation, mesenchymal stem cells and bone regeneration. *Histochem. Cell Biol.* **2018**, *149*, 393–404. [CrossRef] [PubMed]
33. Currey, J.D.; Dean, M.N.; Shahar, R. Revisiting the links between bone remodelling and osteocytes: Insights from across phyla. *Biol. Rev. Camb. Philos. Soc.* **2017**, *92*, 1702–1719. [CrossRef] [PubMed]
34. Katsimbri, P. The biology of normal bone remodelling. *Eur. J. Cancer Care* **2017**, *26*, e12740. [CrossRef] [PubMed]
35. Park, J.Y.; Davies, J.E. Red blood cell and platelet interactions with titanium implant surfaces. *Clin. Oral Implants Res.* **2000**, *11*, 530–539. [CrossRef] [PubMed]
36. Iwata, N.; Nozaki, K.; Horiuchi, N.; Yamashita, K.; Tsutsumi, Y.; Miura, H.; Nagai, A. Effects of controlled micro-/nanosurfaces on osteoblast proliferation. *J. Biomed. Mater. Res. A* **2017**, *105*, 2589–2596. [CrossRef] [PubMed]
37. Franchi, M.; Orsini, E.; Trire, A.; Quaranta, M.; Martini, D.; Piccari, G.G.; Ruggeri, A.; Ottani, V. Osteogenesis and morphology of the peri-implant bone facing dental implants. *Sci. World J.* **2004**, *4*, 1083–1095. [CrossRef] [PubMed]
38. Franchi, M.; Bacchelli, B.; Giavaresi, G.; De Pasquale, V.; Martini, D.; Fini, M.; Giardino, R.; Ruggeri, A. Influence of different implant surfaces on peri-implant osteogenesis: Histomorphometric analysis in sheep. *J. Periodontol.* **2007**, *78*, 879–888. [CrossRef] [PubMed]
39. Lazzara, R.J.; Testori, T.; Trisi, P.M.; Porter, S.S.; Weinstein, R.L. A human histologic analysis of osseotite and machined surfaces using implants with 2 opposing surfaces. *Int. J. Periodontics Restorative Dent.* **1999**, *19*, 117–129. [CrossRef] [PubMed]
40. Fei, L.; Wang, C.; Xue, Y.; Lin, K.; Chang, J.; Sun, J. Osteogenic differentiation of osteoblasts induced by calcium silicate and calcium silicate/β-tricalcium phosphate composite bioceramics. *J. Biomed. Mater. Res. B Appl. Biomater.* **2012**, *100*, 1237–1244. [CrossRef] [PubMed]
41. Ni, S.; Chang, J. In vitro degradation, bioactivity, and cytocompatibility of calcium silicate, dimagnesium silicate, and tricalcium phosphate bioceramics. *J. Biomater. Appl.* **2009**, *24*, 139–158. [CrossRef] [PubMed]
42. Wu, C.; Ramaswamy, Y.; Kwik, D.; Zreiqat, H. The effect of strontium incorporation into CaSiO3 ceramics on their physical and biological properties. *Biomaterials* **2007**, *28*, 3171–3181. [CrossRef] [PubMed]
43. Wang, G.; Lu, Z.; Liu, X.; Zhou, X.; Ding, C.; Zreiqat, H. Nanostructured glass-ceramic coatings for orthopaedic applications. *J. R. Soc. Interface* **2011**, *8*, 1192–1203. [CrossRef] [PubMed]
44. Bezerra, F.; Ferreira, M.R.; Fontes, G.N.; da Costa Fernandes, C.J.; Andia, D.C.; Cruz, N.C.; da Silva, R.A.; Zambuzzi, W.F. Nano hydroxyapatite-blasted titanium surface affects pre-osteoblast morphology by modulating critical intracellular pathways. *Biotechnol. Bioeng.* **2017**, *114*, 1888–1898. [CrossRef] [PubMed]
45. Wei, J.; Heo, S.J.; Kim, D.H.; Kim, S.E.; Hyun, Y.T.; Shin, J.W. Comparison of physical, chemical and cellular responses to nano- and micro-sized calcium silicate/poly(3-caprolactone) bioactive composites. *J. R. Soc. Interface* **2008**, *5*, 617–630. [CrossRef] [PubMed]
46. Herrmann, M.; Verrier, S.; Alini, M. Strategies to stimulate mobilization and homing of endogenous stem and progenitor cells for bone tissue repair. *Front Bioeng. Biotechnol.* **2015**, *3*, 79. [CrossRef] [PubMed]
47. Lampert, F.M.; Kütscher, C.; Stark, G.B.; Finkenzeller, G. Overexpression of Hif-1α in mesenchymal stem cells affects cell-autonomous angiogenic and osteogenic parameters. *J. Cell Biochem.* **2016**, *117*, 760–768. [CrossRef] [PubMed]
48. Lin, W.; Xu, L.; Zwingenberger, S.; Gibon, E.; Goodman, S.B.; Li, G. Mesenchymal stem cells homing to improve bone healing. *J. Orthop. Transl.* **2017**, *9*, 19–27. [CrossRef] [PubMed]

materials

MDPI

Article

Plasma-Induced Crystallization of TiO_2 Nanotubes

Metka Benčina [1], Ita Junkar [1,*], Rok Zaplotnik [1], Matjaz Valant [2,3], Aleš Iglič [4] and Miran Mozetič [1]

1 Department of Surface Engineering and Optoelectronics, Jožef Stefan Institute, Jamova 39, Ljubljana 1000, Slovenia; metka.bencina@ijs.si (M.B.); rok.zaplotnik@ijs.si (R.Z.); miran.mozetic@ijs.si (M.M.)
2 Materials Research Laboratory, University of Nova Gorica, Vipavska 13, Nova Gorica 5000, Slovenia; matjaz.valant@ung.si
3 Institute of Fundamental and Frontier Sciences, University of Electronic Science and Technology of China, Chengdu 610054, China
4 Laboratory of Biophysics, Faculty of Electrical Engineering, University of Ljubljana, Tržaška 25, Ljubljana 1000, Slovenia; ales.iglic@fe.uni-lj.si
* Correspondence: ita.junkar@ijs.si; Tel.: +38-614-773-885

Received: 21 January 2019; Accepted: 18 February 2019; Published: 20 February 2019

check for updates

Abstract: Facile crystallization of titanium oxide (TiO_2) nanotubes (NTs), synthesized by electrochemical anodization, with low pressure non-thermal oxygen plasma is reported. The influence of plasma processing conditions on TiO_2 NTs crystal structure and morphology was examined by X-ray diffraction (XRD) and scanning electron microscopy (SEM). For the first time we report the transition of amorphous TiO_2 NTs to anatase and rutile crystal structures upon treatment with highly reactive oxygen plasma. This crystallization process has a strong advantage over the conventional heat treatments as it enables rapid crystallization of the surface. Thus the crystalline structure of NTs is obtained in a few seconds of treatment and it does not disrupt the NTs' morphology. Such a crystallization approach is especially suitable for medical applications in which stable crystallized nanotubular morphology is desired. The last part of the study thus deals with in vitro biological response of whole blood to the TiO_2 NTs. The results indicate that application of such surfaces for blood connecting devices is prospective, as practically no platelet adhesion or activation on crystallized TiO_2 NTs surfaces was observed.

Keywords: TiO_2 nanotubes; crystallization; gaseous plasma; biological response

1. Introduction

Self-aligned TiO_2 nanotubes synthesized by electrochemical anodization of Ti foil are highly promising materials for biomedical applications due to their advanced biocompatibility in comparison to commonly used plain metal surfaces, and as such can be employed as orthopaedic and dental implants, vascular stents, antibacterial devices or surfaces and smart drug-delivery platforms [1–8]. Improved biological response to the TiO_2 NTs is, among others, the result of increased surface roughness and larger surface area. Moreover, the tunable morphology of NTs that can be controlled by altering the anodization parameters such as voltage and time, allows for selective response of biological material. For instance, proteins, platelets and cell adhesion, cell morphology, proliferation and differentiation are highly affected by the TiO_2 NTs diameter [9]. This is especially important in applications, in which the adhesion of certain cell types is preferred, as in the case of vascular stents.

As-anodized amorphous TiO_2 NTs are, therefore, often subjected to crystallization process in order to enhance their bio-performance. It has been shown that annealed TiO_2 NTs layers with anatase crystal structure have a much better corrosion resistance and bioactivity than amorphous NTs; moreover, they tend to induce more, and faster, hydroxyapatite deposition, which is crucial for successful bone bonding (osteointegration) ability of the body implants [10–12]. In addition, the proliferation and mineralization of osteoblasts cultured on anatase or a mixture of anatase/rutile TiO_2 NTs was significantly higher than on the surface of as-anodized amorphous TiO_2 NTs [13]. Crystallized TiO_2 NTs have also good bacterial resistance; it was found that such layers inhibit the growth of *S. aureus* and *P. aeruginosa* [5], which is important for designing medical devices, since implant-associated infections present a serious health care concern. In addition, the enhanced biocompatibility of crystalline TiO_2 NTs could also be linked with increasing Ti-OH functional groups, responsible for enhancing wettability and apatite deposition [14,15] and increasing stability in bioliquids [16].

Crystallization of TiO_2 NTs induced by annealing in a conventional furnace is a well-established process; the atom rearrangements occur due to elevated temperatures. However, such crystallization requires high temperatures and is time consuming in comparison with the plasma-induced crystallization presented in this contribution. For instance, in order to achieve anatase, mixture of anatase/rutile and rutile crystal structure, TiO_2 NTs should undergo annealing in the furnace for at least 2 h at 450 °C, 550 °C [10] and 800 °C [17], respectively. Besides, changes in the morphology of TiO_2 NTs present a significant drawback of such crystallization process; the annealing in a furnace at 800 °C results in the transition of anatase NTs to a dense rutile layer [17]. By contrast, hydrothermal treatment presents an intriguing low-temperature crystallization method that is catalysed by pressure and mineralizing agents [18]. Liu et al. [19] reported hydrothermal reaction in an autoclave at a temperature of less than 180 °C for 4 h, which led to the amorphous-to-ananase transition of TiO_2 NTs. In spite of low temperature requirements, hydrothermal crystallization adversely alters TiO_2 NTs morphology and leads to the formation of aggregated anatase nano-particles on the surface of NTs at about 200 °C [20]. Lamberti et al. [14] demonstrated another attractive near-room temperature (50 °C) crystallization process of amorphous TiO_2 NTs, which led to anatase crystal phase formation after only 30 min of sample exposure to water vapour [14,20]. However, such a crystallization method also initiates the formation of crystals at the outer and inner walls of the NTs, which finally leads to transformation of nanotubes to nanorods. For this reason, there is an indispensable need to develop novel crystallization processes that would preserve the morphology of the materials.

In the present research, the transformation of rather amorphous TiO_2 NTs to anatase or mixture of anatase/rutile crystal structure by a non-thermal oxygen plasma process, is, according to our best knowledge, reported for the first time. Fast amorphous-to-anatase or -anatase/rutile transition and unaltered NTs morphology are the major advantages of such crystallization before conventional processes, for which the preservation of morphology is still a challenge. Although such a time-saving method is convenient for inducing crystallinity of the nanomaterials/thin films [21–23], there is a lack of information about the mechanisms involved in the crystallization process. In this contribution, the influence of plasma processing conditions on the crystal structure and morphology of the TiO_2 NTs is therefore examined. In addition, the biological aspects of plasma treated TiO_2 NTs are presented. It has already been shown that surfaces treated by highly reactive oxygen plasma improve biological response. For instance, Junkar et al. [24] showed that plasma treatment allows for TiO_2 NTs surface functionalization and enhancement of osteoblast-like cell responses. Authors also confirmed cleaning and sterilizing effects of plasma treatment [25]. Moreover, the wettability of ZrO_2 was noticeably improved by an oxygen plasma treatment, which promoted the attachment, proliferation and differentiation of human osteoblast-like cells [26]. Therefore, simultaneous plasma treatment and crystallization of the material with already tunable morphology offers highly needed improvements for devices used in biomedical applications.

2. Materials and Methods

2.1. Materials

Titanium foil (Advent, 0.1 mm thickness, 99.6%), ethylene glycol (Fluka, ≥99.5%), ammonium fluoride—NH_4F (Sigma Aldrich, 28.0–30.0%), hydrofluoric acid—HF (Sigma Aldrich, ≥40%) acetone

(Honeywell Riedel–de Haen, 99.5%), ethanol (Sigma Aldrich, 96%), phosphate-buffered saline—PBS (Sigma Aldrich), and glutaraldehyde solution (Sigma Aldrich, 25% in H_2O) deionized water (miliQ).

2.2. Synthesis of TiO_2 NTs by Electrochemical Anodization

The TiO_2 NTs were fabricated by an electrochemical anodization method as described in Refs. [24,25]. The synthesis was carried out at room temperature (≈20 °C) in a two electrode system (Pt/Ti) with the size 10×10 mm^2 and working distance of 15 mm. The thickness of the Ti electrode was 0.10 mm. Before anodization, the Ti foil was ultrasonically cleaned in acetone, ethanol and deionized water for 5 min in each and further dried under a nitrogen stream. The electrolyte used in this step was composed of ethylene glycol and NH_4F (0.35 wt.%) and H_2O (1.7 wt.%). The nanotubular layer grown in this step was then detached from the substrate with a successive ultrasonication in H_2O, acetone and ethanol in order to obtain a pre-dimpled surface, which led to enhanced homogeneity of the NTs' surface. In the second step of anodization process, the pre-treated Ti foils were used as a substrate to grow NTs. The electrolyte based on the ethylene glycol, containing water and HF was used and the detailed synthesis parameters of these steps are presented in Table 1. The as-synthesized NTs were kept in ethanol for 2 h in order to remove components from the electrolyte and then dried under a nitrogen stream. The as-prepared NTs were used for further plasma processing (output power 200–800 W for different times) or annealed at 450 °C for 2 h in a furnace with annealing/cooling rate 8 °C/min.

Table 1. Influence of synthesis conditions on the diameter and length of the TiO_2 nanotubes (NTs).

Sample	Voltage (V)	Time (h)	Electrolyte (wt.%)	Diameter (nm)	Length (μm)
TiO_2 NTs–15 nm	10	1	HF (1.1), H_2O (15.3), EG	15	0.21
TiO_2 NTs–100 nm	58	2.5	HF (1.0), H_2O (11.0), EG	100	2.53

2.3. Oxygen Plasma Treatment

Treatment of TiO_2 NTs was performed by oxygen plasma in the plasma reactor designed in house. The system was evacuated with a two-stage oil rotary pump with a nominal pumping speed of 80 m^3 h^{-1}. The discharge chamber was a Pyrex tube with a length of 80 cm and an inner diameter of 3.6 cm. Gaseous plasma was created with an inductively coupled radiofrequency (RF) generator (CESAR 1310, Advanced Energy, Fort Collins, Colorado, USA), operating at a frequency of 13.56 MHz and a nominal power of 1000 W. Generator powers between 200–800 W were used. Commercially available oxygen was leaked into the discharge chamber and the pressure was measured with an absolute vacuum gauge. The pressure during the plasma treatment was fixed at 50 Pa which allows for the highest degree of dissociation of oxygen molecules for this particular plasma reactor. The samples were placed on an object glass and treated for different periods of time and with different output powers, one at a time. Details about the behaviour of reactive gaseous species versus nominal power of the RF generator are presented elsewhere [27].

2.4. Characterization

2.4.1. X-ray Diffraction (XRD) Spectroscopy Analysis

X-ray diffraction (XRD) was performed using MiniFlex 600 Benchtop X-ray diffractometer (Rigaku, Tokyo, Japan) equipped with Cu K-α radiation (1.541 Å) over the 2θ range 10–70°, with a step size of 0.017°, divergence slit of 0.218° and counting step time of 25 s in continuous scanning mode. Carbon tape was used to mount the samples on the glass holder.

2.4.2. Scanning Electron Microscope (SEM) Analysis

Morphology of the materials was analysed with a JSM 7100F scanning electron microscope (SEM, JEOL Ltd., Tokyo, Japan). For biological evaluation of platelets on the surface, the samples were coated with gold/palladium and examined by SEM at an accelerating voltage of 15 kV. The test was done in triplicates and only representative images are shown in this paper.

2.4.3. Temperature Measurements

During the exposure of the samples to plasma, the temperature was measured with a custom-made K-type thermocouple). Chromel and alumel wires (Goodfellow Cambridge Ltd., Huntingdon, UK) were spot welded on the rear of the samples so they were not influenced by plasma. A Keithley Model 2100 digital multimeter (Keithley Instruments/Tektronix, OH, USA) and custom-made software were used to record the time evolution of the samples temperature.

2.5. *In Vitro Biological Response—Interaction With Whole Blood*

All subjects gave their informed consent for inclusion before they participated in the study. The study was conducted in accordance with the Declaration of Helsinki, and the protocol was approved by the Ethics Committee of Slovenia (56/03/10). Whole blood was obtained from healthy volunteers via vein puncture. The blood was drawn into 9 ml tubes already coated with trisodium citrate anticoagulant. The material samples—Ti foil, amorphous and annealed (anatase) TiO_2 NTs of 100 nm in diameter (7 × 7 mm^2)—were incubated with the 250 µL of whole blood for 45 min at room temperature in the 24-well cell culture plates. Afterwards, 250 µL of PBS was added to the incubated samples. The blood with PBS was then removed and the samples were rinsed 3 times with 250 µL of PBS in order to remove weakly adherent platelets and other biological material. The adherent platelets were further fixed by 400 µL of 0.5% glutaraldehyde solution for 2 h at room temperature. Then the materials were rinsed with PBS and dehydrated by using a graded ethanol series (50 vol.%, 70 vol.%, 80 vol.%, 90 vol.%, 100 vol.% and again 100 vol.% of ethanol) for 5 min and in the last stage (100 vol.% ethanol) for 10 min. Afterwards the samples were dried with liquid nitrogen and stored in vacuum.

3. Results and Discussion

3.1. *Crystal Structure Analysis*

The TiO_2 NTs were subjected either to annealing or highly reactive oxygen plasma treatment at different input powers and at different treatment times and the influence of such treatments was further studied in terms of altered surface morphology and crystallization. It has been shown in our previous work [25] and recently also in Ref. [28] that TiO_2 NTs were deformed after exposure to elevated temperatures and that NTs' stability depended on diameter (larger diameter NTs are more resistant to thermal degradation). In the present study, the alteration in crystallinity of TiO_2 NTs was studied by X-ray diffraction spectroscopy, which revealed that the treatment with highly reactive oxygen plasma at different powers significantly influences the TiO_2 NTs crystal structure. The as-anodized TiO_2 NTs samples are amorphous and remain uncrystallized even after 3 min of exposure to highly reactive oxygen plasma at 200 W (Figure 1). The XRD data show that the Ti foil, which is used as a substrate for TiO_2 NTs fabrication, is detected by XRD although NTs of 100 nm in diameter have a length of about 2.5 µm. The higher power of plasma, more than 400 W, seems to already induce crystallization if samples are exposed to plasma for more than 10 s. Actually a mixture of anatase and rutile crystal phase (hardy observable peak at $2\theta = 36.5°$ as seen in Figure 1) appears after the treatment of TiO_2 NTs for 10 s in plasma with the power of 400 W and 600 W. Interestingly, the 1 s plasma exposure at these conditions is not enough to initiate the change of crystal structure, since no anatase nor rutile peaks were detected. However, the plasma treatment at 800 W for 1 s and 10 s results in anatase and a mixture of anatase/rutile crystal phases, respectively (Figures 1 and 2). For comparison, the annealing of TiO_2 NTs in a conventional furnace at 450 °C requires 2 h to induce the transition of amorphous phase to anatase crystal structure.

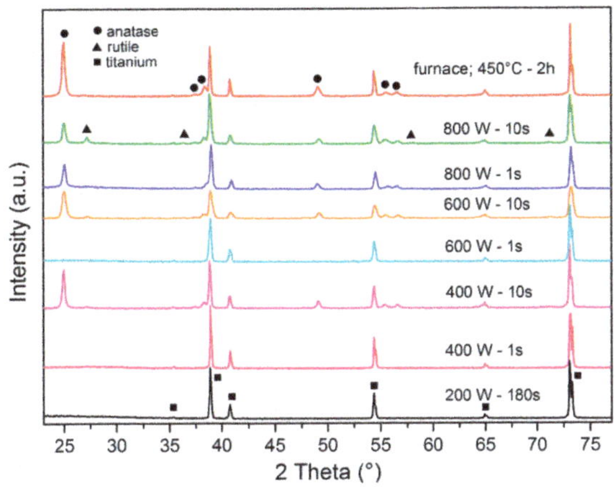

Figure 1. X-ray diffraction (XRD) patterns of TiO_2 NTs of 100 nm diameter measured after plasma treatment at different powers; 200, 400, 600 and 800 W for different times and TiO_2 NTs of 100 nm diameter after annealing in a furnace at 450 °C for 2 h.

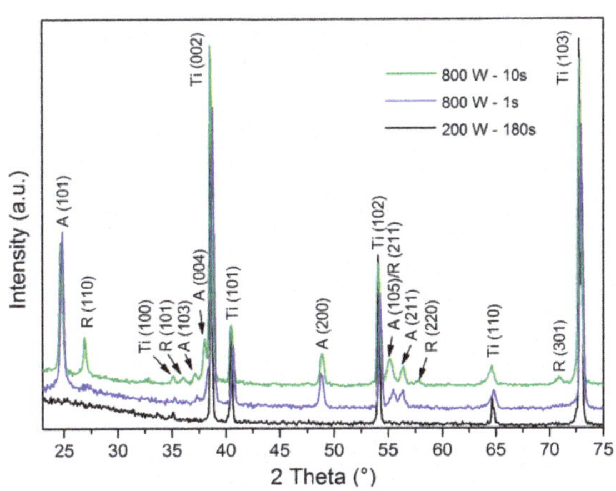

Figure 2. Magnified X-ray diffraction (XRD) patterns of TiO_2 NTs of 100 nm diameter measured after plasma treatment at different powers; 200 and 800 W for different times, in order to show the difference between amorphous, anatase and rutile crystal phase of TiO_2 NTs. A = characteristic XRD peaks for anatase crystal structure, R = characteristic XRD peaks for rutile crystal structure and Ti = characteristic XRD peaks for Ti foil.

In order to study the effect of synthesized TiO_2 layer thickness on the crystallization in plasma, TiO_2 NTs of 15 nm in diameter and length of about 0.2 μm were treated with oxygen plasma at 800 W for 1 s and 10 s. Only one anatase-phase peak at $2\theta = 25°$ (101) was detected for the TiO_2 NTs sample treated with plasma at 800 W for 1 s, indicating that the sample is still mainly amorphous (Figure 3). Interestingly, for TiO_2 NTs of 100 nm in diameter the anatase crystal phase is already observed after 1 s of treatment at 800 W (Figures 1 and 2). However, the rutile crystal phase without any evidence of anatase prevails when plasma treatment at 800 W is prolonged to 10 s for TiO_2 NTs of 15 nm in diameter (Figure 3). For comparison, TiO_2 NTs of 100 nm in diameter crystallize in mixture of anatase/rutile phase after prolonged plasma treatment (10 s). The XRD spectra of plain Ti foil, which is used as a substrate for TiO_2 NTs fabrication, are presented in Figure 3; it can be observed that XRD pattern of the plasma-treated sample at 800 W for 1 s exhibit small rutile and anatase peaks at $2\theta = 36.5°$ and

37.5°, respectively. However, an intense rutile-phase peak at 2θ = 27° is confirmed after the exposure of the Ti foil to plasma of power 800 W for 10 s. These results show that although smaller-sized nanoparticles require less intensive plasma conditions to induce crystallization [21,29], the thermal effects on TiO_2 NTs with 2.5 μm length and 0.21 μm length may be different. It is also noteworthy that plasma treatment induce the formation of crystalline oxide layer on the surface of the Ti foil.

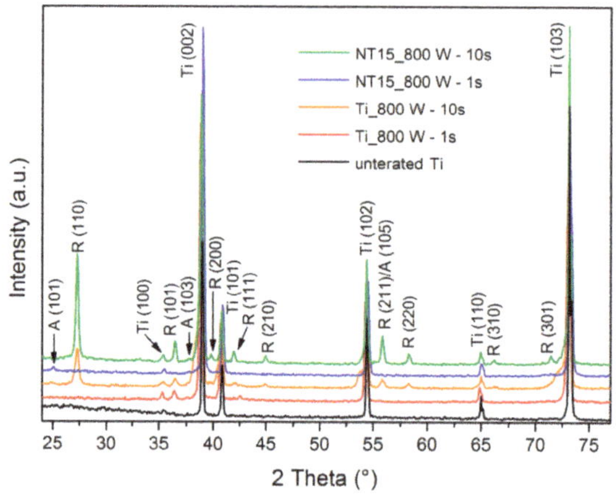

Figure 3. X-ray diffraction (XRD) patterns of untreated Ti foil and Ti foil and TiO_2 NTs of diameter 15 nm (NT15) measured after plasma treatment at 800 W for 1 s and 10 s. A = characteristic XRD peaks for anatase crystal structure, R = characteristic XRD peaks for rutile crystal structure and Ti = characteristic XRD peaks for Ti foil.

3.2. Morphology Analysis

The morphology of as-anodized TiO_2 NTs (100 nm in diameter) is shown in Figure 4, while the morphology of TiO_2 NTs exposed to oxygen plasma is shown in Figure 5. It is noteworthy that TiO_2 NTs formed on the Ti foil are stable at plasma conditions used in the present study, except the NTs treated with plasma of power 800 W for 10 s. SEM analysis of TiO_2 NTs treated at 800 W for 10 s showed that the NTs structure is destroyed as can be seen from Figure 5, where the comparisons between plasma-treated samples at 800 W for 10 s, 800 W for 1 s and 400 W for 10 s are presented. The tops of TiO_2 NTs treated at 800 W for 10 s are partially closed, most probably due to collapsing of NTs walls (Figure 5A) and increase in stoichiometric oxide on the surface. It is also evident form Figure 5A that the bottom layer of the NTs is destroyed or even that there is a formation of oxide layer as already observed in Ref. [30], which could be caused by a longer exposure to high temperatures. Mazare et al. [5] observed the same loss of nanotubular structure after annealing at high temperatures (i.e., 550 °C, 650 °C and 750 °C). It was previously reported that the formation of a rutile oxide layer is initiated at the metal-nanotube interface during the annealing [30], and therefore changes in crystal structure firstly occur at the bottom layer of the NTs. A similar structure, although not destroyed, was formed when the TiO_2 NTs were treated at 800 W for 1 s (Figure 5B). However, the morphology of TiO_2 NTs treated at 400 W for 10 s is not destroyed and the tubes are well defined and opened (Figure 5C).

SEM analysis revealed that TiO_2 NTs of 15 nm in diameter have open tops and their structure is not destroyed after treatment with plasma at 800 W for 1 s (Figure 6A). However, NTs are completely destroyed after plasma treatment at 800 W for 10 s (Figure 6B). Ti foil morphology is unaltered after 1 s plasma treatment at 800 W (Figure 6C), while changes in the morphology of Ti foil after plasma treatment at 800 W for 10 s can be observed; the surface is covered with nanostructures (fused oxide particles) as seen in Figure 6D.

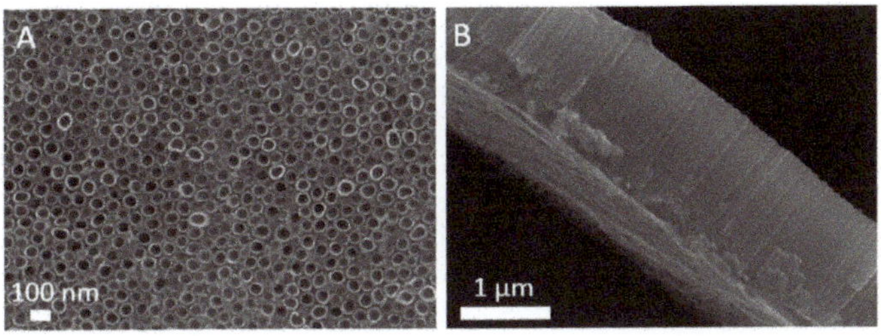

Figure 4. Scanning electron microscope (SEM) images of untreated as-anodized TiO_2 NTs with diameter of 100 nm; (**A**) top view, (**B**) side view.

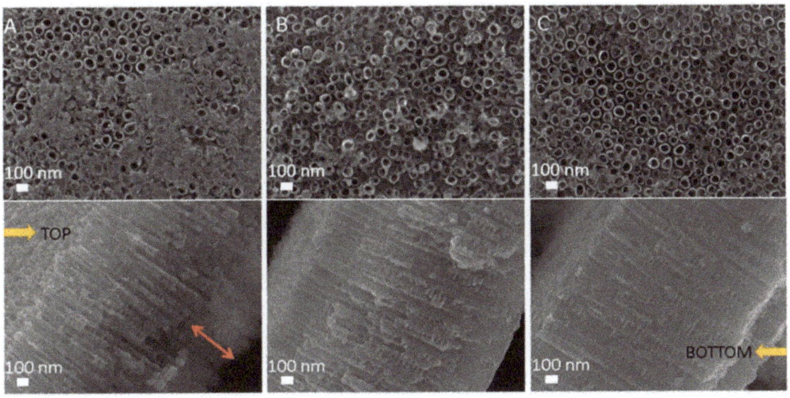

Figure 5. SEM images (above: top view and below: cross sectional view) of TiO_2 NTs of 100 nm in diameter treated with oxygen plasma at (**A**) 800 W for 10 s, (**B**) 800 W for 1 s and (**C**) 400 W for 10 s. The red arrow indicates destroyed bottom layer of NTs.

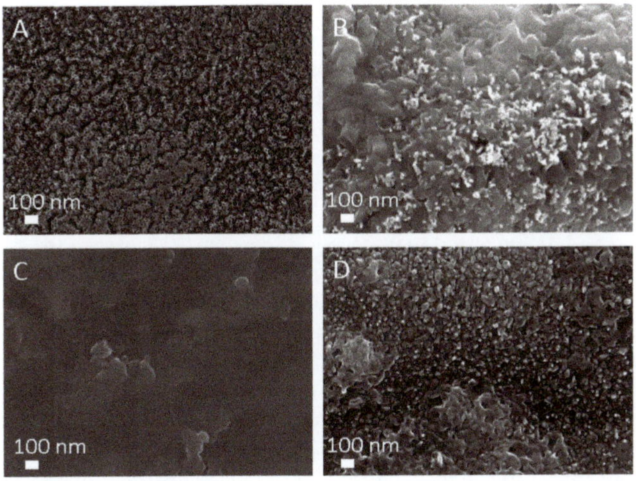

Figure 6. SEM images (top view) of TiO_2 NTs of 15 nm in diameter after plasma treatment at (**A**) 800 W for 1 s, (**B**) 800 W for 10 s and Ti foil after plasma treatment at (**C**) 800 W for 1s and (**D**) 800 W for 10 s.

3.3. Temperature Measurements

Crystallization of the materials in a furnace is governed by the change in temperature. Several studies report on light-induced [31–33], laser-induced [34,35], and microwave-induced [36] crystallization, which are also correlated with sample heating. In the present study we used a plasma treatment and the temperature of the samples during the treatments in correlation with plasma output power were measured. As evident from Table 2, the temperature of the sample increases with increasing output power of the plasma reactor in the time range up to about 10 s. It has been previously reported that plasma used in the present study operates in two different modes, E and H; The E-mode prevails at a low input power and is defined by a relatively low electron density, high electron temperature and low light emission, while the H-mode at higher input power represents a higher electron density, somehow lower electron temperature and higher light emission [37]. Zaplotnik et al. [27] measured the transition from E- to H–mode in the same plasma system as used in this study. It was found that the E–H transition occurs at the generator power of about 400 W at the pressure of 50 Pa. Below the output power of 400 W, the power transmission to plasma is not ideal; for example, at generator forward power of 200 W, the forward minus reflected power is about 65 W, while for example at 400 W, there is much lower reflected power which means the transmission of power is much better. Exposure of TiO_2 NTs to H-mode plasma subjects the sample, among others, to more intense electron density and light emission. Therefore, the samples exposed to H-mode for a sufficient time could crystallize in seconds, while exposure of samples to E-mode, even for 3 min, does not initiate the crystallization.

Table 2. Effect of radiofrequency (RF) forward (output) power and plasma processing time on the temperature and the crystal structure of TiO_2 NTs of 100 nm in diameter.

Forward Power (W)	Time (s)	Temperature (°C)	Crystal Structure
200	180	300	amorphous
400	1	660	amorphous
400	10	1210	anatase + rutile
600	1	970	amorphous
600	10	1240	anatase + rutile
800	1	1030	anatase
800	10	1320	anatase + rutile

Although the process of nanoparticles synthesis by plasma treatment has already been reported [21], there is a lack of reports and detailed mechanisms about the crystallization upon treatment with non-thermal plasmas. Ohsaki et al. [23] claim that the crystallization of sol-gel derived TiO_2 thin films can be achieved within a few minutes by a non-thermal plasma processing. The authors suggest that crystallization of the materials should be derived from the excitation by the RF electromagnetic field and not plasma itself. Similarly, An et al. [38] speculate that formation of the crystalline grains of $BaTiO_3$ is due to the energy provided by ion bombardment. Contrary, Kramer et al. [21] suggested that the silicon nanoparticles exceed the gas temperature in non-thermal plasmas (370–430 K/97–157 °C) to the point of sufficient temperature for crystallization (up to 700 K/427 °C), although measurements showed that the average temperature of the silicon nanoparticles exposed to plasma was close to the gas temperature. According to the authors [21], crystallization of the nanoparticles by non-thermal plasma occurs due to the electron ion recombination and reactions of radicals on the nanoparticle's surface. Similarly, Lopez et al. [22] demonstrated that plasma exposure allowed for crystallization of silicon nanoparticles due to the heating of the nanoparticles to the temperature of 1100 K/827 °C.

In present study, the correlations between TiO_2 NTs crystal structure and plasma processing conditions have been studied. It has been shown that TiO_2 NTs crystallize to anatase and at appropriate conditions also to rutile phase when exposed to highly reactive oxygen plasma for sufficient time and at appropriate plasma power. It was shown that the mixture of anatase/rutile crystal structure was achieved after 10 s at the power of 400 W, where the measured temperature of the sample was above 1210 °C. These operating conditions allow more intense physical processes which occur during interaction of plasma species with the sample. We presume that these interactions contribute to the alteration of the sample's crystal structure, mainly due to rapid heating of the sample induced by surface recombination of neutral oxygen atoms and ion neutralization. However, the contribution of these two mechanisms to the floating sample heating in electrodes low pressure oxygen plasma are not of the same order. The densities of the ions are a few orders of magnitude lower than the densities of the neutral oxygen atoms and, therefore, the heating of the sample due to the ion neutralization can be neglected [39]. Therefore, the heating of samples (P_H) can be expressed with the formula (1):

$$P_H = \gamma \cdot j \cdot W_D/2, \tag{1}$$

where γ is the recombination coefficient of the oxygen atoms for the specific surface, j is the flux of the oxygen atoms onto the sample surface and W_D is the dissociation energy of oxygen molecules. At the same time the sample is being cooled, where two most important mechanisms are radiation and convection. The radiation part (P_r) is described with the Stefan–Boltzmann law (2):

$$P_r = A \cdot \varepsilon \cdot \sigma \cdot T^4, \tag{2}$$

where A is the surface area of the sample, ε is the emissivity, σ is the Stefan–Boltzmann constant and T is the sample temperature. The convection part (P_c) is a linear function of temperature and can be described as (3):

$$P_c = h \cdot A \cdot T, \tag{3}$$

where h is the convection coefficient. The sample immersed in plasma reaches steady temperature when the heating part is equal to the sum of the cooling parts (4):

$$P_H = P_r + P_c, \tag{4}$$

From here we can see that larger flux of the oxygen atoms means a higher temperature of the sample. In Figure 7 the flux of oxygen atoms versus the steady temperature of the samples of NTs of 100 nm in diameter are presented. In this figure the measure points are those presented in Table 2, and the corresponding fluxes were measured with a cobalt catalytic probe. The detailed measurement technique is described elsewhere [27]. It should be noted that in this case only the measured points of samples that reached the steady temperature were used. The theoretical fit is that derived from previous formulas for cooling and heating of samples in plasma [27] taking into account the coefficient $\gamma = 1$.

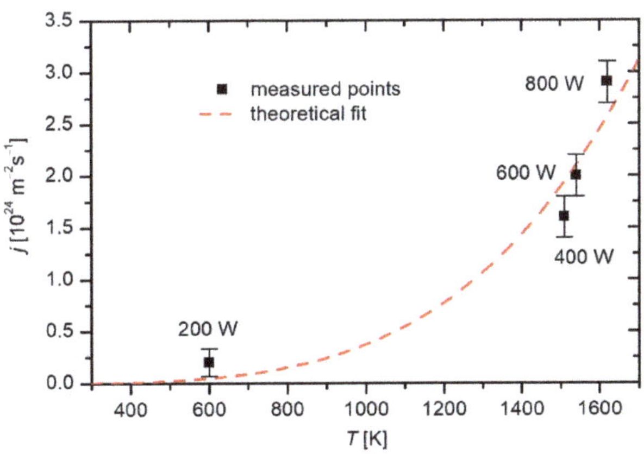

Figure 7. Flux of oxygen atoms and the steady temperature of the NTs samples of 100 nm in diameter.

Destruction of the NTs was observed at higher power plasma (800 W) at prolonged treatment (more than a second). The crystallization mechanisms of TiO_2 in plasma can be correlated with RF electromagnetic field, vacuum ultraviolet radiation, interaction with radicals and ions which induce sample heating, and chemical reactions. However, the prevailing influence of each parameter alone and its possible synergistic influence is not yet understood and definitely beyond the scope of this paper. Further studies in this direction should be conducted. The results of this study confirmed that the temperature, which was measured on the bottom surface of the samples during the treatments, plays an important role in TiO_2 NTs crystallization.

3.4. In Vitro Biological Response—Interaction with Whole Blood

The approach to rapidly alter the crystal structure of TiO_2 NTs without influencing their morphology is an intriguing way for surface modification of NTs used in bio-applications. As mentioned before, the crystal structure of a material affects its performance in biological applications. For instance, it has been shown that platelet adhesion and activation on the crystallized TiO_2 NTs depends of the annealing temperature and crystal phase [40]. More precisely, authors showed that a certain amount of pure anatase in the TiO_2 NTs leads to better platelet adhesion, which is beneficial for successfully integration of dental implants. By contrast, rutile-phase TiO_2 NTs reduced platelet adhesion and activation, perhaps due to the formation of a hydration layer, which reduced intimate contact area between a platelet and the surface [28]. This particular case is advantageous for vascular stents, for which the inhibition of adhesion and activation of platelets is preferred, since platelets' aggregation leads to a blood clot formation and stent thrombosis. Results of platelets performance on the surface of TiO_2 NTs of 100 nm in diameter prepared in present study are shown in Figure 8. Platelets adhere to the surface of Ti foil with filopodia, while adhesion on the amorphous TiO_2 NTs is done via lamelopodia and fillopodia. This indicates that the amorphous TiO_2 NTs provide a better environment for platelet adhesion and activation. By contrast, platelets could not be found on the surface of annealed TiO_2 NTs with anatase crystal structure. No biological material was detected on these surfaces, as platelets seemed not to adhere on these surfaces. These results indicate that anatase crystal structure (obtained by plasma or annealing) of TiO_2 NTs does not provide appropriate conditions for platelets adhesion, which is beneficial from the perspective of vascular stent applications. In previous publications we showed increased hydrophilicity and reduced electrolyte residual (fluorine) on the surface of annealed TiO_2 NTs [26]. Since fluorine is toxic to cells, its lowered content could be the reason for the better performance of platelets. It should be also considered that besides crystal structure also surface chemistry and wettability (surface charge), as well as morphology of the material play important role in platelet response, therefore further studies of TiO_2 NTs interactions with platelets are needed.

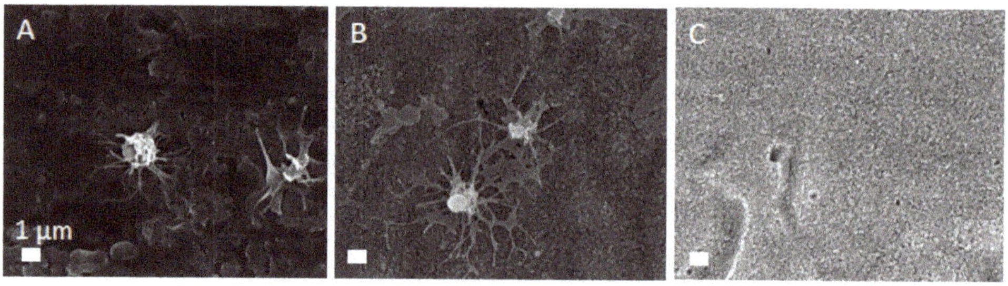

Figure 8. Platelets adhesion and activation on (**A**) Ti foil, (**B**) amorphous and (**C**) annealed TiO_2 NTs with 100 nm in diameter.

4. Conclusions

Non-thermal oxygen plasma induced crystallization of TiO_2 NTs synthesized by electrochemical anodization of Ti foil. The transition of the NTs' amorphous phase to anatase and/or rutile crystal structure was obtained within a few seconds of exposure to oxygen plasma without changing the desired morphology of NTs. Although the gas temperature of non-thermal plasma was well below the thermal crystallization temperature, the sample temperature rose above 1300 °C at a certain plasma treatment condition e.g. the output power of 800 W and 10 s plasma treatment time. Results provide evidence that a rapid temperature increase of samples treated in plasma is correlated with crystallization of TiO_2 NTs. Good correlation between the flux of oxygen atoms and increase in temperature of the sample was found; however, other parameters in plasma may also play a vital role in crystallization and should be further studied. It should be emphasized that crystallization in plasma is also influenced by the NTs diameter, as NTs with 15 nm in diameter seem to crystalize prevalently to the rutile phase, which was at our experimental conditions not the case for NTs of 100 nm in diameter. Moreover, the in vitro biological response of whole blood with crystalline TiO_2 NTs showed that the crystallization reduces adhesion and activation of blood platelets, which is of particular interest for designing medical devices that are likely to contact blood, such as vascular stents.

Author Contributions: Methodology, formal analysis and investigation: M.B., I.J., R.Z., M.M. and M.V., writing—original draft preparation: M.B., I.J., writing—review and editing: I.J., R.Z, M.M. and A.I, conceptualization: M.M.

Funding: This research was funded by the Slovenian Research Agency for financial support [grant numbers Z3-4261, J3-9262, J1-9162, J2-8166, J2-8169, J5-7098, P2-0232] and Slovenian Ministry of Education, Science and Sport [grant "Public call for encouraging young investigators at the beginning of their career 2.0", No. 5442-15/2016/18].

Conflicts of Interest: The authors declare no conflict of interest.

References

1. Wang, Q.; Huang, J.-Y.; Li, H.-Q.; Zhao, A.Z.-J.; Wang, Y.; Zhang, K.-Q.; Sun, H.-T.; Lai, Y.-K. Recent advances on smart TiO_2 nanotube platforms for sustainable drug delivery applications. *Int. J. Nanomed.* **2017**, *12*, 151–165. [CrossRef] [PubMed]
2. Cipriano, A.F.; Miller, C.; Liu, H. Anodic growth and biomedical applications of TiO_2 nanotubes. *J. Biomed. Nanotechnol.* **2014**, *10*, 2977–3003. [CrossRef] [PubMed]
3. Wu, S.; Weng, Z.; Liu, X.; Yeung, K.W.K.; Chu, P.K. Functionalized TiO_2 based nanomaterials for biomedical applications. *Adv. Funct. Mater.* **2014**, *24*, 5464–5481. [CrossRef]
4. Hamlekhan, A.; Butt, A.; Patel, S.; Royhman, D.; Takoudis, C.; Sukotjo, C.; Yuan, J.; Jursich, G.; Mathew, M.T.; Hendrickson, W. Fabrication of anti-aging TiO_2 nanotubes on biomedical Ti alloys. *PLoS ONE* **2014**, *9*, e96213. [CrossRef] [PubMed]
5. Mazare, A.; Totea, G.; Burnei, C.; Schmuki, P.; Demetrescu, I.; Ionita, D. Corrosion, antibacterial activity and haemocompatibility of TiO_2 nanotubes as a function of their annealing temperature. *Corros. Sci.* **2016**, *103*, 215–222. [CrossRef]
6. Ji, M.-K.; Oh, G.; Kim, J.-W.; Park, S.; Yun, K.-D.; Bae, J.-C.; Lim, H.-P. Effects on antibacterial activity and osteoblast viability of non-thermal atmospheric pressure plasma and heat treatments of TiO_2 nanotubes. *J. Nanosci. Nanotechnol.* **2017**, *17*, 2312–2315. [CrossRef]
7. Kulkarni, M.; Flašker, A.; Lokar, M.; Mrak-Poljšak, K.; Mazare, A.; Artenjak, A.; Čučnik, S.; Kralj, S.; Velikonja, A.; Schmuki, P. Binding of plasma proteins to titanium dioxide nanotubes with different diameters. *Int. J. Nanomed.* **2015**, *10*, 1359–1373.
8. Kulkarni, M.; Mazare, A.; Schmuki, P.; Iglič, A. Biomaterial surface modification of titanium and titanium alloys for medical applications. *Nanomedicine* **2014**, *111*, 111–136.
9. Kulkarni, M.; Mazare, A.; Gongadze, E.; Perutkova, Š.; Kralj-Iglič, V.; Milošev, I.; Schmuki, P.; Iglič, A.; Mozetič, M. Titanium nanostructures for biomedical applications. *Nanotechnology* **2015**, *26*, 062002. [CrossRef]
10. Hilario, F.; Roche, V.; Nogueira, R.P.; Junior, A.M.J. Influence of morphology and crystalline structure of TiO_2 nanotubes on their electrochemical properties and apatite-forming ability. *Electrochim. Acta* **2017**, *245*, 337–349. [CrossRef]

11. Liu, Y.; Kim, S.; McLeod, J.A.; Li, J.; Guo, X.; Sham, T.-K.; Liu, L. The effect of crystal structure of TiO_2 nanotubes on the formation of calcium phosphate coatings during biomimetic deposition. *Appl. Surf. Sci.* **2017**, *396*, 1212–1219. [CrossRef]
12. Bai, Y.; Park, I.S.; Park, H.H.; Lee, M.H.; Bae, T.S.; Duncan, W.; Swain, M. The effect of annealing temperatures on surface properties, hydroxyapatite growth and cell behaviors of TiO_2 nanotubes. *Surf. Interface Anal.* **2011**, *43*, 998–1005. [CrossRef]
13. Yu, W.; Zhang, Y.; Jiang, X.; Zhang, F. In vitro behavior of MC3T3-E1 preosteoblast with different annealing temperature titania nanotubes. *Oral Dis.* **2010**, *16*, 624–630. [CrossRef] [PubMed]
14. Lamberti, A.; Chiodoni, A.; Shahzad, N.; Bianco, S.; Quaglio, M.; Pirri, C.F. Ultrafast room-temperature crystallization of TiO_2 nanotubes exploiting water-vapor treatment. *Sci. Rep.* **2015**, *5*, 7808. [CrossRef] [PubMed]
15. Wu, J.-M.; Wang, M.; Li, Y.-W.; Zhao, F.-D.; Ding, X.-J.; Osaka, A. Crystallization of amorphous titania gel by hot water aging and induction of in vitro apatite formation by crystallized titania. *Surf. Coat. Technol.* **2006**, *201*, 755–761. [CrossRef]
16. Mazare, A.; Dilea, M.; Ionita, D.; Titorencu, I.; Trusca, V.; Vasile, E. Changing bioperformance of TiO_2 amorphous nanotubes as an effect of inducing crystallinity. *Bioelectrochemistry* **2012**, *87*, 124–131. [CrossRef] [PubMed]
17. Yang, L.; Zhang, M.; Shi, S.; Lv, J.; Song, X.; He, G.; Sun, Z. Effect of annealing temperature on wettability of TiO_2 nanotube array films. *Nanoscale Res. Lett.* **2014**, *9*, 621. [CrossRef]
18. Kundu, S.; Polshettiwar, V. Hydrothermal crystallization of nano-titanium dioxide for enhanced photocatalytic hydrogen generation. *ChemPhotoChem* **2018**, *2*, 796–800. [CrossRef]
19. Liu, J.; Liu, Z.; Zhang, T.; Zhai, J.; Jiang, L. Low-temperature crystallization of anodized TiO_2 nanotubes at the solid–gas interface and their photoelectrochemical properties. *Nanoscale* **2013**, *5*, 6139–6144. [CrossRef]
20. Yu, J.; Dai, G.; Cheng, B. Effect of crystallization methods on morphology and photocatalytic activity of anodized TiO_2 nanotube array films. *J. Phys. Chem. C* **2010**, *114*, 19378–19385. [CrossRef]
21. Kramer, N.; Anthony, R.; Mamunuru, M.; Aydil, E.; Kortshagen, U. Plasma-induced crystallization of silicon nanoparticles. *J. Phys. D Appl. Phys.* **2014**, *47*, 075202. [CrossRef]
22. Lopez, T.; Mangolini, L. On the nucleation and crystallization of nanoparticles in continuous-flow nonthermal plasma reactors. *J. Vac. Sci. Technol. B Nanotechnol. Microelectron.* **2014**, *32*, 061802. [CrossRef]
23. Ohsaki, H.; Shibayama, Y.; Yoshida, N.; Watanabe, T.; Kanemaru, S. Room-temperature crystallization of amorphous films by RF plasma treatment. *Thin Solid Films* **2009**, *517*, 3092–3095. [CrossRef]
24. Junkar, I.; Kulkarni, M.; Drašler, B.; Rugelj, N.; Recek, N.; Drobne, D.; Kovač, J.; Humpolicek, P.; Iglič, A.; Mozetič, M. Enhanced biocompatibility of TiO_2 surfaces by highly reactive plasma. *J. Phys. D Appl. Phys.* **2016**, *49*, 244002. [CrossRef]
25. Junkar, I.; Kulkarni, M.; Drašler, B.; Rugelj, N.; Mazare, A.; Flašker, A.; Drobne, D.; Humpolíček, P.; Resnik, M.; Schmuki, P. Influence of various sterilization procedures on TiO_2 nanotubes used for biomedical devices. *Bioelectrochemistry* **2016**, *109*, 79–86. [CrossRef] [PubMed]
26. Wu, C.-C.; Wei, C.-K.; Ho, C.-C.; Ding, S.-J. Enhanced hydrophilicity and biocompatibility of dental zirconia ceramics by oxygen plasma treatment. *Materials* **2015**, *8*, 684–699. [CrossRef] [PubMed]
27. Zaplotnik, R.; Vesel, A.; Mozetic, M. Transition from E to H mode in inductively coupled oxygen plasma: Hysteresis and the behaviour of oxygen atom density. *EPL* **2011**, *95*, 55001. [CrossRef]
28. Huang, Q.; Yang, Y.; Zheng, D.; Song, R.; Zhang, Y.; Jiang, P.; Vogler, E.A.; Lin, C. Effect of construction of TiO_2 nanotubes on platelet behaviors: Structure-property relationships. *Acta Biomater.* **2017**, *51*, 505–512. [CrossRef] [PubMed]
29. Kramer, N.; Aydil, E.; Kortshagen, U. Requirements for plasma synthesis of nanocrystals at atmospheric pressures. *J. Phys. D Appl. Phys.* **2015**, *48*, 035205. [CrossRef]
30. Das, S.; Zazpe, R.; Prikryl, J.; Knotek, P.; Krbal, M.; Sopha, H.; Podzemna, V.; Macak, J.M. Influence of annealing temperatures on the properties of low aspect-ratio TiO_2 nanotube layers. *Electrochim. Acta* **2016**, *213*, 452–459. [CrossRef]
31. Feinleib, J.; Moss, S.; Ovshinsky, S. Rapid Reversible Light-Induced Crystallization of Amorphous Semiconductors. In *Disordered Materials;* Springer: Boston, MA, USA, 1991; pp. 34–37.
32. Krylova, G.; Na, C. Photoinduced crystallization and activation of amorphous titanium dioxide. *J. Phys. Chem. C* **2015**, *119*, 12400–12407. [CrossRef]

33. Nishio, T.; Naka, K. Photoinduced crystallization of calcium carbonate from a homogeneous precursor solution in the presence of partially hydrolyzed poly (vinyl alcohol). *J. Cryst. Growth* **2015**, *416*, 66–72. [CrossRef]
34. Janotta, A.; Dikce, Y.; Schmidt, M.; Eisele, C.; Stutzmann, M.; Luysberg, M.; Houben, L. Light-induced modification of a-SiO x II: Laser crystallization. *J. Appl. Phys.* **2004**, *95*, 4060–4068. [CrossRef]
35. Andrä, G.; Bergmann, J.; Falk, F.; Ose, E.; Stafast, H. Laser induced crystallization of amorphous silicon films on glass for thin film solar cells. *Phys. Status Solidi (a)* **1998**, *166*, 629–634. [CrossRef]
36. Shaporev, A.; Ivanov, V.; Baranchikov, A.; Tret'yakov, Y.D. Microwave-assisted hydrothermal synthesis and photocatalytic activity of ZnO. *Inorg. Mater.* **2007**, *43*, 35–39. [CrossRef]
37. Kempkes, P.; Singh, S.; Pargmann, C.; Soltwisch, H. Temporal behaviour of the E to H mode transition in an inductively coupled argon discharge. *Plasma Sources Sci. Technol.* **2006**, *15*, 378–383. [CrossRef]
38. An, J.; Usui, T.; Logar, M.; Park, J.; Thian, D.; Kim, S.; Kim, B.; Prinz, F.B. Plasma processing for crystallization and densification of atomic layer deposition $BaTiO_3$ thin films. *ACS Appl. Mater. Interfaces* **2014**, *6*, 10656–10660. [CrossRef]
39. Zaplotnik, R.; Vesel, A.; Mozetič, M. A fiber optic catalytic sensor for neutral atom measurements in oxygen plasma. *Sensors* **2012**, *12*, 3857–3867. [CrossRef]
40. Zhang, L.; Liao, X.; Fok, A.; Ning, C.; Ng, P.; Wang, Y. Effect of crystalline phase changes in titania (TiO_2) nanotube coatings on platelet adhesion and activation. *Mater. Sci. Eng. C* **2018**, *82*, 91–101. [CrossRef]

materials

MDPI

Article

Effectiveness of Surface Treatment with Amine Plasma for Improving the Biocompatibility of Maxillofacial Plates

Yeon-Woo Jeong [1,†], Seunggon Jung [1,†], Jeong Joon Han [1], Hong-Ju Park [1], Rok Young Kim [2], Byung-Hoon Kim [3] and Min-Suk Kook [1,*]

1 Department of Oral and Maxillofacial Surgery, School of Dentistry, Chonnam National University, Gwangju 61186, Korea
2 KJ Meditech, Gwangju 61009, Korea
3 Department of Dental Materials, College of Dentistry, Chosun University, Gwangju 61452, Korea
* Correspondence: omskook@jnu.ac.kr
† These authors equally contributed to this work.

Received: 18 July 2019; Accepted: 12 August 2019; Published: 13 August 2019

check for updates

Abstract: To date, no products have been presented for the surface treatment of metal plates used for repairing maxillofacial defects caused by trauma. Plasma surface treatment is a useful technique for chemically modifying the surfaces of biomaterials. Amine plasma-polymerization is an efficient way to prepare bioactive thin film polymers terminated with nitrogen-containing functional groups. The purpose of this study was to investigate the improvement in biocompatibility of titanium (Ti) plates treated with amine plasma-polymerization, and analyze their surfaces characteristics. To compare biocompatibility levels, in vitro test and animal study were performed using an amine plasma-polymerized Ti plate and an untreated Ti plate. After amine plasma-polymerization, the hydrophilicity of the Ti surface was remarkably improved. Biocompatibility was also improved for the Ti plates treated with amine plasma. The clinical application of this technique will not only shorten the time required for osseointegration, but will also improve the regeneration of bone.

Keywords: titanium plate; amine plasma; surface modification; hydrophilicity; biocompatibility; new bone formation

1. Introduction

In the oral and maxillofacial region, various trauma-related bone defects can occur. Therefore, various efforts have been made to develop treatments for bone defect recovery and reconstruction. [1–4]. For the reconstruction of bone defects, various bone graft materials and bone substitutes are used in combination with frequent applications of metal plate fixation [5–9]. In the case of the maxillofacial titanium plates used in bone defect site(s), the characteristics of plate surfaces may determine initial osseointegration. If the surface roughness of a plate is non-uniform, it may lead to extended time for osseointegration or failure of osseointegration. Various surface treatments can be used to modify the physical properties of titanium surfaces, such as titanium plasma spraying, hydroxyapatite coating, sandblasted with large acid-etched grits (SLA), and resorbable blast media [10–15].

Thus far, no products have been proposed for the surface treatment of metal plates used to treat maxillofacial defects caused by trauma without dental implants. This is because the metal plates used in the maxillofacial area are fabricated for the purpose of fixing reduced bone and being removed in the future, instead of supporting significant loads, unlike dental implants [16,17]. However, if a surface-treated plate is developed for the maxillofacial area, it may be possible to promote faster recovery and regenerate higher quality bone [18–23]. Among available surface modification techniques, plasma surface modification is a useful technique for chemically modifying the surfaces of biomaterials [24,25]. In particular, amine molecules are very useful for immobilization because they form covalent bonds with bioactive substances and proteins, which can improve biocompatibility compared to plates without surface modification. Additionally, amine plasma is very inexpensive and easy to mass-produce based on its simple treatment process. Furthermore, the generation of by-products, such as ozone, is minimized during plasma generation, meaning uniform surface treatment can be achieved [26–30]. A lot of studies on amine plasma surface treatment for biological implants have been reported [31–34]. The plasma-polymerization of amine-based monomers is an efficient way to prepare bioactive thin film polymers terminated with nitrogen containing functional groups [35]. These surfaces are characterized by a positive surface charge in aqueous environments and thus prone to interaction, especially with hyaluronan, a negatively charged substance of the extracellular matrix, playing a key role in the initial adhesion of osteoblasts to artificial surfaces [35]. However, no studies have been reported of the amine plasma-polymerization technique applied to titanium plate surface coating.

The purpose of this study was to investigate the improvement in biocompatibility of titanium plates treated with amine plasma through in vitro/in vivo evaluation and analyze their surfaces. Furthermore, after implanting different plates (plates treated with amine plasma and untreated plates) into leporine crania with bone defects, bone regenerative capacity was measured on the bone-side surfaces.

2. Materials and Experimental Methods

2.1. Materials

In this study, two kind of titanium (Ti) disks (diameter = 16 mm, thickness = 3 mm) and Ti plates (length = 10.85 mm, thickness = 1 mm, hole diameter = 2 mm) were used as samples for cell response and animal study, respectively. Samples were supplied by KJ Meditech Company, Gwangju in Korea. One is an untreated Ti surface (control group) and the other is sandblasted with an SLA-treated Ti surface (experimental group). In general, commercially Ti maxillofacial plate was treated with SLA for improving the osseointegration. Allylamine monomer was purchased from Sigma-Aldrich. All other chemicals and bioreagents were used without further purification. Figure 1 shows the samples used in this study.

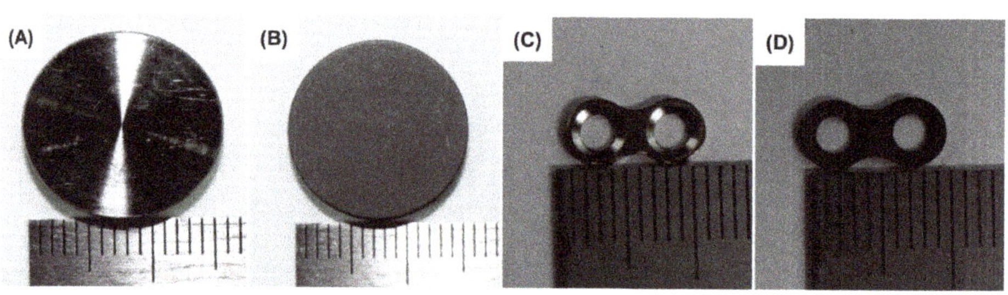

Figure 1. A Ti disk with no treatment (**A**) and sandblasted with large acid-etched grits (SLA) treated Ti disk (**B**). Ti disks were used as sample for surface analysis and in vitro evaluation. Untreated Ti mini plate (**C**) and SLA treated Ti mini plate (**D**). Ti mini plates were used as sample for animal study.

2.2. Amine Plasma-Polymerization on Sandblasted with Large Acid-Etched Grits (SLA)-Treated Ti Surface

The amine plasma-polymerization of SLA treated Ti disks and plates were carried out by depositing an ultra-thin polymeric film containing amine groups through plasma-polymerization of allylamine monomer. The equipment used for amine plasma-polymerization was a low-pressure radio frequency (RF) glow discharge plasma deposition system (model UM1000134-001A Miniplasma-station, PLASMART, Daejeon, Korea). Amine plasma-polymerization was performed under the process conditions listed in Table 1. Here, preconditioning process was performed to activate and clean on Ti surface, and post-processing was performed to enhance the degree of polymerization of the polyallylamine ultra-thin film.

Table 1. Process conditions for amine plasma-polymerization.

Treatment Conditions	Pre-Treatment	Polymerization	Post-Treatment
Plasma gas	Argon	Monomer-Allylamine	Argon
Radio Frequency discharge power	100 W	30 W, 50 W, 70 W	30 W
Deposition time	300 s	300 s	60 s

2.3. *Tiatanium Surface Characterization after Amine Plasma-Polymerization*

2.3.1. Hydrophilicity Evaluation

Following amine plasma-polymerization, the contact angles of the modified titanium surface were measured to evaluate its hydrophilicity. The contact angle was examined by the sessile drop method using a goniometer (GS, Surface Tech Co. Ltd., Gwangju, Korea). A water droplet (7 μL) was dropped using a syringe mounted vertical to the sample surface. The measurement was repeated five times for each experimental and control group.

2.3.2. Surface Analysis of Polyallylamine Thin Film Deposited by Amine Plasma-Polymerization

Following amine plasma-polymerization, Fourier transform infrared spectrometry (FT-IR, Perkin Elmer, Spectrum Two^{TM} FT-IR Spectrometer, Buckinghamshire, UK) was used to analyze the polymer thin film on the titanium surface [36,37]. The resolution was 4 cm^{-1} and eight scans were accumulated.

2.3.3. Evaluation of Amine Concentration

To determine the amine concentration on the amine plasma-polymerized titanium disk surface, a methyl orange (MO) assay was used. The MO colormetric measurement can be used to determine the number of amino groups on polymer surfaces. In brief, amino functionalized Ti samples were treated with the acidic methyl orange solution for 1 h. The absorbance of the methyl orange solution desorbed in a desorbing agent was measured at 465 nm [36]. The $-NH_2$ content was obtained from a calibration plot of the optical density versus dye concentration with a 1:1 ratio assumed between the dye and amino groups [21].

2.3.4. Surface Chemistry Analysis

The amine plasma-polymerized Ti surfaces were also characterized by X-ray photoelectron spectroscopy (XPS, VG Multilab 2000, Thermo VG Scientific, London, UK) to obtain the elemental composition and chemical state. For each specimen, a compositional survey scan was acquired using a pass energy of 50 eV and the core level spectra were acquired with a pass energy of 20 eV and a step size of 0.5 eV. All surface compositions reported are expressed in atomic percent (at.%) values.

2.4. *In Vitro Evaluation of Titanium Plates following Amine Plasma Treatment*

2.4.1. Culture of Cell Lines

The cell line used in this experiment consisted of osteoblastic cells (MC3T3-E1) from a mouse skull. These cells were purchased from ATCC.

Cell cultures were supplemented with growth factors (10% (w/v)) fetal bovine serum (FBS, PAA Laboratoris, Inc. A15-751) and antibiotics (100 units/mili-liter (mL) penicillin and 100 μg/mL streptomycin) in an α-MEM medium (alpha minimum essential medium with ribonucleosides, deoxyribonucleosides, 2 mM L-glutamine, and 1 mM sodium pyruvate, but without ascorbic acid/GIBCO, Custom Product, Catalog NO. A1049001, Thermo Fisher Scientific, Waltham, MA, USA). The cells were cultivated in a CO_2 incubator (Sanyo Electric, Osaka, Japan) maintained at 100% humidity and 37 °C with 5% CO_2. Subcultured third generation cell lines were used for cell proliferation assay at 80% confluency every three days.

2.4.2. Evaluation of the Osteoblast Proliferation

The proliferation rate of osteoblastic cell lines was evaluated using a MTT (3-(4,5-Dimethyl-2-thiazolyl)-2,5-diphenyl-2H-tetrazolium bromide)) assay based on the following process. After removing all excess α-MEM medium, the cultured MC3T3-E1 cells were washed with phosphate buffered saline (PBS, Sigma-Aldrich, St. Louis, MO, USA) and separated from the culture dish by adding trypsin/ethylenediaminetetraacetic acid (EDTA, Sigma-Aldrich, St. Louis, MO, USA). FBS was then added to the separated cells to halt any reactions and the cells were collected using a centrifuge. After the cells were suspended by adding medium, they were seeded at 5×10^4 cells/well in a 12-well plate containing the prepared sample. On the first, third, and fifth days, 100 µL of MTT (M2128, Sigma-Aldrich, St. Louis, MO, USA) reagent was added to each well and it was confirmed that blue-violet crystals were formed. After 4 h, 1,000 µL/well of dimethyl sulfoxide (35535-0350, Junsei Chemical Co., Ltd., Tokyo, Japan) was added, followed by incubation at room temperature for 30 min. To measure absorbance, 200 µL of each reaction mixture was dispensed into a 96-well plate and the absorbance was measured at 540 nm using an EPOCH device (BioTek Instruments, Inc., Winooski, VT, USA).

2.4.3. Observations of Cytoskeletons and Nuclei of the Osteoblast

MC3T3-E1 cells were seeded at a concentration of 2×10^5 cells/mL in a 12-well plate for each sample. After 48 h of cell seeding, the cells were fixed with a 4% paraformaldehyde (Electron Microscopy Sciences 15714) solution and washed with PBS for 5 min. The cells were then treated with 0.1% Triton X-100 (T1020, Bio-Sesang, Seongnam, Korea) and 1% bovine serum albumin(BSA, SIGMA A9647, Sigma-Aldrich, St. Louis, MO, USA) for 5 min. A Rhodamine-Phalloidin (R415, Thermo Fisher Scientific, Waltham, MA, USA) reagent was applied for 15 min to stain the cells. After washing the cells with PBS three times for 5 min each, the samples were then attached to cover glasses using a fluorescence mounting media (VECTOR H-1200, Vector Laboratories, Burlingame, CA, USA) for nuclei staining and stored at 4 °C for further observation using a fluorescence microscope [38].

2.4.4. Evaluation of Osteoblasts Differentiation

The differentiation of osteoblasts was confirmed by measuring alkaline phosphatase (ALP) activities. The cells were cultured on an untreated titanium plate and amine plasma-treated titanium plate at a concentration of 2×10^5 cells/mL. The original medium was replaced with a mixed medium containing α-MEM (GIBCO), 50 µg/mL of ascorbic acid (JUNSEI), and 10 mM of β-glycerophosphate (Sigma-Aldrich) the following day. To evaluate the level of differentiation, p-nitrophenol (p-NP) release from p-NP phosphate was measured after 5 d. The medium containing MC3T3-E1 cells was washed twice with Dulbecco's PBS (Welgene) and again with a 0.9% NaCl solution containing 0.2% of Triton X-100 (Sigma-Aldrich) for 10 min. The cells were sonicated for 1 min at 65 W on ice using a Vibra-Cell instrument (SONICS, Newtown, CT, USA). The washed lysate was centrifuged at 4 °C and 2500 × *g* for 10 min. The supernatant was cultivated using a p-NP phosphate solution for 30 min at 37 °C. The reaction was halted by adding 600 µL of 1.2-N NaOH. ALP activities were confirmed by measuring absorption at 405 nm and normal protein concentration using an enzyme-linked immunosorbent assay (ELISA) reader (BioTek Instruments, Inc. Winooski, VT, USA). Protein concentrations were determined using Bradford protein quantification (Bio-Rad Protein Assay, Bio-Rad Laboratories, Inc., Hercules, CA, USA) and the results were expressed as µmole of p-NP/min/µg protein.

2.4.5. Evaluation of Bone Calcification

Bone calcification was measured using Alizarin Red S staining. Differentiated cells were washed twice using a Dulbecco's PBS solution and fixed with a 10% formalin solution for 15 min. Following fixation, the cells were washed twice with distilled water. Shortly thereafter, an Alizarin Red S solution was mixed with distilled water and an NH_4OH (ammonium hydroxide) solution was added to titrate the pH level to 4.1. The treated Alizarin Red S solution was added to the fixed cell mixtures to cover all cells and was covered with foil for approximately 45 min. The Alizarin bonded to the Ca^{2+} in the cell matrix exhibited a reddish-orange color.

2.5. Animal Study

Domestic male rabbits weighing 2 kg were used as the experimental subjects. Experiments were conducted on a total of 20 rabbits. Two-hole straight plates with no treatment and two-hole straight plates with amine plasma treatment were tested.

The study protocol was approved by the Institutional Review Board at Chonnam National University Hospital. The authorization number is CNU IACUC-H-2018-32. All surgical procedures were performed under sterile conditions. The experimental procedure was as follows. The day before the experiment, sterilization using an autoclave was performed on all surgical instruments used during animal experimentation (121 °C. 15 psi, 15 min). The rabbits were weighed and 3 mL of Zolethyl (Zolethyl 50, Virbac Co, Carros, France) per kg bodyweight was injected intramuscularly into the thigh to induce sedation and anesthesia. The rabbits were placed in the prone position and shaved from the bregma to the supraorbital margin. Povidone iodine, chlorhexidine gluconate, and isopropanol were then used to disinfect the surgical site. Infiltration anesthesia was administered with lidocaine containing 1 ppm of epinephrine at the surgical site. The cranium was exposed after incision and dissection of subcutaneous tissue and the periosteum using a No. 15 blade. After exposing the cranium, a round bur was used under irrigation to form a shallow groove in which a small metal plate could be placed. Groove formation also causes moderate bleeding to promote osteogenesis. Each rabbit received one untreated titanium plate and one treated with amine plasma. These two-hole miniplates were fixed with screws for both test groups. The periosteum was sutured using 3-0 vicryl, followed by the skeletal muscle, subcutaneous tissue, and skin using 3-0 vicryl, 5-0 vicryl, and 6-0 nylon, respectively. The surgical sites were disinfected using povidone iodine, chlorhexidine gluconate, and isopropanol, mirroring the pre-operation procedure. Finally, Zenocef® (first-generation cephalosporin antibiotic) and Keromin® (non-steroidal anti-inflammatory drug) were injected. The systemic conditions of the rabbits, as well as the healing and infection conditions at the surgical sites, were monitored daily and the sites were disinfected. After four weeks, the rabbits were euthanized using a KCl solution and bone fragments with a margin of 5 mm around the small metal plates were extracted using an osteotome. Figure 2 shows the extracted Ti mini plate samples used in animal study.

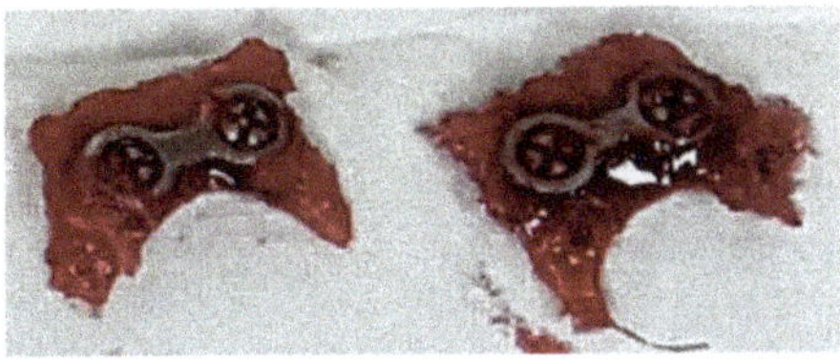

Figure 2. Samples extracted four weeks after the operations.

2.6. Statistical Analysis

In vitro data were statistically analyzed using a Kolmogorov–Smirnov (KS) test to check for normality. A student's t-test was then used based on the normal distribution identified by the KS test.

A threshold of $p \leq 0.05$ was considered to be significant (*) and a threshold of $p \leq 0.01$ was considered to be highly significant (**).

Histomorphometric analysis of newly formed bone was carried out using an Image J® Version 1.52 device (National Institutes of Health, Wayne Rasband, Bethesda, MD, USA). After tracing the new bone growth, the pixels in the traced area were counted. The mean results were compared through a paired t-test using the SPSS 23.0 TM software (IBM, Armonk, NY, USA). The differences were considered to be statistically significant at a threshold of $p < 0.05$.

3. Results and Discussion

3.1. Changes of Ti Surface Characterizations after Amine Plasma-Polymerization

3.1.1. Determination of Amine Concentrations

Figure 3 shows the optical density of methyl orange solution desorbed and amine concentrations as a function of plasma RF powers. As the shown in Figure 3, amine concentrations on the Ti surface were 41 nmol/cm^2 at 30 W, 64 nmol/cm^2 at 50 W, and 73 nmol/cm^2 at 70 W, indicating that as power increases, the amine concentration on the surface also increases. The highest degree of NH_2 density on the surface is 73 nmol/cm^2, was detected at the higher RF power 70 W. The hydrophilicity is influenced by the $-NH_2$ grafted density on the modified Ti surface, so we selected the optimized condition for the plasma-polymerization process (RF power 70 W). Therefore, we treated all plasma-polymerized Ti samples used in the in vitro and animal studies with this condition.

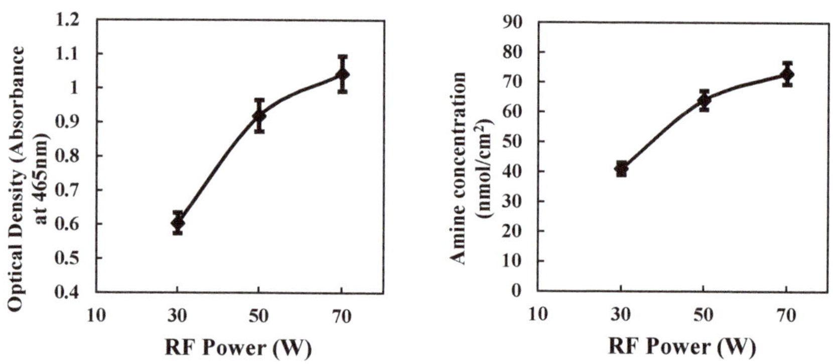

Figure 3. Optical density values of methyl orange solution (**left**) and amine concentrations on the titanium surface (**right**) with increasing radio frequency (RF) plasma powers.

3.1.2. Identification of Polyallylamine Ultra-Thin Film Deposited on the Ti Surface

To confirm the presence of polyallylamine thin film after the amine plasma-polymerized Ti surface, FT-IR analysis was performed. Figure 4 shows the FT-IR spectra of the SLA-treated Ti surface and amine plasma-polymerized Ti surfaces. As the shown in Figure 4, the FT-IR spectrum of the SLA-treated Ti surface did not shown the amines absorption bands, on the other hand, amine plasma-polymerized Ti surface clearly shows several absorption bands characteristic of the presence of primary (1600 cm^{-1}) and/or secondary amines (3380 cm^{-1}) groups. Based on this result, the presence of a polyallylamine ultra-thin film on the surface of amine plasma-polymerized titanium disk was confirmed. Therefore, it is clear that the plasma surface modification of Ti surface through amine plasma-polymerization was successful.

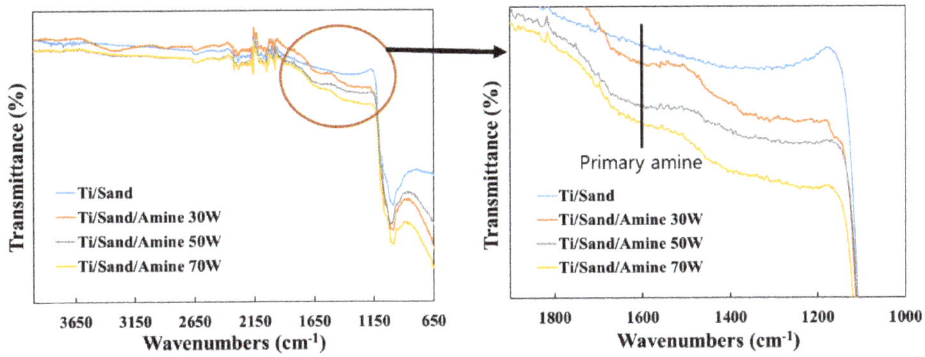

Figure 4. Fourier transform infrared (FT-IR) spectra of the SLA-treated Ti surface and amine plasma-polymerized Ti surfaces under three different plasma powers.

3.1.3. Surface Chemistry Analysis

To examine the surface chemistry of the Ti sample after amine plasma-polymerization, X-ray photoelectron spectroscopy (XPS, VG Multilab 2000, Thermo VG Scientific, London, UK) analysis was performed. The surface compositions determined from the wide scan XPS spectra are shown in Figure 5. The peaks at 284.8 and 531.2 eV were present in untreated Ti samples and assigned to C 1s and O 1s, respectively. The peak at 399.9 eV, corresponding to N 1s, was detected in amine plasma-polymerized Ti surface only. From the XPS analysis results, we confirmed that amine functional group containing polyallylamine thin film was successfully deposited on Ti surface thorough plasma-polymerization process.

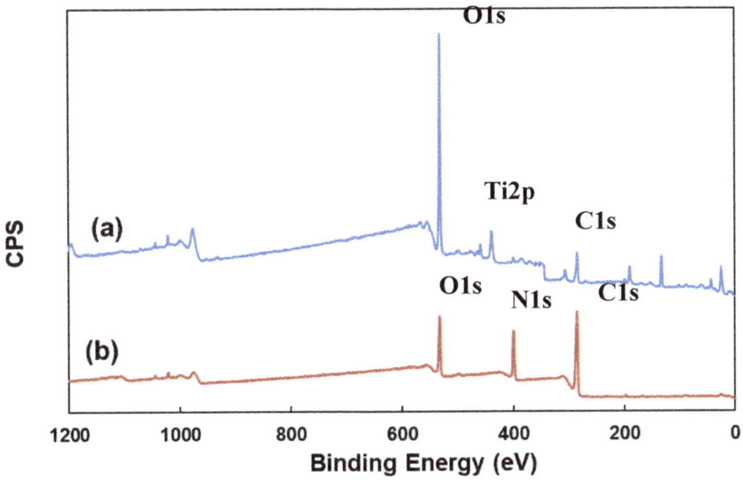

Figure 5. X-ray photoelectron spectroscopy (XPS) spectra of (**a**) untreated Ti surface and (**b**) amine plasma-polymerized Ti surface.

3.1.4. Hydrophilicity

The hydrophilicity was determined by measuring the water contact angle (WCA) on the plasma-polymerized Ti surface. Figure 6 shows the contact angle values and representative images of the corresponding water droplets. It can be seen that there are obvious differences in the wetting properties before and after plasma-polymerization. The contact angle of the untreated titanium surface was 77.18°, representing strong hydrophobicity. In contrast, the contact angle of the surface treated with amine plasma was 28.14°, representing strong hydrophilicity. These results confirm that changes in surface chemical compositions can affect the wettability of the materials, which is a useful parameter that has often been strongly correlated with the cell-biomaterial interfacial interactions [38].

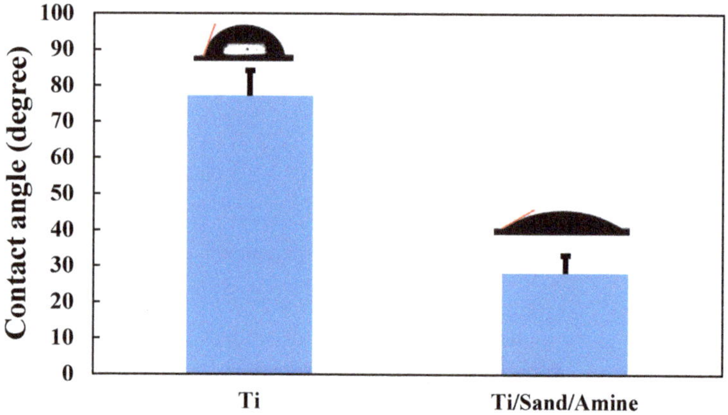

Figure 6. Contact angles of the untreated titanium surface and titanium surface treated with amine plasma-polymerization.

3.2. *In Vitro Evaluations*

3.2.1. Proliferation Rate of Osteoblastic Cell Lines

Cell proliferation was measured using an MTT assay. Figure 7 presents the cell proliferation rates after one, three, and five days of cell culturing. As a control, the cell proliferation rate on the surface of the untreated titanium plate was defined as 100%. The Ti samples with amine plasma-polymerization showed higher MC3T3-E1 proliferation than untreated Ti samples. All the experimental groups for MTT assay showed statistically significant differences ($p < 0.01$, in comparison to untreated Ti sample).

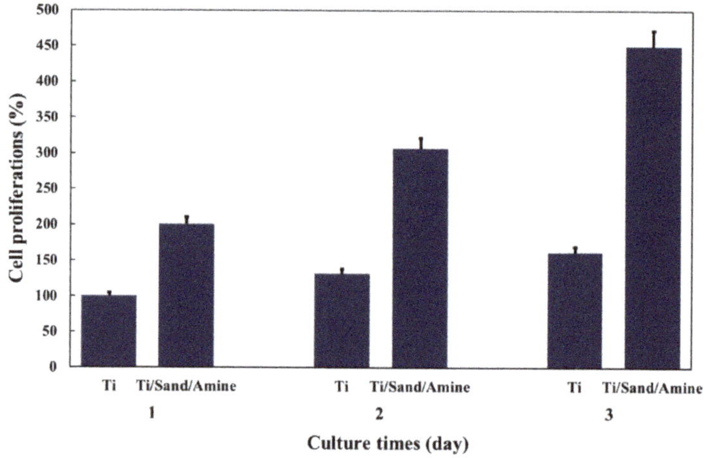

Figure 7. Untreated titanium surface and titanium surfaces treated with amine plasma. Proliferation rates of cultured osteoblasts on the surfaces after one, three, and five days.

3.2.2. Observations of Cytoskeletons and Nuclei of the Osteoblastic Cell Lines

Rhodamine-phalloidin and 4′,6-diamidino-2-phenylindole (DAPI) fluorescence staining were performed to observe the cell morphologies attached to the titanium plates (Figure 8). Four h after cell seeding, fluorescence microscopy revealed that the number of cells was greater on the titanium surface treated with amine plasma than on the untreated titanium surface. Additionally, cytoskeletons were spread widely on the titanium surface treated with amine plasma, confirming the enhanced biocompatibility of the treated surface.

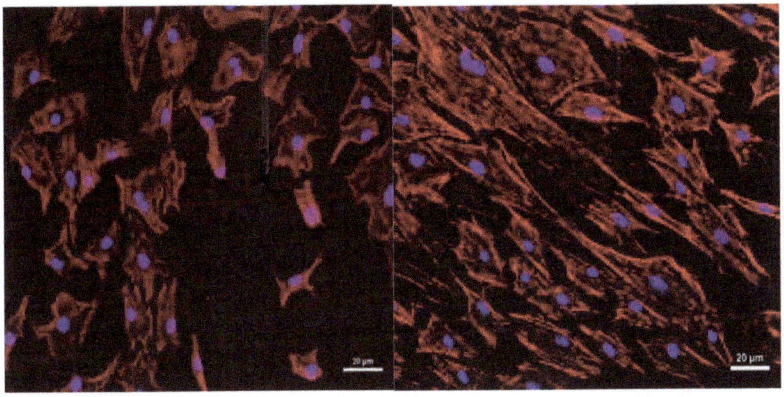

Figure 8. Cell morphologies on the surfaces of plates: untreated titanium plate (**left**) and titanium plate treated with amine plasma (**right**). Nuclei color is blue.

3.2.3. Differentiation of Osteoblastic Cell Lines

ALP activities were measured to analyze the differentiation of cells attached to the titanium plates. After seven days of osteoblast cultivation, ALP activities were measured on the untreated titanium surface and the titanium surface treated with amine plasma. It was apparent that the level of differentiation increased significantly in the sample treated with amine plasma (Figure 9). All the experimental groups for ALP activity showed statistically significant differences ($p < 0.01$, in comparison to untreated Ti sample).

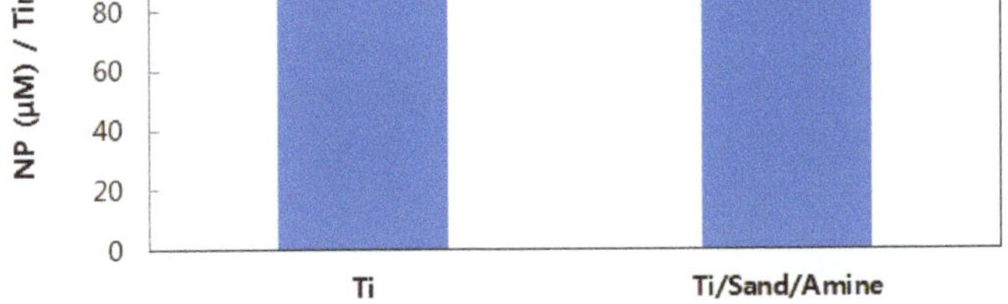

Figure 9. Alkaline phosphatase (ALP) activities of osteoblasts cultured for seven days on the untreated titanium surface and the titanium surface treated with amine plasma.

3.2.4. Bone Calcification

Alizarin Red S staining was performed to examine the calcification of bone generated on the titanium plates. Osteoblasts were cultivated on both the untreated titanium surface and the titanium surface treated with amine plasma for 14 d. As the shown in Figure 10, amine plasma-polymerized Ti disk shows the strong reddish color compare to the untreated Ti disk. Therefore, the sample treated with amine plasma exhibited excellent bone calcification.

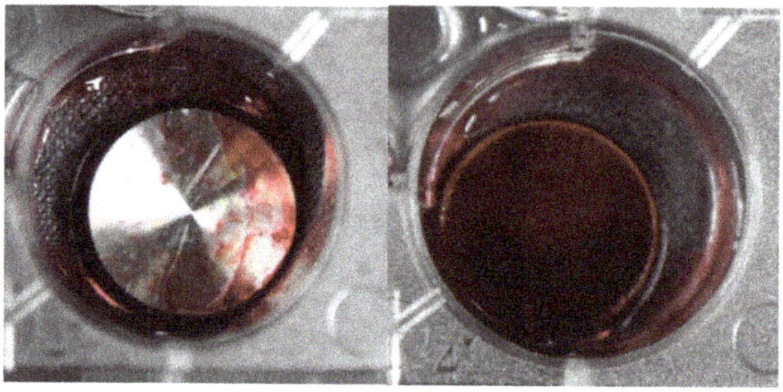

Figure 10. Levels of calcification on the surfaces of plates: bone calcification on the untreated titanium plate (**left**) and on the titanium plate treated with amine plasma (**right**).

3.3. *Animal Study*

3.3.1. Clinical Evaluation

On the day of euthanizing the rabbits, the following visual observations were made. On the soft-tissue-side surfaces of the Ti plates in the experimental group where polishing was performed during dissection, the dissection was relatively clean compared to the control group (Figure 11). Additionally, the amount of periosteum and inflammation remaining around the metal plate was relatively small and the smooth surface of the bone could be observed.

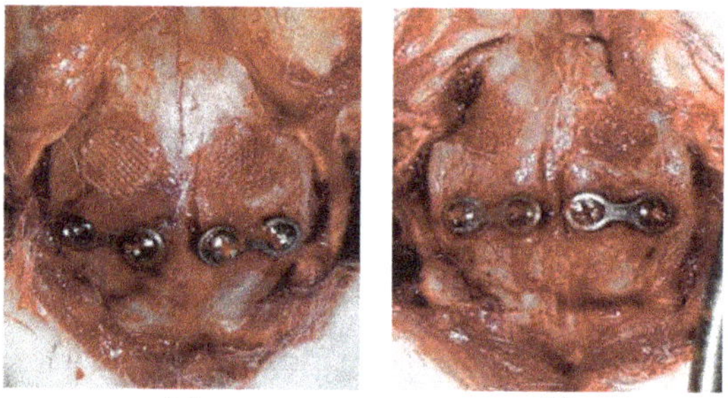

(**a**) (**b**)

Figure 11. Four weeks after the initial operation, the untreated two-hole straight plate (**a**) and two-hole straight plate treated with amine plasma (**b**) were dissected. The periosteum and surrounding tissue remained around the untreated two-hole straight plate. In contrast, a relatively clear separation of tissue around the two-hole straight plate treated with plasma can be observed.

3.3.2. Histological Findings

Figure 12 shows the hematoxylin and eosin (H&E) staining of the tissue around the plates after four weeks for the untreated two-hole straight plates and amine plasma-polymerized plates. The histological findings from photographs of the collected samples in Figure 12 are discussed below. Compared to the control group, dissection was cleaner on the soft-tissue-side surfaces of the experimental group, which showed little or no inflammatory tissue. The experimental group treated with amine plasma had closer contact between the small metal plate and newly formed bone layer compared to the control group. Additionally, there was very little fatty tissue intercalating with the soft tissue.

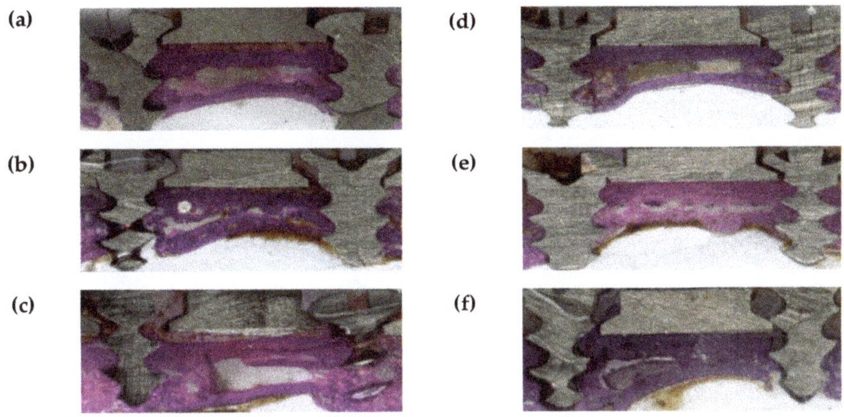

Figure 12. (**a**–**c**) Hematoxylin and eosin (H&E) staining of the tissue around the plates after four weeks for the untreated two-hole straight plates. A foreign body can be observed around the plates and the bone is tightly connected in some sections. Additionally, inflammatory tissue can be observed on the bone-side surfaces. (**d**–**f**) H&E staining of the tissue around the plates after four weeks for the two-hole straight plates treated with amine plasma. A small amount of foreign material can be observed around the plate and the newly formed bone on the bone-side surface is more tightly connected to the plates compared to the untreated two-hole straight plate group.

3.3.3. Histomorphometirc Findings

The ratio of the length of the newly formed bone to the length of the entire Ti plate on the bone-side surface was greater for the plasma-treated group compared to the control group (Figure 13).

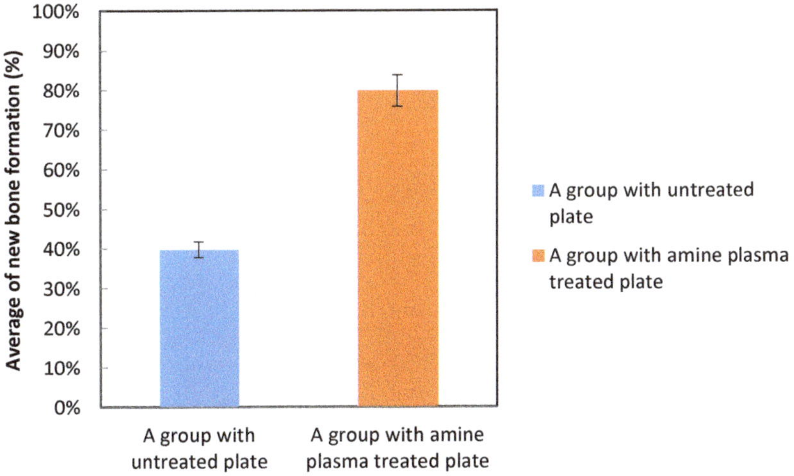

Figure 13. Comparison of the average percentage of new bone formation for each specimen in the animal experiment.

The mean values of the lengths of the six bone-contacting tissues for the control and experimental groups are listed in Table 2.

Table 2. Average lengths of bone-to-plate contact areas (the distance between the two screws is 15 mm).

-	Average Length of Bone-to-Plate Contact Area (mm)	Average Percentage of New Bone Formation (%)
Group with Untreated Plates	6.06	39.70
Group with Amine Plasma-Treated Plates	12.18	79.76

($p = 0.005$).

3.3.4. Micro-Computed Tomography (Micro-CT) Findings

Figure 14 shows the micro-computed tomography (micro-CT) images of the untreated two-hole straight plates and amine plasma-polymerized plates after four weeks. As the shown in Figure 14, micro-CT data revealed that the amount of newly formed bone on the bone-side surfaces of the experimental group was significantly higher than that on the corresponding surfaces of the control group. Additionally, the distance between the metal plate and layer of newly formed bone was relatively small for the experimental group. These results match the results of microscopic evaluation.

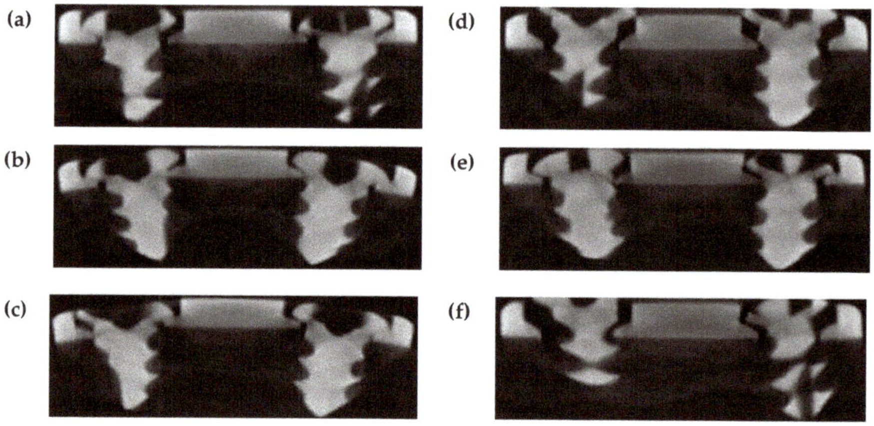

Figure 14. (**a–c**) Micro-computed tomography (micro-CT0 images of the untreated two-hole straight plates after four weeks. One can see that weak bone formed on the bone-side surface of the metal plate. (**d–f**) Micro-CT images of the two-hole straight plates treated with amine plasma after four weeks. One can see that dense bone formed on the bone-side surface of the titanium plate.

Our experimental results confirmed that titanium treated with amine plasma is biocompatible.

As reported by Kristen R. Kull et al., the hydrophilicity of amine plasma treated titanium plate was increased compared to that of untreated titanium plate [39]. This may lead to a significant difference in terms of biocompatibility when considering the hydrophilic environment of the bone surface. In addition, FT-IR analysis showed that the amine group was actually present on the surface of the titanium plate, and the concentration of the amine group on the surface of the titanium plate also increased with an increase in the intensity of the polymer [28]. Amine plasma treatment is not so complicated, so it is possible to fabricate a titanium plate with higher biocompatibility by polymerizing at stronger intensity in clinical practice. Animal studies artificially induced some bleeding to further activate bone regeneration. In order to minimize the differences between individuals, a titanium plate treated with amine plasma and a titanium plate treated with no treatment were plated for each individual. The proliferation rate and degree of differentiation of osteoblasts measured in this same environment showed a great difference in two types of titanium plates. As time passed, the difference in the proliferation rate of the osteoblasts between the titanium plate treated with no treatment and the titanium plate treated with the amine plasma was found to be large, and the difference was more than twice as high as on the fifth day of culture. Fluorescence microscopy showed that the number of nuclei in the titanium plate treated with amine plasma was much higher than that of the untreated titanium plate and that the cytoskeleton was more developed [40]. These results accurately reflect the results obtained from the specimens collected from animal studies. Histological findings showed that the titanium plate treated with amine plasma showed more uniformly formed new bone than the untreated titanium plate. And this is the same in micro CT. These results suggest that amine plasma treatment on titanium plate may increase biocompatibility and osteogenesis.

However, in the case of sprayed plasma-coated titanium, chemical changes on the surface may cause changes in physical properties. Therefore, additional studies on the tensile load, compressive load, bending load, torsion load, and amount of deformation of titanium plates treated with amine plasma are required [41].

Additionally, titanium powder is very reactive, making it difficult to reproduce accurate samples using sprayed plasma coating. Therefore, it is necessary to study safe methods for the stable and accurate surface treatment of titanium with plasma [42–44].

Nutritional supply may also be important for the formation of new bone. Therefore, if nutrient materials such as polyglycolic acid, one of the bioreactors, are provided with amine treatment, the regeneration of bone will be improved [45].

4. Conclusions

In this study, we found that titanium surface treatment through amine plasma-polymerization was successfully performed on maxillofacial titanium plates used in clinical surgery. We characterized the amine plasma-treated Ti surface, and investigated the effectiveness of amine plasma-polymerization in improving the biological activities in both in vitro and in vivo tests. After plasma surface treatment, polyallylamine ultra-thin film was deposited on Ti surface and hydrophilicity of Ti surface was improved. Amine groups contained polyallylamine on the Ti surface positively influenced osteoblast cell behaviors, such as proliferation, and differentiation. In addition, results of animal test showed the improvement of new bone formation. Therefore, the clinical application of this technique can not only shorten the time required for osseointegration, but also improve the regeneration of bone. Based on the results, the superior bone regenerative capacity of plates treated with amine plasma is expected to increase their therapeutic effects in clinical practice.

Author Contributions: Conceptualization, Y.-W.J. and M.S.K.; Methodology, Y.-W.J., B.-H.K. and M.-S.K.; Software, B.-H.K.; Validation, J.J.H. and R.Y.K. and Z.Z.; Formal Analysis, S.J.; Investigation, Y.-W.J., S.J. and B.-H.K.; Data Curation, Y.-W.J., S.J. and R.Y.K.; Writing—Original Draft Preparation, Y.-W.J. and S.J.; Writing—Review and Editing, J.J.H., H.-J.P., B.-H.K. and M.-S.K.; Visualization, R.Y.K.; Supervision, H.-J.P., M.-S.K.; Project Administration, M.-S.K.

Funding: This research was funded by National Research Foundation of Korea, grant number NRF-2017R1D1A1A02017564 and it was also supported by a grant (BCRI 19023) from Chonnam National University Hospital Biomedical Research Institute, Republic of Korea.

Conflicts of Interest: The authors declare no conflict of interest.

References

1. Brown, A.; Zaky, S.; Ray, H.; Sfeir, C. Porous magnesium/PLGA composite scaffolds for enhanced bone regeneration following tooth extraction. *Acta Biomater.* **2015**, *11*, 543–553. [CrossRef] [PubMed]
2. Tevlin, R.; McArdle, A.; Atashroo, D.; Walmsley, G.G.; Senarath-Yapa, K.; Zielins, E.R.; Paik, K.J.; Longaker, M.T.; Wan, D.C. Biomaterials for craniofacial bone engineering. *J. Dent. Res.* **2014**, *93*, 1187–1195. [CrossRef] [PubMed]

3. Neovius, E.; Engstrand, T. Craniofacial reconstruction with bone and biomaterials: Review over the last 11 years. *J. Plast. Reconstr. Aesthet. Surg.* **2010**, *63*, 1615–1623. [CrossRef] [PubMed]
4. Porter, J.R.; Ruckh, T.T.; Popat, K.C. High calcium bioglass enhances differentiation and survival of endothelial progenitor cells, inducing early vascularization in critical size bone defects. *PLoS ONE* **2009**, *25*, 1539–1560.
5. Kim, D.G. Finite element analysis of stress and deformation according to the shape of plates for internal bone fixation. *J. Korean Orthop. Assoc.* **2012**, *47*, 257–263.
6. Kweon, H.Y.; Yoo, M.K.; Park, I.K.; Kim, T.H.; Lee, H.C.; Lee, H.S.; Oh, J.S.; Akaike, T.; Cho, C.S. A novel degradable polycaprolactone network for tissue engineering. *Biomaterials.* **2003**, *24*, 801–808. [CrossRef]
7. Suh, S.W.; Shin, J.Y.; Kim, J.; Kim, J.; Beak, C.H.; Kim, D.I.; Kim, H.; Jeon, S.S.; Choo, I.W. Effect of different particles on cell proliferation in polymer scaffolds using a solvent-casting and particulate leaching technique. *ASAIO J.* **2002**, *48*, 460–464. [CrossRef]
8. Kook, M.S.; Roh, H.S.; Kim, B.H. Effect of oxygen plasma etching on pore size-controlled 3D polycaprolactone scaffold s for enhancing the early new bone formation in rabbit calvaria. *Dent. Mater. J.* **2018**, *37*, 599–610. [CrossRef]
9. Won, J.Y.; Park, C.Y.; Bae, J.H.; Ahn, G.; Kim, C.; Lim D., H.; Cho, D.W.; Yun, W.S.; Shim, J.H.; Huh, J.B. Evaluation of 3D printed PCL/PLGA/β-TCP versus collagen membranes for guided bone regeneration in a beagle implant model. *Biomed. Mater.* **2016**, *11*, 055013. [CrossRef]
10. Shim, J.H.; Won, J.Y.; Park, J.H.; Bae, J.H.; Ahn, G.; Kim, C.H.; Lim, D.H.; Cho, D.W.; Yun, W.S.; Bae, E.B.; et al. Effects of 3D-printed polycaprolactone/-tricalcium phosphate membranes on guided bone regeneration. *Int. J. Mol.* **2017**, *18*, E899. [CrossRef]
11. Lavik, E.; Langer, R. Tissue engineering: Current state and perspectives. *Appl. Microbiol. Biotechnol.* **2004**, *65*, 1–8. [CrossRef] [PubMed]
12. Johnson, P.C.; Mikos, A.G.; Fisher, J.P.; Jansen, J.A. Strategic directions in tissue engineering. *Tissue Eng.* **2007**, *13*, 2827–2837. [CrossRef] [PubMed]
13. Mooney, D.J.; Baldwin, D.F.; Suh, N.P.; Vacanti, J.P.; Langer, R. Novel approach to fabricate porous sponges of poly (D,Llactic co-glycolic acid) without the use of organic solvents. *Biomaterials* **1996**, *17*, 1417–1422. [CrossRef]
14. Yamoso-Scholl, K.; Jacobson, J.A.; Bradica, G.; Lerner, A.L.; O'Keefe, R.J.; Schwarz, E.M.; Zuscik, M.J.; Awad, H.A. Evaluation of dense polylactic acid/beta-tricalcium phosphate scaffolds for bone tissue engineering. *J. Biomed. Mater. Res. A* **2010**, *95*, 717–726.
15. Holliser, S.J.; Lin, C.Y.; Saito, E.; Schek, R.D.; Taboas, J.M.; Williams, J.M.; Partee, B.; Flanagan, C.L.; Diggs, A.; Wilke, E.N.; et al. Engineering craniofacial scaffolds. *Orthod. Craniofac. Res.* **2005**, *8*, 162–173. [CrossRef] [PubMed]
16. Lourenço, M.N.; Sartori, E.M.; Padovan, L.E.M.; Thomé, G.; Faeda, R.S.; Marcantonio, J.E.; Claudino, M. Bone apposition and surface treatment in dental implants: Histomorphometric pilot evaluation in rabbits. *RSBO* **2013**, *10*, 326–334.
17. Cheng, Z.Y.; Teoh, S.H. Surface modification of ultra thin poly (epsilon-caprolactone) films using acrylic acid and collagen. *Biomaterials* **2004**, *25*, 1991–2001. [CrossRef] [PubMed]
18. Sarasam, A.R.; Krishnaswamy, R.K.; Madihally, S.V. Blending chitosan with polycaprolactone: Effects on physicochemical and antibacterial properties. *Biomacromolecules* **2006**, *7*, 1131–1138. [CrossRef] [PubMed]
19. Tiaw, K.S.; Goh, S.W.; Hong, M.; Wang, Z.; Lan, B.; Teoh, S.H. Laser surface modification of poly(ε-caprolactone) (PCL) membrane for tissue engineering applications. *Biomaterials* **2005**, *26*, 763–769. [CrossRef]
20. Katti, D.; Vasita, R.; Shanmugam, K. Improved biomaterials for tissue engineering applications: Surface modification of polymers. *Curr. Top. Med. Chem.* **2008**, *8*, 341–353. [CrossRef]
21. Ma, Z.; Mao, Z.; Gao, C. Surface modification and property analysis of biomedical polymers used for tissue engineering. *Colloid. Surface. B* **2007**, *60*, 137–157. [CrossRef] [PubMed]
22. Nanci, A.; Wuest, J.D.; Brunet, P. Chemical modification of titanium surfaces for covalent attachment of biological molecules. *J. Biomed. Mater. Res. B* **1998**, *40*, 324–335. [CrossRef]
23. Rossini, P.; Colpo, P.; Cecone, G. Surface engineering of polymeric films for biomedical applications. *Mater. Sci. Eng. C* **2003**, *23*, 353–358. [CrossRef]
24. Christophorou, L.G.; Olthoff, J.K. *Fundamental Electron Interactions with Plasma Processing Gases*; Springer: New York, NY, USA, 2004.
25. Conrads, H.; Schmidt, M. Plasma generation and plasma sources. *Plasma Sources Sci. Technol.* **2000**, *9*, 441–454. [CrossRef]

26. Yameen, B.; khan, H.U.; Knoll, W.; Förch, R.; Jonas, U. Surface initiated polymerization on pulsed plasma deposited polyallylamine: A polymer substrate-independent strategy to soft surfaces with polymer brushes. *Macromol. Rapid Commun.* **2011**, *32*, 1735–1740. [CrossRef] [PubMed]
27. Hook, A.L.; Thissen, H.; Quinton, J.; Voelcker, N.H. Comparison of the binding mode of plasmid DNA to allylamine plasma polymer and poly(ethylene glycol) surfaces. *Surf. Sci.* **2008**, *602*, 1883–1891. [CrossRef]
28. Choukourov, A.; Biederman, H.; Kholodkov, I.; Slavinska, D.; Trchova, M.; Hollander, A. Properties of amine-containing coatings prepared by plasma polymerization. *J. Appl. Polym. Sci.* **2004**, *92*, 979–990. [CrossRef]
29. Ho-Shui-Ling, A.; Bolander, J.; Rustom, L.E.; Johnson, A.W.; Luyten, F.P.; Picart, C. Bone regeneration strategies: Engineered scaffolds, bioactive molecules and stem cells current stage and future perspectives. *Biomaterials* **2018**, *180*, 143–162. [CrossRef]
30. Gaelle, A.; Nathalie, D.G.; Rino, M. Incorporation of primary amines via plasma technology on biomaterials. *IntechOpen* **2015**, *10*, 59691.
31. Finke, B.; Hempel, F.; Testrich, H.; Artemenko, A.; Rebl, H.; Kylián, O.; Meichsner, J.; Biederman, H.; Nebe, B.; Weltmann, K.D.; et al. Plasma processes for cell-adhesive titanium surfaces based on nitrogen-containing coatings. *Surf. Coat. Technol.* **2011**, *205*, S520–S524. [CrossRef]
32. Testrich, H.; Rebl, H.; Finke, B.; Hempel, F.; Nebe, B.; Meichsner, J. Aging effects of plasma polymerized ethylenediamine (PPEDA) thin films on cell-adhesive implant coatings. *Mater. Sci. Eng. C* **2013**, *33*, 3875–3880. [CrossRef] [PubMed]
33. Seo, H.S.; Kim, B.H.; Ko, Y.M. Fabrication of anodized titanium with immobilization of hyaluronic acid to improve biological performance. *Prog. Org. Coat.* **2010**, *69*, 38–44. [CrossRef]
34. Heller, M.; Kämmerer, P.W.; Al-Nawas, B.; Luszpinski, M.-A.; Förch, R.; Brieger, J. The effect of extracellular matrix proteins on the cellular response of HUVECS and HOBS after covalent immobilization onto titanium. *J. Biomed. Mater. Res. Part A* **2015**, *103A*, 2035–2044. [CrossRef] [PubMed]
35. Pietro, M.; Federico, M.; Paola, R.; Stefano, C. Surface treatments and functional coatings for biocompatibility improvement and bacterial adhesion reduction in dental implantology. *Coatings* **2016**, *6*, 7.
36. Hartwig, A.; Mulder, M.; Smolders, C.A. Surface amination of poly(acrylonitrile). *Adv. Colloid Interface Sci.* **1994**, *52*, 65–78. [CrossRef]
37. Ciapetti, G.; Ambrosio, L.; Savarino, L.; Granchi, D.; Cenni, E.; Ba ldini, N.; Paqani, S.; Guizzardi, S.; Gausa, F.; Giumti, A. Osteoblast growth and function in porous poly-ε-caprolactone matrices for bone repair: A preliminary study. *Biomaterials* **2003**, *24*, 3815–3824. [CrossRef]
38. Lopez-Perez, P.M.; Marques, A.P.; Silva, R.M.P.D.; Pashkuleva, I.; Reis, R.L. Effect of chitosan membrane surface modification via plasma induced polymerization on the adhesion of osteoblast-like cells. *J. Mater. Chem.* **2007**, *17*, 4064–4071. [CrossRef]
39. Kull, K.R.; Steen, M.L.; Fisher, E.R. Surface modification with nitrogen-containing plasmas to produce hydrophilic, low-fouling membranes. *J. Membr. Sci.* **2005**, *246*, 203–215. [CrossRef]
40. Schmitz, J.P.; Hollinger, J.O. A preliminary study of the osteogenic potential of a biodegradable alloplastic osteoinductive alloimplant. *Clin Orthop.* **1988**, *237*, 245–255. [CrossRef]
41. Jones, A.C.; Arns, C.H.; Hutmacher, D.W.; Milthorpe, B.K.; Sheppard, A.P.; Knackstedt, M.A. The correlation of pore morphology, interconnectivity and physical properties of 3D ceramic scaffolds with bone ingrowth. *Biomaterials* **2009**, *30*, 1440–1451. [CrossRef]
42. Morent, R.; De Geyter, N.; Desmet, T.; Dubruel, P.; Leys, C. Plasma surface modification of biodegradable polymers: A review. *Plasma Process. Polym.* **2011**, *8*, 171–190. [CrossRef]
43. Desmet, T.; Morent, R.; De Geyter, N.; Leys, C.; Schacht, E.; Dubruel, P. Nonthermal plasma technology as a versatile strategy for polymeric biomaterials surface modification: A review. *Biomacromolecules* **2009**, *10*, 2351–2378. [CrossRef]
44. Chu, P.K.; Chen, J.Y.; Wang, L.P.; Huang, N. Plasma-surface modification of biomaterials. *Mat. Sci. Eng. R.* **2002**, *36*, 143–206. [CrossRef]
45. Karande, T.S.; Ong, J.L.; Agrawal, C.M. Diffusion in musculoskeletal tissue engineering scaffolds: Design issues related to porosity, permeability, and nutrient mixing. *Ann. Biomed. Eng.* **2004**, *32*, 1728–1743. [CrossRef]

materials

MDPI

Communication

Titanium Plasma-Sprayed Coatings on Polymers for Hard Tissue Applications

Artur Wypych [1], Piotr Siwak [2], Daniel Andrzejewski [3] and Jaroslaw Jakubowicz [1,*]

1 Institute of Materials Science and Engineering, Poznan University of Technology, Jana Pawla II 24, 61-138 Poznan, Poland; artur.wypych@put.poznan.pl

2 Institute of Mechanical Technology, Poznan University of Technology, Piotrowo 3, 60-965 Poznan, Poland; piotr.siwak@put.poznan.pl

3 Metal Forming Institute, Jana Pawla II 14, 61-139 Poznan, Poland; daniel.andrzejewski@inop.poznan.pl

* Correspondence: jaroslaw.jakubowicz@put.poznan.pl; Tel.: +48-61-665-3781

Received: 19 November 2018; Accepted: 11 December 2018; Published: 13 December 2018

check for updates

Abstract: The paper presents the results of titanium plasma spraying (TPS) on polymer substrates. Polyethylene (PE300), polyamide PA6, and fiber glass-reinforced polyamide (PA6.6-GF30) were used as substrates. The PE300 and PA6.6-GF30 substrates exhibited appropriate behavior during the TPS process, whereas the PA6 substrate did not "accept" Ti during plasma spraying, and the coating did not form. The TPS coatings exhibited low porosity and high homogeneity, and they had a typical multilayer structure composed of Ti and its oxides. The nanoindentation test showed good mechanical properties of the coatings and demonstrated a hardness and a Young's modulus of approximately 400 HV and 200 GPa, respectively. The bending test confirmed the good adhesion of the titanium coatings to the polymer substrates. The Ti coatings did not fall off the substrate after its significant bending deformation.

Keywords: plasma spraying; Ti coating; polymers; biomaterials

1. Introduction

Metallization of polymer materials is a potentially good choice in terms of the improvement of their mechanical properties, elastic modulus, stiffness, hardness, and wear resistance. Elements made from polymer covered with metallic coatings are lighter than bulk metallic biomaterials and stronger than biopolymers without metallic coatings. The plasma spraying process is useful for the deposition of different types of coatings on biomaterials [1,2]. Plasma spraying of titanium powders (TPS—titanium plasma spraying) is commonly used for the enhancement of osseointegration through the formation of a rough, biocompatible surface [3]. In the process, the Ti powders are introduced into the stream of high-temperature plasma, melted, and deposited with high velocity on the substrate. Usually, the process is used for metallic substrates [4,5]. Recently, some attempts have been made to deposit Ti on polymer substrates, such as polyether ether ketone (PEEK) [6,7], although most of the attempts focused on the cold spraying process [8] at a significantly lower temperature than that of high-temperature plasma spraying. During plasma spraying, the Ti powder particles are liquid when flowing from the plasma gun nozzle onto the surface. However, their high velocity and energy, and their relatively large area of particles flattened during contact with the substrate, result in a great ability to dissipate energy. Hence, the substrate usually softens but does not melt.

The properties of many polymer substrates may be incompatible with the sprayed material. During plasma spraying, the substrate should have comparable thermal properties with the sprayed coating material, which would provide proper coating integrity with the substrate without cracks and delamination at the substrate–coating interphase.

The relatively poor knowledge related to the TPS process on polymer substrates and a very high application potential of the Ti coatings on polymers have motivated the authors to study this field. In this work, the authors show the results of the TPS coating formation on selected synthetic biopolymer substrates that may be potential candidate materials for hard tissue applications. In some applications, polymer materials with Ti coatings could provide sufficient strength at a low weight and hence could phase out metallic ones of greater weight.

2. Materials and Methods

In this work, titanium particles (average size 100 mesh = 149 μm, purity 99.7%; Sigma-Aldrich, St. Louis, MO, USA) were used for the plasma spraying process for coating preparation. The TPS coatings were made on polymer substrate rods (ϕ15 mm): high-density polyethylene PE300, polyamide PA6, and polyamide reinforced with 30% glass fibers (PA6.6-GF30) (Table 1).

Table 1. Properties of the polymer substrates and the coating material.

Properties	PE300	PA6	PA6.6-GF30	Ti
Density (g/cm^3)	0.96	1.14	1.36	4.51
Yield strength (MPa)	25	76	180	280
Tensile strength (MPa)	36	79	210	350
Elongation (%)	50	50	5	20
Modulus of elasticity (GPa)	1.0	3.3	7.0	105
Hardness (MPa)	45	155	200	970
Thermal conductivity (W/mK)	0.41	0.28	0.23	17
Coefficient of linear thermal expansion ($\times 10^{-6}$ K^{-1})	200	90	30	8.4

In order to prepare the polymer surface to provide adhesion of the coating, abrasive blasting was applied. Owing to the softness of the polymer substrate, the energy of the abrasive Al_2O_3 particles (100 mesh, Alfa Aesar, Ward Hill, MA, USA) was reduced to prevent merging into the substrate. The particles were transported in air at a pressure of approximately 0.3 MPa. To generate the plasma stream, the AP-50 Plasma Spray System (Flame Spray Technologies, Duiven, The Netherlands) was used. The process parameters were automatically selected and controlled by programmable logic controllers (PLC) when the plasma generator operates at a current of up to 400 A. The JP-5000 plasma gun (Praxair Surface Technologies, Indianapolis, IN, USA) was used for the TPS process with a mixture of He (0.8 MPa) and H_2 (0.15 MPa). The distance of the plasma gun to the surface was fixed at 480 mm. The diameter of the spraying field was 25 mm. The plasma spraying parameters were optimized for polymer substrates in our preliminary experiments (not shown here).

The structure analysis was performed using a PANalytical X-Ray diffractometer (Cu Kα radiation, Empyrean model, Royston, UK). The ICDD-JCPDS (International Centre for Diffraction Data-Joint Committee on Powder Diffraction Standards) crystallographic database was used for the phase identification. The microscopic observations were conducted using a GX51 optical microscope (Olympus, Tokyo, Japan) and Vega 5135 SEM (Tescan, Brno, Czech Republic). The coating porosity was estimated based on the SEM (Scanning Electron Microscope) images analyzed using Corel software (Corel Photo-Paint X5). The black spots in the coatings were qualified as the pores and the percentage porosity in the considered area was calculated.

The roughness was measured using an SRT-6210 roughness tester (Landtek, Guanghzou, China). In order to obtain the statistics to evaluate average roughness, three measurements of 2.5 mm scan length on different spots of the sample were performed. A Gaussian type filter was used for the profiles, and the cut-off length for the waviness-roughness separation was 0.25 mm. Uncertainties of the roughness measurements result from the accuracy of the testing device and are less than 10% of the indication.

The mechanical properties were measured using a Picodentor HM500 (Fischer, Windsor, CO, USA) nanoindentation tester at the indentation force load of 300 mN that was applied for 20 s. The load–displacement curves were recorded for the substrate and the coating. The Vickers Hardness (HV), the indentation modulus (E_{IT}), the indentation creep (C_{IT1}), the total mechanical work of indentation (W_t), and the plastic deformation portion (N_{plast}) were all measured.

The three-point bending test was performed using a 4483 mechanical testing machine (Instron, Norwood, MA, USA) with a constant crosshead speed of 4 mm/min. The distance between the supports was 50 mm and the bending mandrel was 10 mm in diameter. The bending was carried out up to a displacement of the bending mandrel by 19–20 mm, which corresponds to approximately 45°, or to the point when the sample breaks. The mechanical properties were measured for five samples of each series.

3. Results and Discussion

The titanium plasma spraying process applied to polymer rods resulted in a formation of relatively thick coatings of high roughness. The core material and its properties (Table 1) affected the deposition of the Ti particles, the formation of the coating, and its final properties.

Among the three different substrate materials, the PA6 was found to be inadequate for the formation of the Ti coating (Figure 1). The surface of PA6 (Figure 1c) was incoherent with the sprayed Ti particles at the process setup, whereas for PE300 (Figure 1a) and PA6.6-GF30 (Figure 1b), the Ti coatings had sufficient adhesion, thickness, and roughness. The PA6 substrate material underwent surface transformation through significant softening and formed polymer blisters under the stream of melted Ti particles, which resulted in a failure to form the coating (Figure 1c). For both the PE300 and the PA6.6-GF30 substrate materials, the deposited Ti particles formed coatings.

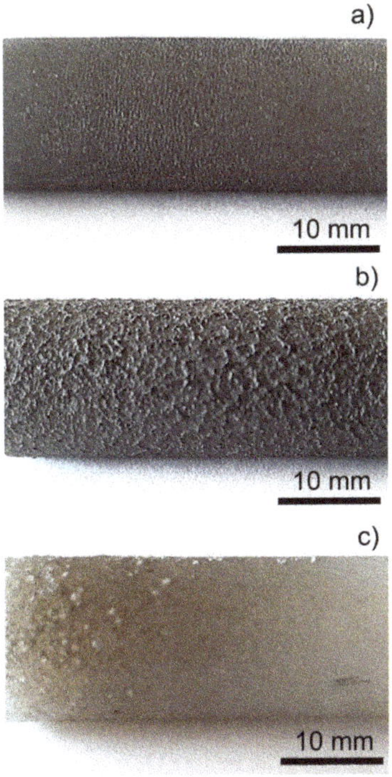

Figure 1. Titanium plasma spraying (TPS) coatings on the (**a**) PE300, (**b**) PA6.6-GF30, and (**c**) PA6 rods.

The roughness of the Ti coatings was sufficiently high for the potential osseointegration process. The average values of the roughness parameters R_a, R_z, R_q, and R_t for Ti deposited on PE300 were

25.5, 72.2, 26.1, and 72.9 µm, respectively, whereas for PA6.6-GF30, the roughness parameters R_a, R_z, R_q, and R_t assumed the values of 25.4, 71.8, 22.7, and 77.0 µm, respectively. Given the macroscale topography of the coating, the more developed surface was created for PA6.6-GF30 in comparison to PE300. However, the roughness parameters, averaged from the given measurement area, were comparable for both samples. The roughness parameters R_a, R_z, R_q, and R_t for the PA6 rod with the failed Ti coating assumed the values of 8.0, 22.6, 7.4, and 23.6 µm, respectively. For comparison, the typical average roughness R_a of the plasma-sprayed titanium coating is around 26 µm [9].

The TPS coatings deposited on the polymer rods, i.e., PE300 and PA6.6-GF30, had a typical microstructure regardless of the substrate material used (Figure 2). Coatings deposited on metallic substrates, for example, Ti alloys, have a comparable microstructure and morphology [9]. The coatings are composed of multilayers because the plasma gun moves repeatedly along the surface, spraying the particles to obtain a coating of the desired thickness. The first sprayed Ti particles are mechanically embedded into the polymer substrate that is softened under high temperature. When the small liquid Ti particles hit the substrate, they undergo fast solidification and cooling, which does not increase the polymer substrate temperature beyond 80-100 °C. This leads to a mechanical connection of the coating with the substrate. The Ti coatings described in this study had a uniform morphology with an internal porosity of <4.8% and <3.4% on PE300 and PA6.6-GF30, respectively. The porosity is heavily dependent on the TPS process optimization [10] and can reach up to 15% [9]. The thickness of the applied coatings fell into the range of 0.6 to 1.0 mm, which should be sufficiently high to provide good mechanical and physicochemical properties of the polymer elements for hard tissue application, for example, its coating integrity, its fixation with substrate, its continuity, and its density [7]. The typical thickness of plasma-sprayed titanium or hydroxyapatite coatings on Ti alloy substrates is in the approximate range of 0.2–0.6 mm [11,12]. Thicker coatings, however, provide better insulation of the substrate from the tissue [11]. Cold-sprayed metals on polymer substrates have a thickness of up to 1 mm [8]. The main phase in the coatings is pure Ti; however, oxides such as TiO_2 and TiO are also present.

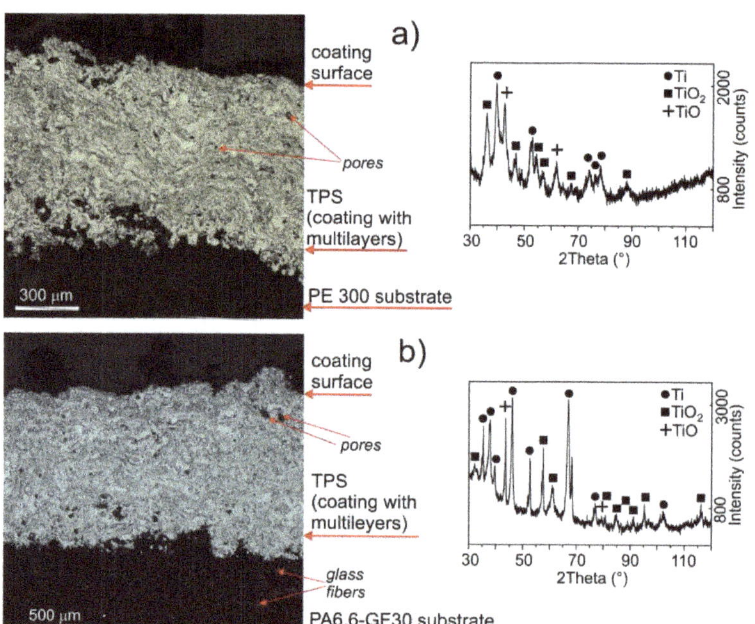

Figure 2. Microstructure as seen with an optical microscope and XRD spectra for the TPS coatings on (**a**) PE300 and (**b**) PA6.6-GF30.

The mechanical properties were measured by nanoindentation (Table 2, Figure 3) and bending tests (Figure 4). The nanoindentation was performed on polymer substrates, coatings, and bulk Ti for comparison. The load-displacement curves for the polyethylene substrate (Figure 3a) overlapped each other, exhibiting good material homogeneity. For the polyamide–glass fiber composite substrate (Figure 3c), the curves had a wide spread, which was a result of the two-phase polyamide–glass fiber substrate composition. The curves for the coatings (Figure 3b,d) had some spread that resulted from the interlayered morphology of the coatings, oxide distribution, and porosity. No delamination was observed at the diamond tip loading. The mechanical properties of the coatings were confirmed by the XRD analysis. The high hardness and Young's modulus of the Ti coatings resulted from the presence of oxygen (TiO, TiO_2), which significantly boosts the mechanical properties [13]. The mixture of oxides and pure Ti, as well as the multilayer morphology, acts against coatings delamination, as the layers and oxides facilitate plastic deformation through the slip mechanism, and brittle cracking.

Table 2. Mechanical properties of the investigated polymer substrates, Ti coatings, and bulk Ti.

Material	HV	E_{IT} (GPa)	W_t (μJ)	N_{plast} (%)	C_{IT} (%)
PE300 substrate	5.43 ± 0.04	1.28 ± 0.02	1.67 ± 0.02	65.0 ± 0.3	6.8 ± 0.2
Ti coating (on PE300 substrate)	396.7 ± 34.5	204.6 ± 27.3	0.38 ± 0.06	30.0 ± 5.8	5.6 ± 1.0
PA6.6-GF30 substrate	20.3 ± 2.1	4.82 ± 0.59	0.74 ± 0.07	55.3 ± 1.3	4.9 ± 0.8
Ti coating (on PA6.6-GF30 substrate)	420.3 ± 39.2	214.4 ± 21.5	0.39 ± 0.07	32.0 ± 4.6	5.1 ± 0.8
Bulk Ti (for comparison)	189.3 ± 11.7	115.2 ± 3.1	0.28 ± 0.03	86.2 ± 1.2	3.3 ± 0.3

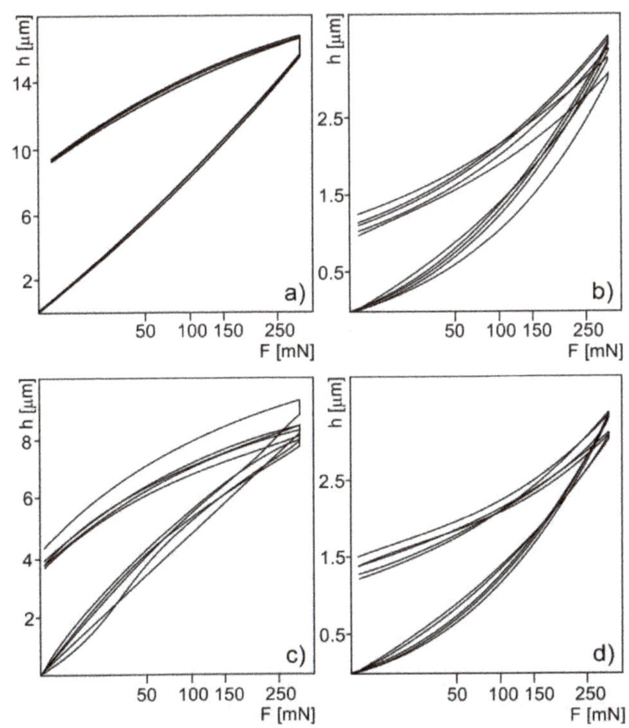

Figure 3. Load-displacement curves for (**a**) PE300 substrate, (**b**) TPS coating on PE300, (**c**) PA6.6-GF30 substrate, and (**d**) TPS coating on PA6.6-GF30.

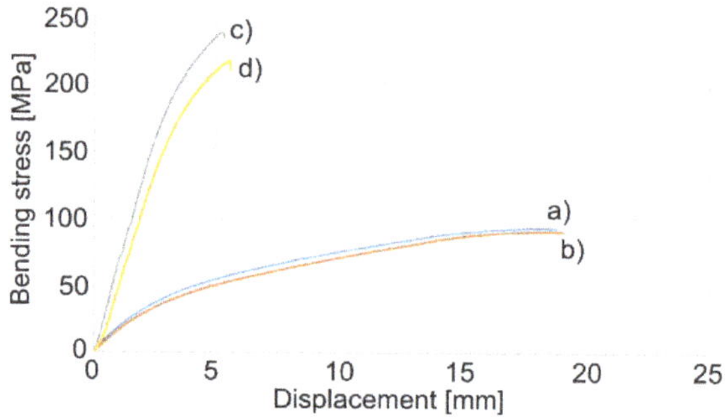

Figure 4. Bending test of the substrate polymer rods (**a**, **c**) and those with the TPS coating (**b**, **d**), (**a**, **b**) PE300, and (**c**, **d**) PA6.6-GF30.

The bendability of the PE300 type samples was higher in comparison to the PA6.6-GF30 ones (Figure 4). The PE300 samples did not crack at the displacement set on the machine (approximately 19 mm), whereas the PA6.6-GF30 samples, owing to the fiber glass reinforcement and thus greater stiffness, broke at the displacement of approximately 5.4 mm. The measured maximum bending stress at a given displacement for PE300 and PE300 with TPS was 92 MPa and 90 MPa, respectively. For PA6.6-GF30 and PA6.6-GF30 with TPS, the maximum bending stress (strength) was 236 MPa and 215 MPa, respectively. The samples with the TPS coatings exhibit a lower bending stress in comparison to the substrate rods without the coatings. This behavior correlated with the coating properties. During the bending test, the TPS coating deformed, cracked, and easily slid on the machine test supports—better than the polymer substrate—and the coating was pressed into the polymer substrate by the mandrel and the supports of the testing machine. The coatings bent together with the polymer substrate without delaminating. Additionally, the relatively low thickness and the multilayer morphology of the coatings resulted in lower stress during the bending test for the TPS samples.

The Ti coatings sprayed on the polymer substrate exhibited good adhesion qualities. In the bending tests, the coatings cracked under a high level of substrate deformation. However, they did not delaminate but adhered to the polymer surface (Figure 5). Due to the bending behavior, one side of the sample was compressed (the part of the sample affected by the mandrel) and the opposite side (from the supports side) was extended. The extended side revealed more cracks (Figure 5a,b) compared to the compressed side (Figure 5c). In both cases, however, the coatings did not fall off the substrate.

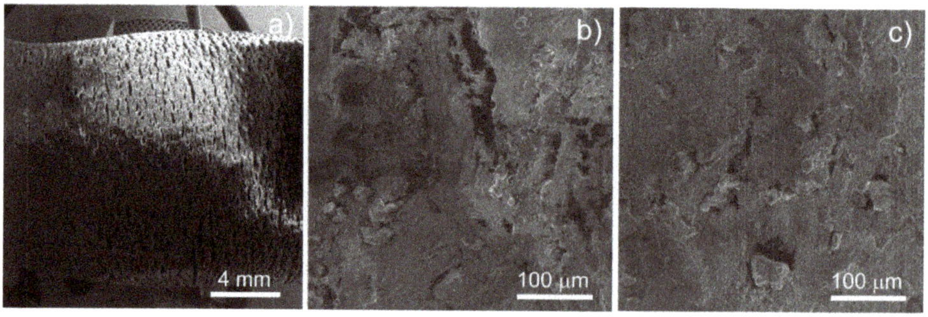

Figure 5. SEM images of the TPS coating on the PE300 substrate after the bending test; (**a**) low magnification sample view, (**b**) the extended side of the sample, and (**c**) the compressed side of the sample.

4. Conclusions

In this work, the authors investigated the method of titanium plasma spraying on polymer substrates. The Ti coatings exhibit a typical multilayer morphology of low porosity and good adhesion to polyethylene and glass fiber-reinforced polyamide. For pure polyamide, the TPS process was unsuccessful. The coatings are composed of Ti and its oxides, which resulted in the coatings' greater hardness of approximately 400 HV and a Young's modulus of approximately 200 GPa. The coating did not improve the bending strength, yet the bending tests have shown good adhesion of the coatings.

The preliminary results shown in this paper are an introduction to stimulate further research into the TPS process with polymer substrates with a view to its optimization and practical application.

Author Contributions: Conceptualization and writing, J.J. and A.W.; methodology, A.W. and J.J.; samples preparation, A.W.; mechanical measurements, P.S. and D.A.; structure and microstructure analysis, J.J. and A.W.

Funding: The work has been financed by the Polish Ministry of Science and Higher Education within its statutory activity.

Conflicts of Interest: The authors declare no conflict of interest.

References

1. Ong, J.L.; Carnes, D.L.; Bessho, K. Evaluation of titanium plasma-sprayed and plasma-sprayed hydroxyapatite implants in vivo. *Biomaterials* **2004**, *25*, 4601–4606. [CrossRef] [PubMed]
2. Morks, M.F. Plasma spraying of zirconia–titania–silica bio-ceramic composite coating for implant application. *Mater. Lett.* **2010**, *64*, 1968–1971. [CrossRef]
3. Le Guehennec, L.; Soueidan, A.; Layrolle, P.; Amouriq, Y. Surface treatments of titanium dental implants for rapid osseointegration. *Dent. Mater.* **2007**, *23*, 844–854. [CrossRef] [PubMed]
4. Quek, C.H.; Khor, K.A.; Cheang, P. Influence of processing parameters in the plasma spraying of hydroxyapatite/Ti–6Al–4V composite coatings. *J. Mater. Proc. Technol.* **1999**, *89–90*, 550–555. [CrossRef]
5. Sathish, S.; Geetha, M.; Aruna, S.T.; Balaji, N.; Rajam, K.S.; Asokamani, R. Studies on plasma sprayed bi-layered ceramic coating on bio-medical Ti–13Nb–13Zr alloy. *Ceram. Int.* **2011**, *37*, 1333–1339. [CrossRef]
6. Gonzalez, R.; Ashrafizadeh, H.; Lopera, A.; Mertiny, P.; McDonald, A. A Review of thermal spray metallization of polymer-based structures. *J. Therm. Spray Technol.* **2016**, *25*, 897–919. [CrossRef]
7. Walsh, W.R.; Bertollo, N.; Christou, C.H.; Schaffner, D.; Mobbs, R.J. Plasma-sprayed titanium coating to polyetheretherketone improves the bone-implant interface. *Spine J.* **2015**, *15*, 1041–1049. [CrossRef] [PubMed]
8. Gardon, M.; Latorre, A.; Torrell, M.; Dosta, S.; Fernández, J.; Guilemany, J.M. Cold gass pray titanium coatings onto a biocompatible polymer. *Mater. Lett.* **2013**, *106*, 97–99. [CrossRef]
9. Fousova, M.; Vojtech, D.; Jablonska, E.; Fojt, J.; Lipov, J. Novel approach in the use of plasma spray: Preparation of bulk titanium for bone augmentations. *Materials* **2017**, *10*, 987. [CrossRef] [PubMed]
10. Vogel, D.; Dempwolf, H.; Baumann, A.; Bader, R. Characterization of thick titanium plasma spray coatings on PEEK materials used for medical implants and the influence on the mechanical properties. *J. Mech. Behav. Biomed.* **2018**, *77*, 600–608. [CrossRef] [PubMed]
11. Sun, L.; Berndt, C.C.; Gross, K.A.; Kucuk, A. Material fundamentals and clinical performance of plasma-sprayed hydroxyapatite coatings: A review. *J. Biomed. Mater. Res.* **2001**, *58*, 570–592. [CrossRef] [PubMed]
12. Vercaigne, S.; Wolke, J.G.C.; Naert, I.; Jansen, J.A. The effect of titanium plasma-sprayed implants on trabecular bone healing in the goat. *Biomaterials* **1998**, *19*, 1093–1099. [CrossRef]
13. Oh, J.M.; Lee, B.G.; Cho, S.W.; Lee, S.W.; Choi, G.S.; Lim, J.W. Oxygen effects on the mechanical properties and lattice strain of Ti and Ti-6Al-4V. *Met. Mater. Internat.* **2011**, *17*, 733–736. [CrossRef]

materials

MDPI

Article

Electromagnetic Analysis, Characterization and Discussion of Inductive Transmission Parameters for Titanium Based Housing Materials in Active Medical Implantable Devices

Waldemar Gruenwald [1,*], Mayukh Bhattacharrya [1], Dirk Jansen [1] and Leonhard Reindl [2]

[1] Institute for Applied Research, Offenburg University, 77652 Offenburg, Germany; mayukh.bhattacharyya@hs-offenburg.de (M.B.); dirk.jansen@hs-offenburg.de (D.J.)

[2] Department of Microsystems Engineering, Institute of Albert-Ludwigs University, 79098 Freiburg, Germany; reindl@imtek.uni-freiburg.de

* Correspondence: waldemar.gruenwald@hs-offenburg.de; Tel.: +49-179-1424830

Received: 3 September 2018; Accepted: 19 October 2018; Published: 25 October 2018

check for updates

Abstract: The growing demand for active medical implantable devices requires data and or power links between the implant and the outside world. Every implant has to be encapsulated from the body by a specific housing and one of the most common materials used is titanium or titanium alloy. Titanium thas the necessary properties in terms of mechanical and chemical stability and biocompatibility. However, its electrical conductivity presents a challenge for the electromagnetic transmission of data and power. The proposed paper presents a fast and practical method to determine the necessary transmission parameters for titanium encapsulated implants. Therefore, the basic transformer-transmission-model is used with measured or calculated key values for the inductances. Those are then expanded with correction factors to determine the behavior with the encapsulation. The correction factors are extracted from finite element method simulations. These also enable the analysis of the magnetic field distribution inside of the housing. The simulated transmission properties are very close to the measured values. Additionally, based on lumped elements and magnetic field distribution, the influential parameters are discussed in the paper. The parameter discussion describes how to enhance the transmitted power, data-rate or distance, or to reduce the size of the necessary coils. Finally, an example application demonstrates the usage of the methods.

Keywords: titanium; low frequency; inductive transmission; metallic housing; hermetic sealing; longevity; FEM model; active implantable medical devices

---

1. Introduction

Active implantable medical devices (AIMDs) have been on the market since the first fully implanted pacemaker in the 1960s. Due to the miniaturization of the electronics and the growing capability of micro-processing units, new markets and applications have been developed. Those include implants for cochlear, retina, muscle-stimulation and body-function-sensing. Additionally to system functionality, the AIMDs require a contactless communication interface. This interface can be realized via optical, ultrasound, capacitive, inductive or radio link [1] (Ch. 1).

To ensure the longevity of the implant and to protect the implant electronics from the corrosive body- fluids,while at the same time the body from the partially poisonous compounds of the implant, a hermetically sealed housing is required [2] (Ch. 1.6). While there are several materials in usage, most of the physicians prefer titanium and titanium alloys. This is due to its mechanical strength, the chemical stability, watertightness and biocompatibility [2,3].

Because of the great importance of inductive- interfaces and titanium housings, this paper aims to describe a simple method to estimate the general behavior of inductive energy and data transmission through a hermetically sealed titanium housing for active implantable medical devices. Similar to [4], the standard transformer model is expanded and used to determine the influence of the housing, which is also separated from the transmission without housing as far as possible. The characteristics of the housing are determined through finite element method (FEM) simulations, which then can be applied to the standard two coil transmission in resonant tuned mode. In addition, the physical boundaries, depending on the physical properties of the housing, are derived from simulations. The results shall enable a pre-prototype determination of the housing possibilities and constraints.

While [5] develops theoretical expressions and [4] uses simulated transfer functions, this paper aims to introduce a new method by determining correction factors. Those can be used directly in a transformer model or given coil setup to estimate the effects of conductive materials for inductive links. Therefore, this paper is structured in the following way. Firstly, the FEM simulation is explained with all necessary parameters with a simple example. Secondly, a real coil model with the main parasitic parameters is presented to determine the parts most impacted by the conductive object. This is shown by FEM simulations and impedance measurements. Thirdly, this model is expanded for transmission parameter estimation and validation for a standard two coil transmission in a resonance tuned mode. Additionally, the magnetic vector field distribution is analyzed to identify the eddy current concentration spots, which may allow optimization of the geometric properties. Furthermore, the identified relevant parameters are discussed to highlight the modification and optimization possibilities. Finally, an application of an implantable infusion pump system demonstrates the implementation of the method. These enable a high data rate transmission system for conductive housings at low frequencies.

2. FEM Impedance Parameter Estimation

To calculate magnetic field problems, tools like finite element methods are applied. The Finite Element Method Magnetics (FEMM) [6] used here solves two-dimensional planar and axially symmetric problems for low frequency magnetics, with interfaces to Octave [7], for interactive simulations.

For an axially symmetric problem, as shown in Figure 1a, two coils L_1 and L_2 are used (depicted in red), with a titanium tube T around L_2 (depicted in blue). This physical model is translated into FEMM, according to Figure 1b with the necessary geometric dimensions, shown in the parameter box. The problem is enclosed by a spheric boundary (indicated with the green semicircle), where the Robin boundary condition, after [8] (Ch. A.3.2) of the form

$$\frac{\partial \vec{A}}{\partial n} + c_0 \cdot \vec{A} = 0 \quad \text{with} \quad c_0 = \frac{n}{\mu_0 \cdot r_0} \tag{1}$$

is applied, to approximate an unbounded solution region. In Equation (1), $\partial \vec{A}/\partial n$ is the normal derivative of the magnetic vector potential $\vec{A}$ and c_0 describes the condition at the boundary, with the harmonic $n = 1$, the magnetic permeability $\mu_0 = 4\pi \cdot 10^{-7}\mathrm{H/m}$ and the radius of the boundary circle r_0. Additional properties for the coils are the number of windings N, which are evenly distributed over the cross section of the coil with the wire diameter $\varnothing_W$. The conductivity of the coil is $\sigma_{Cu} = 58\,\mathrm{MS/m}$ for copper. For the titanium tube, the conductivity is $\sigma_{Ti2} = 2.08\overline{3}\,\mathrm{MS/m}$ and the relative permeability is $\mu_{r\,Ti2} = 1.000178$.

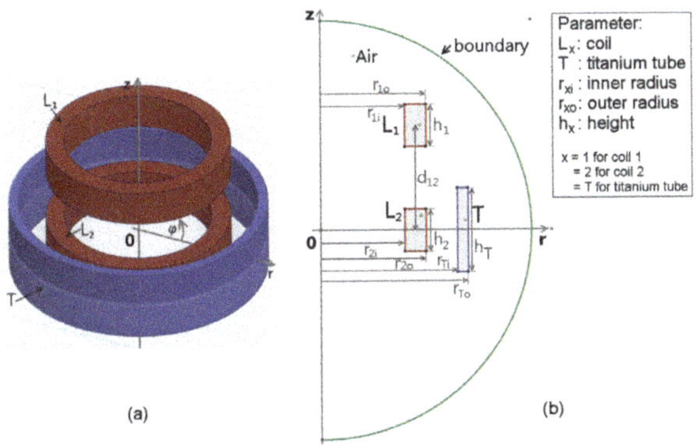

Figure 1. FEM model of an axially symmetric magnetic problem setup, with two coils, L_1 and L_2 (red) and a titanium tube T around L_2 (blue). (**a**) shows the three-dimensional model of the setup in a polar coordinate system where z is the symmetry axis; (**b**) shows the the setup in FEMM, where the green semicircle is the boundary region and the legend describes the used parameters.

The simulation is performed while only one coil is excited with the current $I_i = 1$ A and the other coil remains open. The results, taken from the simulation, are basically the impedance parameters. First, the total magnetic flux $\Psi = N \cdot \Phi$, which is the magnetic flux Φ over all windings N, can be derived from the magnetic field energy W_m [9] (Ch. 9.5), which is

$$W_{m\,i\,j} = \frac{1}{2} \iiint\limits_{V_j} (\vec{A}_i \cdot \vec{J}_j) \cdot dV = \Psi_{i\,j} \cdot I_i \tag{2}$$

with the volume of the coil V, the vector potential $\vec{A}$ and the current density $\vec{J}$.

Case coil 1 is excited with the current I_1, the magnetic flux in the same coil is $\Psi_{11} = W_{m\,11}/I_1$, while the magnetic flux in coil 2 is $\Psi_{12} = W_{m\,12}/I_1$ and vice versa. Equation (2) leads to the self inductance L_i

$$L_i = \frac{N_i \cdot \Re\{\Phi_{i\,i}\}}{I_i} = \frac{\Re\{\Psi_{i\,i}\}}{I_i} \tag{3}$$

and the mutual inductance $M_{i\,j}$

$$M_{i\,j} = \frac{N_j \cdot \Phi_{i\,j}}{I_i} = \frac{\Psi_{i\,j}}{I_i}. \tag{4}$$

Next, the induced voltage U_i, which is the time derivative of Φ, is determined by

$$U_i = -N_i \cdot \Re\left\{\frac{d\Phi_{i\,i}}{dt}\right\} = -\Re\left\{\frac{d\Psi_{i\,i}}{dt}\right\} \tag{5}$$

and used to calculate the coil resistance R_i

$$R_i = \frac{U_i}{I_i} = -\frac{N_i}{I_i} \cdot \Re\left\{\frac{d\Phi_{i\,i}}{dt}\right\} = -\frac{1}{I_i} \cdot \Re\left\{\frac{d\Psi_{i\,i}}{dt}\right\}. \tag{6}$$

Equations (3) and (4) are based on [10] (Ch. 8.10). However, (3) is slightly modified for this case, because only the real part of Ψ is relevant for the inductance, whereas the imaginary part would contribute to the coil resistance. Equation (4) is required to be a complex value, to reproduce the real behavior of the mutual inductance. In addition, Equation (6), which is based on [10] (Ch. 9.1), is modified so that only the real part of Ψ is considered for the resistance calculation.

Finally, the coupling factor $k_{i\,j}$ between two coils can be calculated from Ψ, after [11] (Ch. 3.4.7.3)

$$k_{ij} = \frac{\Psi_{ij}}{\Psi_{ii}}. \tag{7}$$

Based on the FEM simulation without the housing, where the titanium tube is ignored, the basic inductive parameters can be obtained from Equations (3), (4), (6) and (7). Including the titanium tube in the simulation, the primed inductive parameters can be obtained. These provide the necessary values for the further analysis of the impedance and transmission parameters in the following chapters.

3. Physical Coil Impedance Model

To describe the frequency dependent properties of a coil, one of the common models is the Resistor Inductor Capacitor (RLC) model [1] (Ch. 2.4). Therein, L_1 is the main inductance, R_1 describes the resistive losses and C_p describes the parasitic capacitance, which leads to the first resonant frequency. However, this model does not describe the quality factor at the self resonant frequency. Therefore, this model is extended by an additional parallel resistance R_p. Thus, the electrical behavior from direct current (DC) up to and above the first resonant frequency of the coil can be expressed as a complex impedance Z, like shown in Figure 2. This is a slightly modified model of the described large inductor model in [12] (Ch. 5.2). This model is selected because of the practical usability for fast parameter estimation by measurements, which will be explained in Appendix A. However, there is a lot of literature and research on theoretical models, for the estimation of the self-capacitance in [13], the calculation of proximity effects in [14] or the distributive equivalent model of a coil in [15,16].

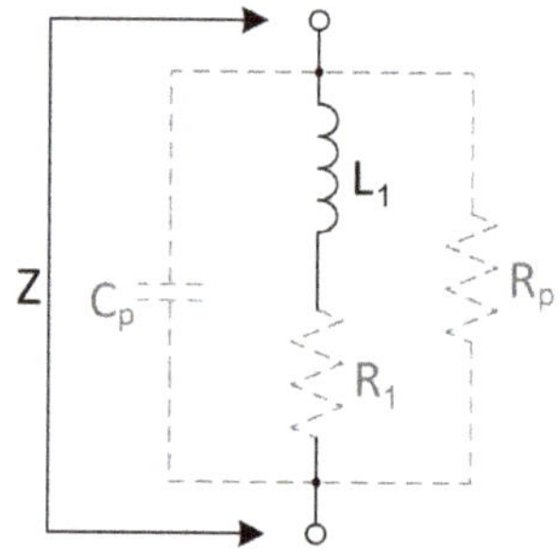

Figure 2. Lumped element model of a simple coil impedance Z, consisting of the main inductance L_1, the equivalent series resistance R_1, the parallel resistance R_p and the parasitic parallel capacitance C_p.

For the analysis of the impact of the metallic housing or any conductive object in the proximity of the coil, the influence of the impedance of a single coil is analyzed first. Therefore, the housing can be seen as an inductively coupled additional short circuited winding, like in [1,17]. This virtual impedance $Z_h = j\omega L_h + R_h$ consists of L_h for the inductance and R_h for the resistive losses. Thus, the coupling between the coil and the housing can be treated like a standard transformer with two windings. Those are coupled by the complex coupling factor k or the complex mutual inductance M (see [18] (Ch. 6)). In the transformer model calculations, it is assumed that only the components L_1 and R_1 of the coil are relevant for the coupling. The basic set of equations for the transformer is

$$\begin{bmatrix} U_1 \\ U_h \end{bmatrix} = \begin{bmatrix} R_1 + j\omega L_1 & j\omega M_{1h} \\ j\omega M_{1h} & R_h + j\omega L_h \end{bmatrix} \cdot \begin{bmatrix} I_1 \\ I_h \end{bmatrix}. \tag{8}$$

Because eddy currents run in closed loops, the voltage $U_h = 0$ allows for eliminating I_h and simplify (8), which then can be resolved for the complex impedance Z' to:

$$Z'(j\omega) = \frac{U_1}{I_1} = R_1 + j\omega L_1 + \frac{\omega^2 M_{1h}^2}{R_h + j\omega L_h}, \tag{9}$$

where $M_{1h} = k_{1h} \cdot \sqrt{L_1 \cdot L_h}$ is the mutual inductance and k_{1h} is the mutual coupling between the coil and the housing. The impedance can then be separated into the components $Z' = R' + j\omega L'$ with

$$\begin{aligned} Z'(j\omega) = \frac{U_1}{I_1} &= R_1 + j\omega L_1 \cdot \left[1 - k_{1h}^2 \cdot \frac{j\omega L_h}{R_h + j\omega L_h}\right] \\ &= R_1 + j\omega L_1 \cdot \left[1 - k_{1h}^2 \cdot \chi\right] \end{aligned} \tag{10}$$

into

$$L_1' = L_1 \cdot \left[1 - \Re\left\{k_{1h}^2 \cdot \chi\right\}\right] \tag{11}$$

and

$$R_1' = R_1 + \omega L_1 \cdot \Im\left\{k_{1h}^2 \cdot \chi\right\} \tag{12}$$

with

$$\chi = \frac{j\omega L_h}{R_h + j\omega L_h}. \tag{13}$$

The combined housing parameter χ is a complex value. In combination with the complex coupling factor k_{1h}, it contributes to the inductance variation in (11) and the resistance variation in Equation (12). Because of the continuous distribution of the eddy currents and the strong dependency of the physical properties of the housing, it is not possible to simply separate χ and k_{1h}. In addition, the housing parameters L_h and R_h cannot be separated (see also [17]).

However, by extraction of the impedance results from FEM simulations, with and without housing, correction factors can be calculated, which further can be applied to the measured coil values and thus allow an estimation of the influence of the housing. Assuming that the coil parameters L_1 and R_1 are constant values, the correction factors can be calculated by

$$\xi_L(\omega) = \frac{L_1'(\omega)}{L_1} = 1 - \Re\left\{k_{1h}^2 \cdot \chi\right\} \tag{14}$$

and

$$\xi_R(\omega) = \frac{R_1'(\omega)}{R_1} = 1 + \frac{\omega L_1}{R_1} \cdot \Im\left\{k^2{}_{1h} \cdot \chi\right\}. \tag{15}$$

According to Equations (14) and (15), the self inductance L_1 of the coil is reduced by the conductive object, while the series resistance is increased. These effects are also stated in [5,19]. These calculated factors are displayed in Figure 3, where (a) shows the variation of the inductance and (b) shows the variation of the equivalent series resistance, compared to DC values (shown in bold blue). Additionally, in (b), the frequency dependent variation of a simple air coil is shown as $\xi_{Rac} = R_{1ac}/R_{1dc}$ (represented in dashed green). The parameters of coil 'A' are: number of turns $N_A = 200$, wire diameter $\varnothing_{WA} = 0.1\,\text{mm}$, coil diameter $\varnothing_C = 71\,\text{mm}$, coil height $h_C = 3\,\text{mm}$ and coil width $w_C = 1\,\text{mm}$. See also Table A1 and Chapter B (The values $\varnothing_C$, h_C and w_C are also used for all other coils in this paper.). The skin-effect and the proximity effect inside the coil are dominant above $f_{ac} = 1\,\text{MHz}$ and significantly increase the resistance of the coil. Methods for the reduction of the skin- and proximity-effects of the coil are given in [20]. However the focus here is on a frequency of $f_0 = 125\,\text{kHz}$, where these effects are insignificant for the given coil setup. The estimated values of the correction factors are $\xi_L(f_0) = 0.412$ and $\xi_R(f_0) = 4.66$. Furthermore, the skin depth for the used material reaches the critical thickness $\delta_{crit} = 0.3\,\text{mm}$ of the housing wall at the frequency $f_{\delta\,crit} = 1.35\,\text{MHz}$. Above $f_{\delta\,crit}$, a penetration of the housing by magnetic fields is impractical.

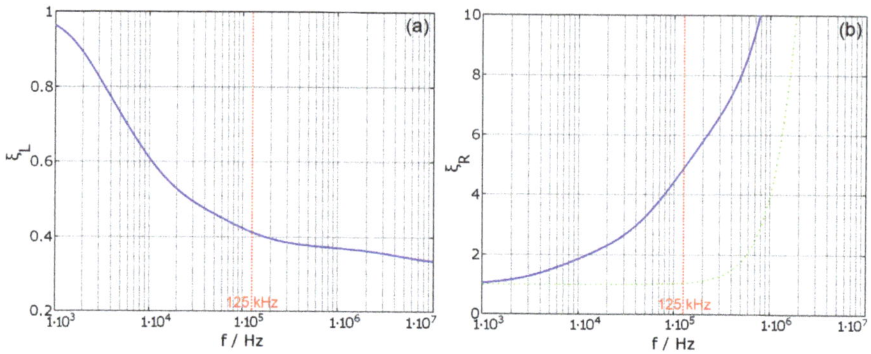

Figure 3. Correction factors ξ_L for the main inductance (**a**) and ξ_R for the equivalent series resistance (**b**). Additionally, in (**b**), the alternating current (AC) resistance factor ξ_{Rac} without housing is plotted in dashed green, which shows the dominance of the skin- and proximity effect above the frequency $f_{ac} = 1\,\text{MHz}$. The parameters, belonging to coil 'A' are $N_A = 200$ and $\varnothing_{WA} = 0.1\,\text{mm}$ (see also Table A1).

Adapted to the coil model from Figure 2, the correction factors can be applied to the values L_1 and R_1. Thus, the impedance of the coil inside the housing can be calculated as:

$$Z'(j\omega) = \frac{(R_1 \cdot \xi_R + j\omega L_1 \cdot \xi_L(\omega))}{1 + \left(\frac{1}{R_p} + j\omega C_p\right) \cdot (R_1 \cdot \xi_R(\omega) + j\omega L_1 \cdot \xi_L(\omega))}. \tag{16}$$

Since one coil is usually tuned to a specific resonance frequency, rather then used in self resonance, Equation (16) is expanded with additional tuning capacitors in series C_s and in parallel C_{p2} to

$$Z_s'(j\omega) = \frac{1}{j\omega C_s} + \frac{(R_1 \cdot \xi_R + j\omega L_1 \cdot \xi_L)}{1 + \left(\frac{1}{R_p} + j\omega C_p\right) \cdot (R_1 \cdot \xi_R + j\omega L_1 \cdot \xi_L)} \tag{17}$$

and

$$Z_p'(j\omega) = \frac{(R_1 \cdot \xi_R + j\omega L_1 \cdot \xi_L)}{1 + \left(\frac{1}{R_p} + j\omega (C_p + C_{p2})\right) \cdot (R_1 \cdot \xi_R + j\omega L_1 \cdot \xi_L)}. \tag{18}$$

Figure 4 shows the impedance characteristic for coils 'D', 'K' and 'B', which are tuned to a resonance frequency of $f_0 = 125\,\text{kHz}$. In (a), the tuning is in serial mode, according to Equation (17), showing the typical impedance drop at the tuned resonance frequency, but also an impedance peak at higher frequencies. The peak results from the natural resonance of the coil $f_{L1} = (2 \cdot \pi \cdot \sqrt{L_1 \cdot C_p})^{-1}$. In (b), the tuning is in parallel mode, according to Equation (18), with the typical impedance peak at the tuned frequency. The impedance drop at higher frequencies and thus the deviation between calculated and measured values is due to additional parasitic components of the system, which are not included in the used coil model. The graphs show that the absolute value of the impedance can be estimated with the parameters. Only at resonance do the calculated values slightly diverge from the measured values. However, the resonance frequency is estimated correctly.

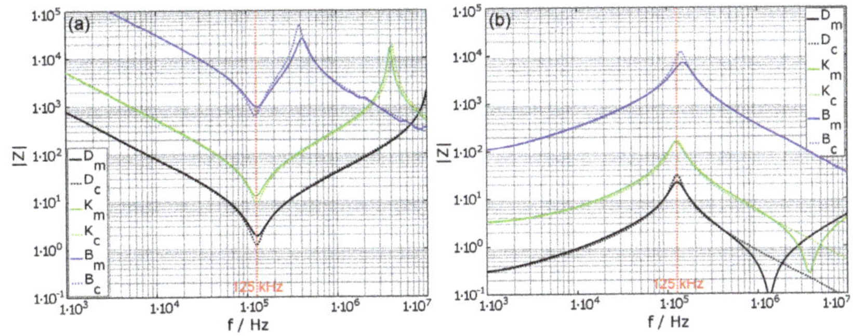

Figure 4. Comparison of the calculated (- -) and measured (–) absolute impedance frequency response for serial (**a**) and parallel (**b**) tuned coils to 125 kHz. The different parameters of the coils 'B', 'D', 'K' are $N_B = 200$, $\varnothing_{WB} = 0.1\,\text{mm}$, $N_D = 10$, $\varnothing_{WD} = 0.4\,\text{mm}$, $N_K = 25$ and $\varnothing_{WK} = 0.2\,\text{mm}$ (see also Table A1).

4. Transfer Function with Housing

Like the impedance of a single coil, the transfer function between two coils can also be expanded with the simulated housing values. Thus, the transmission of energy and data between two coils in proximity to a conductive material can be estimated; therefore, the standard transformer model for three coils is used, which is based on the system of equations in (19)

$$\begin{bmatrix} U_1 \\ U_2 \\ U_h \end{bmatrix} = \begin{bmatrix} R_1 + j\omega L_1 & j\omega M_{12} & j\omega M_{1h} \\ j\omega M_{12} & R_2 + j\omega L_2 & j\omega M_{2h} \\ j\omega M_{1h} & j\omega M_{2h} & R_h + j\omega L_h \end{bmatrix} \cdot \begin{bmatrix} I_1 \\ I_2 \\ I_h \end{bmatrix}. \tag{19}$$

Now, the third winding is used for the housing and is indexed with the character h. Again, a symmetric coupling is assumed between the coils and also the housing and so are the mutual inductances. In addition, the voltage of the housing is assumed to be $U_h = 0$ because of the induced eddy currents. Thus, the current I_h can be expressed in terms of I_1 and I_2 as

$$I_h = \frac{-j\omega M_{1h}}{R_h + j\omega L_h} \cdot I_1 + \frac{-j\omega M_{2h}}{R_h + j\omega L_h} \cdot I_2 \tag{20}$$

and Equation (19) can be reduced to Equation (21)

$$\begin{bmatrix} U_1 \\ U_2 \end{bmatrix} = \begin{bmatrix} \left(R_1 + j\omega L_1 + \dfrac{\omega^2 M_{1h}^2}{R_h + j\omega L_h}\right) & \left(j\omega M_{12} + \dfrac{\omega^2 M_{1h} M_{2h}}{R_h + j\omega L_h}\right) \\ \left(j\omega M_{12} + \dfrac{\omega^2 M_{1h} M_{2h}}{R_h + j\omega L_h}\right) & \left(R_2 + j\omega L_2 + \dfrac{\omega^2 M_{2h}^2}{R_h + j\omega L_h}\right) \end{bmatrix} \cdot \begin{bmatrix} I_1 \\ I_2 \end{bmatrix}. \tag{21}$$

The previously discussed impedance values from Equations (11) and (12) are still valid for the first coil, but also for the second coil if the index is changed from 1 to 2. In addition, the mutual inductance M'_{12} between the first and the second coil can now be expressed as

$$M'_{12} = M_{12} - \left(\frac{j\omega \cdot M_{1h} \cdot M_{2h}}{R_h^2 + \omega^2 L_h^2}\right). \tag{22}$$

M'_{12} consists of the mutual inductance between the two coils M_{12}, which is affected by the conductive object. By using the coupling coefficient instead of the mutual inductance, Equation (22) can be reformulated in a way, which better shows the influential parts to the coupling:

$$M'_{12} = \sqrt{L_1 \cdot L_2} \cdot (k_{12} - k_{1h} \cdot k_{2h} \cdot \chi)\,. \tag{23}$$

Because M'_{12} is a complex value, the real part contributes to the inductive coupling and is reduced by the housing. Likewise, the imaginary part contributes to the resistive losses, which are increased by the housing. Again the correction factor for the housing can be estimated for the mutual coupling. It relies on the coupling coefficients between the housing and the first, respectively, the second coil, divided by the coupling coefficient between the first and the second coil times the complex housing parameter:

$$\xi_M = \frac{M'_{12}(\omega)}{M_{12}} = 1 - \frac{k_{1h} \cdot k_{2h}}{k_{12}} \cdot \chi. \tag{24}$$

Since the mutual inductance is the main parameter for the transmission, the plot of the correction factor over the frequency is displayed in Figure 5, where (a) shows the real part and (b) shows the imaginary part of the calculated value. The deviation of the values for different distances between the coils increases with decreasing distance of L_1 to the housing. The value for the frequency of interest is $\xi_M(f_0) = 0.023 - 0.115i$.

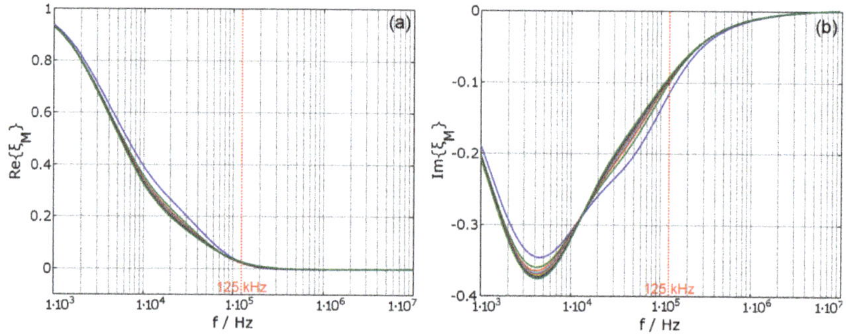

Figure 5. Correction factor for the mutual inductance χ_M, split into real part (**a**) and imaginary part (**b**) for coil configuration 'A' ($N_A = 200$ and $\varnothing_{W\,A} = 0.1$ mm, see also Table A1).

With these values given, the forward voltage transfer function (FVTF) $v_f = u_2/u_1$ can be calculated by using the corrected values of R'_1, L'_1 and M'_{12}. They are composed of the measured or calculated original values multiplied with the frequency dependent values ξ_Z. In addition, here the dependency is on the geometric arrangement of the housing and the coils. The typical transmission with two coils, where the sending coil L_1 is tuned in series resonance and the receiving coil is tuned in parallel resonance is depicted in Figure 6.

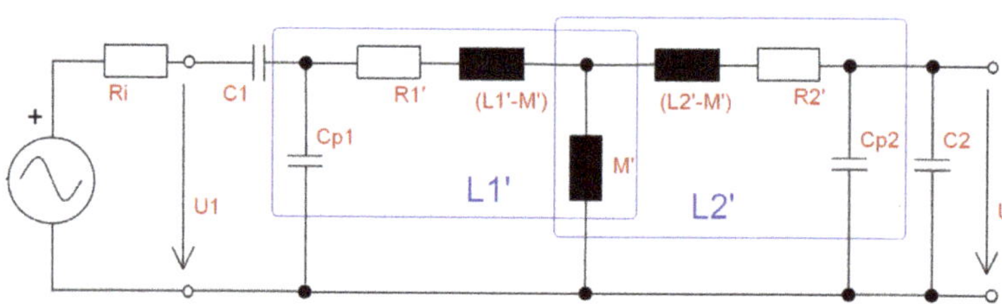

Figure 6. Schematic of typical inductive resonant transmission for the estimation of the transfer function, by using a modified transformer. L'_1 and L'_2 represent the coils, which include the correction factors of the housing.

With the known basic parameters of the coils and the remaining circuit components, the calculated and the measured transfer functions can be compared to each other. Figure 7a shows the FVTF from outside the implant into the implant. Figure 7b shows the backward voltage transfer function (BVTF) $v_b = U_1/U_2$ from inside the implant to outside the implant. The transfer functions are shown for the case without housing (black graphs, nH) and also for the case with the housing (blue graphs, wH). The distance between the coils is 10 mm, 50 mm and 90 mm. The resonance frequency of the system is tuned to $f_0 = 125\,\text{kHz}$ on both sides, by additional capacitors in series to the sending coil and in parallel to the receiving coil.

For the configuration without housing, the results show that the resonant peak spreads into two peaks after $\omega_{01/02} = \omega_0 \cdot (1 \pm k)^{0.5}$ [4] (Equation (2)), as the coils have a high quality factor. This effect vanishes with increased distance between the coils which also means a reduced coupling factor. The comparison between the measured and the calculated results show good correlation up to about $f \approx 1.7 \cdot f_0 \approx 200\,\text{kHz}$. Above 200 kHz, the curves start to deviate because the used model only includes the primary parasitic capacitance and neglects additional parasitic components of the system.

In the configuration with conductive housing, the results are obtained by using the transmission parameters without housing and modifying them with the estimated correction factors from Equations (14), (15) and (24). Since the correction factors only include magnetic field effects up to the first self-resonant frequency, the calculated transfer function starts to deviate from the measurements shortly after the resonant frequency of the system. For most implementations, this is sufficient. For distances of 50 mm and 90 mm, the calculations are precise; however, for 10 mm, the coupling of the outer coil to the housing is not estimated correctly. This results in a smaller value at the resonant peak, but the frequency shift is predicted correctly. In general, the damping of the transfer function increases with the frequency. Furthermore, the resonant frequency increases for small coil distances, due to the additional coupling of the outer coil to the housing. The reduction of the inductance and therefore the resulting resonant frequency shift are compensated by adjustment of additional capacitors.

Since the transfer functions are ratios of received voltage to transmitted voltage in the case with housing, it looks like $v_f > v_b$, but only because the influence on U_2 is stronger than the influence on U_1.

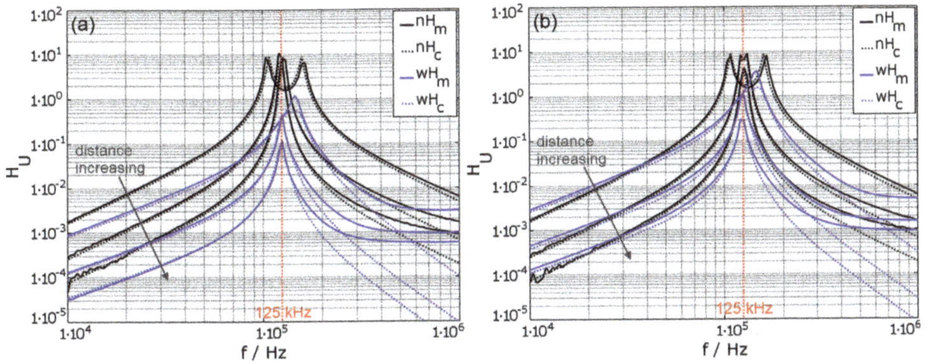

Figure 7. Comparison of the forward voltage transfer function into the implant (**a**) and the reverse voltage transfer function out of the implant (**b**) in a typical inductive transmission without housing (nH) in black and with conductive housing (wH) in blue, – for the measured (m) and - - for the calculated (c) results. The used coil parameters for coils 'A' and 'B' are $N_B = N_A = 200$, $\varnothing_{W\,A} = \varnothing_{W\,B} = 0.1\,\text{mm}$ (see also Table A1) and the distance between the coils increases from 10 mm over 50 mm to 90 mm.

5. Eddy Current Distribution in the Housing

In the previous sections, the housing was considered in the form of the influence to the lumped element model of the coil transmission, which is used to tune the transmission and the resonance frequency. However, during the transmission, the magnetic field of the coils induce eddy currents in the housing. Those eddies create an opposing magnetic field, which weakens the original field. Hence, this chapter focuses on the distribution of these magnetic fields. The aim is to find hotspots of the field distribution, which further can be used to redesign the housing or to modify the transmission coils.

Because magnetic fields can be calculated by superposition of different field sources, one can also reverse this effect. The field which results from the eddies in the housing material is simply the subtracted field vectors of the field with housing from the field without, which only consists of two explicit coils. This resulting field distribution for an excited coil L_2 at a frequency of about 125 kHz is displayed in Figure 8. Since the real part is responsible for the influence on the inductance and the imaginary part is responsible for the influence on the resistance of the coils, the real and the imaginary parts of the distributed field are separated. The strength of the field is normalized to the maximum occurring value and represented by a color according to the color bar on the right. The real part of the B field clearly indicates a concentrated magnetic source directly beneath the excited coil. A high concentration of the resulting B field is curling around it with the axis in $-\varphi$ direction, whereas the current flow in the coil is in the $+\varphi$ direction. For the imaginary part of the B field, the hotspot appears to be inside the base plate of the housing in the $+r$ direction, directly underneath the excited coil. Both fields oppose the original field and thereby attenuate it, which results in reduced inductance by factor ξ_L, increased resistance by factor ξ_R of the excited coil and also a reduced mutual inductance between the coil by factor ξ_M. The parameters of the used housing model are described in Appendix B in Figure A1.

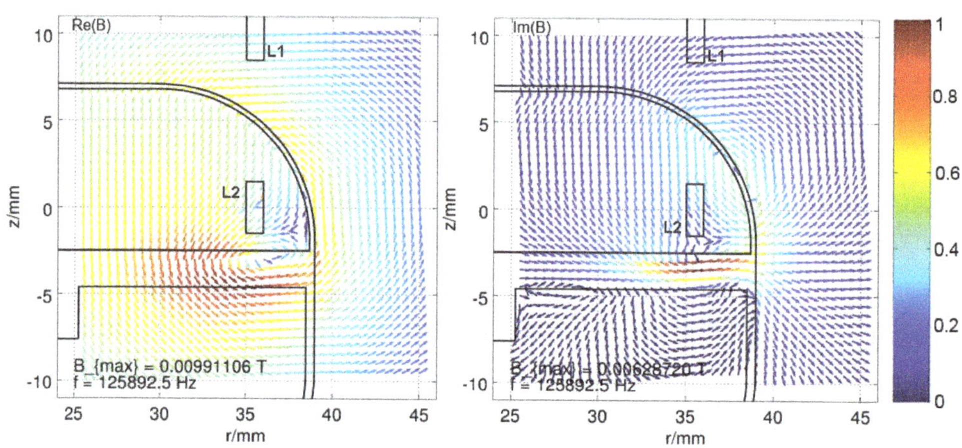

Figure 8. Resulting magnetic field vector distribution, caused by eddy currents in the housing, while exciting L_2, split into real part (Re(B)) and imaginary part (Im(B)) of the field. The field strength is normalized to the maximum field strength B_{max} and shown in color, according to the colorbar.

In case the coil L_1 is excited, the resulting B field distribution looks completely different, since now the top cover of the implant housing is the closest conductive part to the coil and since the absolute value of the B field declines with respect to $1/r$, where r is the distance from the source. The upper cover of the housing has the strongest eddy currents, which reduces the penetration of the B field into the housing and therefore also reduces the eddy currents in the base plate (see Figure 9).

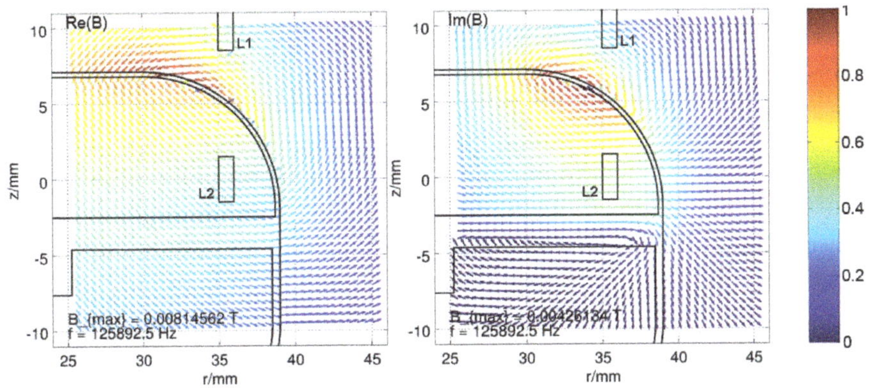

Figure 9. Resulting magnetic field vector distribution, caused by eddy currents in the housing, while exciting L_1, split into real part (Re(B)) and imaginary part (Im(B)) of the field. The field strength is normalized to the maximum field strength B_{max} and shown by the color, according to the colorbar.

6. Parameter Discussion

After analyzing the effects of the conductive housing for the transmission, the results can be used to discuss the parameters and further to optimize the inductive link. For eddy currents, the skin effect describes the behavior of the housing on the basis of the skin depth δ, which contains the operational frequency f, as well as the magnetic permeability μ and the electric conductivity σ of the used material. It is given by:

$$\delta = \sqrt{\frac{1}{\pi f \mu \sigma}}. \tag{25}$$

According to Equation (25), for example, if the conductivity σ of the housing increases, the same δ is achieved at a lower frequency f, which means that the curve progression of the correction factors from Equations (14), (15) and (24) is similarly shifted to lower frequencies with increasing conductivity and vice versa. The same effect takes place, if, instead of the conductivity, the thickness t_h increases because the resistance of the housing wall depends on the thickness of the material. These and further parameters are summarized and discussed in Table 1, which is categorized into electrical, coil and housing modifications.

Table 1. Discussion of advantages and disadvantages of different parameters for titanium encapsulated implantable devices with inductive data and power link.

Type	Parameter	Advantages	Disadvantages
electrical	higher frequency f	increased data rate	smaller skin depth δ, therefore higher attenuation and smaller range
	lower frequency f	bigger skin depth δ, therefore smaller attenuation and bigger range	decreased data rate
	higher transmission power	increased range	hard to realize inside the implant, because of the limited power source
	higher modulation rank	increased data rate	requires higher bandwidth, or reduces the Signal to Noise Ratio (SNR)

Table 1. *Cont.*

Type	Parameter	Advantages	Disadvantages
coil	increased geometric coil size	increased range	usually limited by the overall size of the implant and the space inside the housing
	higher quality factor	increased range	reduced bandwidth which leads to lower data rate
	lower quality factor	increased bandwidth which allows higher data rate	smaller range
	using litz wire instead of solid wire for coil windings	reduces proximity and skin effect, which reduces the ohmic losses and increases power transmission	usually more expensive
	intermediate resonant coil outside the housing, under the skin	increases range	more expensive, complicated setup, additional encapsulation necessary
housing	reduced conductivity of the material	increases skin depth, which reduces the attenuation and increases power transmission and range	approval for material usage necessary, eventually reduced mechanical properties
	reduced wall thickness	decrease attenuation by eddy currents, which increases power transmission and range	reduced mechanical strength
	non conductive material	no eddy currents, hence no negative effects of the housing	smaller longevity, biocompatibility and mechanical strength
	reducing hot spots of eddy currents	reduced attenuation, which improves power transmission and range	more complexity in housing and big effort to find the hot spots
	ferrite plate below the coil inside housing	deviation of the magnetic field and reduction of eddy currents, which improves power transmission and range	problematic for Magnetic Resonance Imaging (MRI) scans
	coil outside the housing	improved power transmission and range	lead through and additional encapsulation of coil with non conductive material necessary

7. Realization of a Communication Link for an Implantable Infusion Pump

In a third party project, the proposed analysis techniques were used for the design of a communication link for an implantable infusion pump. The requirements for the link were to manage the functionality of the implanted infusion pump (IIP) with an external controlling device (ECD). In practice, the IIP is located in the abdominal cavity below the skin. The ECD is placed in the vicinity of the IIP and allows a bidirectional communication with the ECD as a master device, as shown in Figure 10.

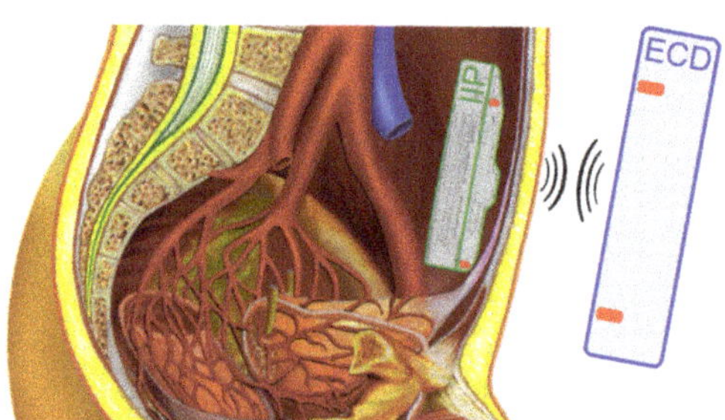

Figure 10. Schematic representation of the inductive link application, with the External Controlling Device (ECD) outside the body (**blue**), the Implantable Infusion Pump (IIP) inside the abdominal cavity (**green**) and the indicated transmission coils (**red**).

The purpose of the IIP is to deliver a continuous and specific amount of medication to the patient. Therefore, the implant has a drug bellow, surrounded by a tank with pressurized gas, which enables the drug to release. The flow of the medication is controlled by a valve that opens and closes in a timed pattern. Additionally, two pressure sensors allow measuring and adjusting the medication flow and a piezo buzzer can warn the patient in case of any system warning or error. For the data transmission, a DQPSK-ASIC (see [21]) is used in combination with a simple analog circuit and an inductive coil. The carrier frequency of the transmission is $f_0 = 125\,\text{kHz}$. The implant housing is made of grade 2 titanium and hermetically seals the complete interior of the implant, which includes the transmission coil, as shown in the model in Figure A1. This ensures a good receptiveness of the implant inside the patient's body, since the outer material is biocompatible. Furthermore, the design with the inlaying coil simplifies the development of the implant and also reduces the costs. Although there have been a few attempts for this design, the established data rate and also transmission distance have not been accomplished before. The complete system with all its components is depicted in Figure 11.

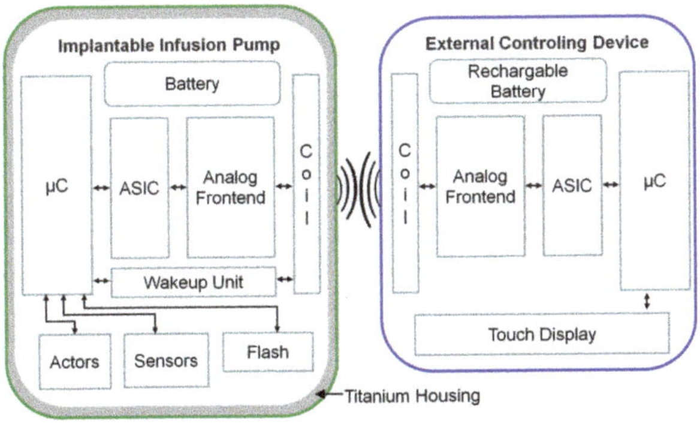

Figure 11. Block diagram of the complete electronics of the implantable infusion pump system, with the IIP on the left side and the ECD on the right side. The main transmission components are the micro-controller (μC), the QPSK-ASIC, the analog frontend circuit and the coil. Additionally, the IIP has sensors, actors, a flash memory and a wake-up unit. The ECD has a touch display, where all commands are handled. Both devices are powered by batteries.

The implant electronics are powered by a non rechargeable battery with a capacity of $C_N = 1600\,\text{mAh}$ and designed for a long lifespan by using low power components for the electric circuits and also keeping the μC in a deep sleep mode for most of the time. Only the real-time clock (RTC) is running permanently, which is responsible for the scheduling of the valve operation. This allows for operating the complete implant electronics at a standby current of $I_{stby} < 4\ \mu\text{A}$ and hence achieving an operation time of more than eight years. However, the ECD can wake up the implant by sending a specific wake-up packet. The wake-up unit in the IIP is also permanently powered and activates the micro-controller, if it receives this signal. The subsequently exchanged protocol ensures that only authorized devices can access the IIP; otherwise, it goes back to sleep. For transmission, the detuning of the resonance frequency by the metallic housing is compensated according to simulated results.

The shielding effect of the housing also has several advantages. It shields all signals, which are higher than approx. 1 MHz due to the skin effect. This enhances the electromagnetic compatibility of the device. It also dampens the signal strength of lower frequency signals, which only permits a short range wake-up and communication to the IIP. The transmission parameters for the IIP are listed in Table 2.

Table 2. Transmission parameters of the Implantable Infusion Pump (IIP).

Parameter	Description	Value
U_{in}	input voltage	1.8 V to 3.6 V
f	operating frequency	12 MHz
f_0	carrier frequency	125 kHz
B	bandwidth	≈10 kHz
R	data rate	≈10 kBit/s
	modulation	CP-DQPSK
I_{TX}	current for sending	<6 mA
I_{stby}	current in standby	<4 μA
d	transmission range	10 mm to 60 mm

8. Results

The paper presents simple methods to analyze the behavior of a conductive housing for active medical implantable devices with inductive transmission. Section 2 explains the usage of the FEMM tool, in order to simulate the conductive housing influences on the impedance and transmission parameters to the intersection point between magnetic field analysis and lumped element model. Section 3 describes the determination of the essential coil parameters in the form of a modified RLC model with additional parallel resistance for the quality factor. Based on this model, the correction factors ξ_L and ξ_R for the coil impedance in the vicinity of conductive objects are introduced, which result from the abstracted conductive object χ as an additional short circuited winding. In Section 4, the previous analysis is extended to an actual inductive transmission, based on a standard transformer model, expanded with χ. A new correction factor ξ_M is introduced, which enables for determining the transmission parameters, with attenuation and shifting of the resonance frequency. The comparison between measurement of a transmission and simulated results show a good agreement for configurations, where only one coil is affected by the conductive object. In this case, it means that the coil L_1 is not closer than 10% of the coil diameter to the conductive object. Section 5 focuses on the magnetic field distribution, caused by the housing, which is achieved by the superposition principle. For the analyzed housing, hotspots are in the base plate of the housing directly beneath the implant coil L_2 for the outwards transmission. For the inwards transmission, by L_1, the hotspot is in the top cover of the housing. From the results, parameter modifications are derived in Section 6 to discuss the advantages and disadvantages of the transmission of data and power. The discussion is categorized in electrical, coil geometry and housing parameters.

The results are also put in practice in Section 7, where a realization of an implantable infusion pump with external controlling device is described. This system achieves data rates of 10 kBits/s for distances of up to 60 mm and a longevity of more than eight years and went successfully to market in the year 2014. Until now, these parameters have not been achieved by other products in this sector.

9. Conclusions

The proposed analysis techniques enable a fast and easy way to determine the effects of a conductive object in the vicinity of inductively coupled coils for the impedance and transmission parameters. Conductive objects affect the impedance and therefore the transmission between the two coils, which strongly depend on the geometry of the complete transmission system and also on the working frequency. This is due to the frequency dependent eddy current distribution in the object.

The measurements and also the simulations show that the correction factor ξ_L for the inductance remains the same for the different coil configurations, as long as the geometry does not change. For the coil, this means that the outer dimensions have to be the same, while the number of turns and also the wire diameter may vary. This specifically allows for generalizing ξ_L for certain geometries.

However, this is not the case for ξ_R, which belongs to the equivalent series resistance (ESR) of the coils due to the strong dependency of the ESR from the current distribution in every winding, which relates to the skin effect, but also due to the inter-winding dependencies in form of proximity effects. Since both effects are frequency dependent, it is not possible to generalize the correction factor ξ_R for a certain geometry. Hence, all the effects have to be considered at once and, for every variation in the coil, a new calculation or simulation has to be performed.

The correction factor for the mutual inductance ξ_M is a complex value, whereas ξ_L and ξ_R can be seen as real values, which applies to both the inductance and the resistance of the coils. Therefore, it depends not only on the outer geometry, but also on the number of windings, the width of the coil wire, the space between the wires and also the winding configuration. Hence, the value influences the transmission of the inductance and also of the resistance and has to be simulated specifically for every coil configuration. A generalization is not possible in terms of lumped elements. Instead, it has to be considered on the magnetic field level.

The additional parameter discussion is categorized in three groups: the coil, electrical and housing variations. Therein, principle methods are described to enhance the data rate, distance or the transmitted power and thereby give new perspectives for fast development and implementation.

While the coupling coefficient between two distinct coils purely depends on the geometry of the coils and their arrangement, the coupling between a conductive object and a coil also depends on the distribution of the induced eddy currents in the object, which are strongly dependent on the frequency. Therefore, it is not possible to give general expressions for the coupling coefficient or for the object parameters like R and L. Nevertheless, the given model is a useful tool for the understanding of the behavior of a conductive object, since it allows for separating the resistive losses in terms of R and also the inductive interaction between the coil and the object in terms of L.

The FEM model, together with all simulation and calculation scripts are provided in the supplementary files. This FEM model can be expanded with details like the electronics and the printed circuit board (PCB), the medication fluid, gas and the tissue surrounding the housing. However, first measurements show no big deviations from the existing model and also the focus in this paper is on the general behavior of the housing and not specific details, which undergo much stronger variation for each application.

Additionally, this model also allows for modifying the electrical and magnetic properties of the housing material and also for modifying the coil parameters for further investigation and optimization.

Supplementary Materials: The complete simulation scripts, as well as the FEM models and the measurement data are available online at www.mdpi.com/1996-1944/11/11/2089/s1. Refer to README.txt.

Author Contributions: Conceptualization, W.G.; Data Curation, M.B.; Formal Analysis, W.G.; Funding Acquisition, D.J.; Investigation, W.G., M.B., D.J. and L.R.; Project Administration, D.J.; Resources, D.J.; Supervision, D.J. and L.R.; Writing—Original Draft, W.G.; Writing—Review & Editing, W.G., M.B. and D.J.

Funding: This research received no external funding.

Acknowledgments: This research was supported by the Albert-Ludwigs University, Freiburg, and Offenburg University, especially the Institute of Applied Research. Additional thanks goes to Tricumed Medizintechnik GmbH in Kiel, Germany, who helped to finance the research and who are also manufacturing the implantable infusion pumps, which enabled for verifying the theoretical methods. Further thanks go to xHain hack & make space Berlin, where I had the time and space to finish this article.

Conflicts of Interest: The authors declare no conflict of interest. The founding sponsors had no role in the design of the study; in the collection, analyses, or interpretation of data; in the writing of the manuscript, and in the decision to publish the results.

Abbreviations

The following abbreviations are used in this manuscript:

AC	Alternating Current
AIMD	Active Medical Implantable Device
ASIC	Application-Specific Integrated Circuit
BVTF	Backward Voltage Transfer Function
CP	Continuous Phase
DC	Direct Current
DQPSK	Differential Quadrature Phase Shift Keying
ECD	External Controlling Device
ESR	Equivalent Series Resistance
FEM	Finite Element Method
FEMM	Finite Element Method Magnetics[6]
FVTF	Forward Voltage Transfer Function
IIP	Implantable Infusion Pump
MRI	Magnetic Resonance Imaging
PCB	Printed Circuit Board
QPSK	Quadrature Phase Shift Keying
RTC	Real Time Clock
SNR	Signal to Noise Ratio

Appendix A. Estimation of Coil Parameters by Measurement

By measuring the frequency response of a coil, the parameters for the coil model in Figure 2 can be estimated. The DC resistance of the coil is $R_1 = \Re\{Z(f_{low})\}$, the inductance of the coil is $L_1 = \Im\{Z(f_{low})\} \cdot (2\pi \cdot f_{low})^{-1}$, where $f_{low} \ll f_0$. At the resonance frequency f_0, the parallel resistance $R_p = Z(f_0)$ can be measured because the inductance and the capacitance of the coil cancel out at resonance. The capacitance is then $C_p = \left(4\pi^2 \cdot f_0^2 \cdot L_1\right)^{-1}$. The coils used for the simulation as well as for the experiments are circular air coils with a diameter of $\varnothing_C = 71$ mm and a height of $h_C = 3.0$ mm with different wire diameters $\varnothing_W$ and number of turns N. The parameters, measured via Bode100 analyzer (OMICRON electronics GmbH, Klaus, Austria) [22], are described in Table A1.

Table A1. Coil properties.

Name	C	D	H	K	A	B
N	10	10	25	25	200	200
$\varnothing_W$/mm	0.4	0.4	0.2	0.2	0.1	0.1
L_s/μH	18.36	18.55	111.4	120.7	6909	6829
R_s/Ω	0.2845	0.2818	3.223	3.212	101.1	98.36
R_p/kΩ	55.94	55.44	42.44	62.30	312.8	354.8
C_p/pF	33.18	31.38	51.22	34.88	75.03	67.14
f_0/kHz	6448	6598	2107	2453	221.0	235.0

Appendix B. Description of the FEM Housing Model

The FEM model for the simulation of the housing, which is based on a housing prototype for an implantable infusion pump, is depicted in Figure A1. This model is axially symmetric to the z-axis. It consists of the active coils L_1 outside the housing and L_2 inside the housing (highlighted in red) and the passive housing (highlighted in blue). The housing itself is made of three parts: (A), (B) and (C). (B) is the base plate, which holds all parts of the implant together. (A) is the top cover, which encloses all electronic components. (C) is the bottom cover, which encloses the drug below and the pressurized gas. The surrounding medium is air. All dimensions are given in the table below the figure. For the simulation, the distance d_{12} between the coils is varied to estimate the coupling behavior. The material, used for the prototype and also the FEM model is grade 2 titanium, with an electrical conductivity $\sigma = 2.08\overline{3}$ MS/m (see [23]). The simulation was performed in FEMM 4.2 (see [6]) and controlled and analyzed in Octave (version 4.0.0, written by John W. Eaton and many other) (see [7]).

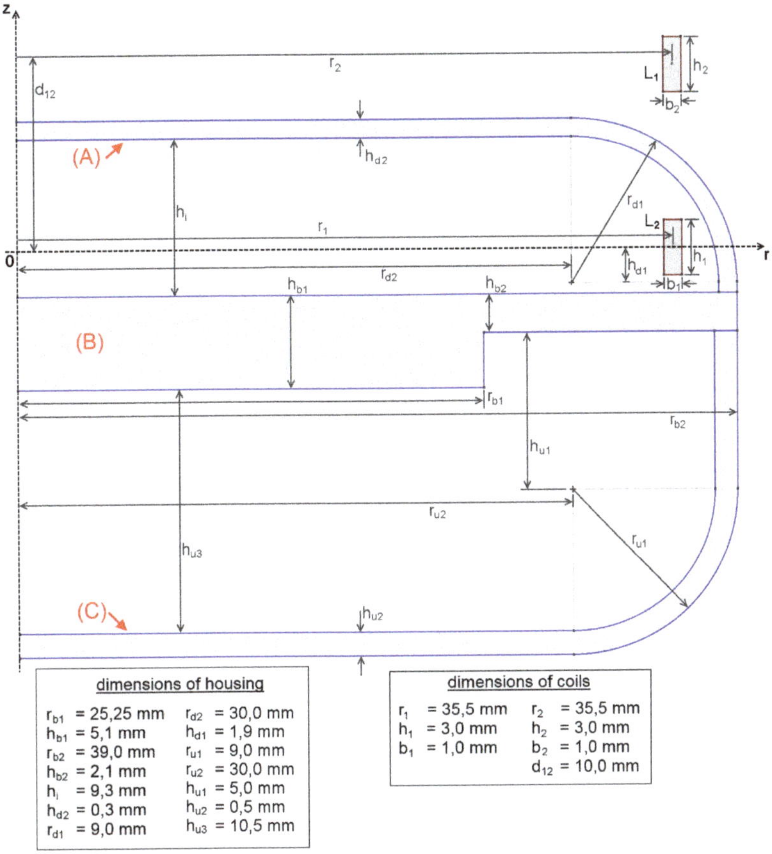

Figure A1. FEM model of the housing and the coils for the estimation of transmission parameters with dimensions, based on a prototype housing of an implantable infusion pump. This model is axially symmetric to the *z*-axis .

Appendix C. Complete Reformulation of the Transformer Equations for Three Windings

$$\begin{bmatrix} U_1 \\ U_2 \end{bmatrix} = \begin{bmatrix} \left(R_1 + j\omega L_1 + \dfrac{\omega^2 k_{1h}^2 L_1 L_h}{R_h + j\omega L_h}\right) & \left(j\omega k_{12}\sqrt{L_1 L_2} + \dfrac{\omega^2 k_{1h}\sqrt{L_1 L_h} k_{2h}\sqrt{L_2 L_h}}{R_h + j\omega L_h}\right) \\ \left(j\omega k_{12}\sqrt{L_1 L_2} + \dfrac{\omega^2 k_{1h}\sqrt{L_1 L_h} k_{2h}\sqrt{L_2 L_h}}{R_h + j\omega L_h}\right) & \left(R_2 + j\omega L_2 + \dfrac{\omega^2 k_{2h}^2 L_2 L_h}{R_h + j\omega L_h}\right) \end{bmatrix} \cdot \begin{bmatrix} I_1 \\ I_2 \end{bmatrix}, \tag{A1}$$

$$\begin{bmatrix} U_1 \\ U_2 \end{bmatrix} = \begin{bmatrix} (R_1 + j\omega L_1 \cdot (1 - k_{1h}^2 \cdot \chi)) & (j\omega\sqrt{L_1 L_2} \cdot (k_{12} - k_{1h}k_{2h} \cdot \chi)) \\ (j\omega\sqrt{L_1 L_2} \cdot (k_{12} - k_{1h}k_{2h} \cdot \chi)) & (R_2 + j\omega L_2 \cdot (1 - k_{2h}^2 \cdot \chi)) \end{bmatrix} \cdot \begin{bmatrix} I_1 \\ I_2 \end{bmatrix}. \tag{A2}$$

References

1. Lenaerts, B.; Puers, R. *Omnidirectional Inductive Powering for Biomedical Implants*, 3rd ed.; Springer: Berlin, Germany, 2009; ISBN 978-1-4020-9075-2.
2. Neves, H.P. Materials for implantable systems. In *Implantable Sensor Systems for Medical Applications*; Woodhead Publishing Series in Biomaterials; Imnann, A., Hodgins, D., Eds.; Woodhead Pub. Ltd.: Cambridge, UK; Philadelphia, PA, USA, 2013; Volume 52, pp. 3–38. [CrossRef]

3. Niinomi, M. Recent metallic materials for biomedical applications. *Metall. Mater. Trans. A* **2002**, *33*, 477–486. [CrossRef]
4. Jacobi, R.C.; Hennig, A.K.D. Simulation methods for inductively coupled sensor systems in varying environments. In *PRIME 2012*; IEEE: Piscataway, NJ, USA, 2012; pp. 1–4. ISBN 978–3–8007–3442–9.
5. Geselowtiz, D. B.; Hoang, Q.G.R.P. The effects of metals on a transcutaneous energy transmission system. *Biomed. Eng. IEEE Trans.* **1992**, *39*, 928–934. [CrossRef] [PubMed]
6. Finite Element Method Magnetics: HomePage. Available online: http://www.femm.info/wiki/HomePage (accessed on 24 August 2015).
7. GNU Octave. Available online: https://www.gnu.org/software/octave/ (accessed on 21 December 2017).
8. Meeker, D. Finite Element Method Magnetics. 2018. Available online: www.femm.info/wiki/Files/files.xml?action=download&file=manual.pdf (accessed on 1 October 2018).
9. Ida, N. *Engineering Electromagnetics*, 3rd ed.; Springer International Publishing: Cham, Germany, 2015; ISBN 978-3-319-07805-2, doi:10.1007/978-3-319-07806-9. [CrossRef]
10. Hayt, W.H.; Buck, J.A. *Engineering Electromagnetics*, 8th ed.; McGraw-Hill: New York, NY, USA, 2012; ISBN 978-0-07-338066-7.
11. Weißgerber, W. *Elektrotechnik für Ingenieure 1*; Springer Fachmedien Wiesbaden: Wiesbaden, Germany, 2013; ISBN 978-3-8348-0903-2, doi:10.1007/978-3-8348-2065-5. [CrossRef]
12. Agilent Technologies. Agilent Impedance Measurement Handbook: A Guide to Measurement Technology and Techniques. Available online: http://home.deib.polimi.it/svelto/didattica/materiale_didattico/materialedidattico_MRF/appnote/handbook_imp-meas.pdf (accessed on 1 August 2018).
13. Massarini, A.; Kazimierczuk, M.K. Self-capacitance of inductors. *Power Electron. IEEE Trans.* **1997**, *12*, 671–676. [CrossRef]
14. Lotfi, A.W.; Gradzki, P.M.L.F.C. Proximity effects in coils for high frequency power applications. *Magn. IEEE Trans.* **1992**, *28*, 2169–2171. [CrossRef]
15. Yang, Z.; Liu, W.; Basham, E. Inductor modeling in wireless links for implantable electronics. *IEEE Trans. Magn.* **2007**, *43*, 3851–3860. [CrossRef]
16. Huang, K.; Zhou, Y.; Wu, X.; Liu, W.; Yang, Z. Design and optimization of inductive power link for biomedical applications. In *Applied Biomedical Engineering*; Chapter 10; Gargiulo, G.D., McEwan, A., Eds.; IntechOpen: Rijeka, Croatia, 2011; pp. 221–242, doi:10.5772/18562. [CrossRef]
17. Donaldson, N.d.N. Effect of the metallic seal of a hermetic enclosure on the induction of power to an implant. *Med. Biol. Eng. Comput.* **1992**, *30*, 63–68. [CrossRef] [PubMed]
18. Weißgerber, W. *Elektrotechnik für Ingenieure 2: Wechselstromtechnik, Ortskurven, Transformator, Mehrphasensysteme*; 8th ed.; Springer Fachmedien Wiesbaden: Wiesbaden, Germany, 2013; ISBN 978-3-8348-1031-1, doi:10.1007/978-3-8348-2072-3. [CrossRef]
19. Cheng, D.H.S. The reflected impedance of a circular coil in the proximity of a semi-infinite medium. *IEEE Trans. Inst. Meas.* **1965**, *IM-14*, 107–116. [CrossRef]
20. Shinagawa, H.; Suzuki, T.; Noda, M.; Shimura, Y.; Enoki, S.; Mizuno, T. Theoretical analysis of AC resistance in coil using magnetoplated wire. *Magn. IEEE Trans.* **2009**, *45*, 3251–3259. [CrossRef]
21. Gruenwald, W.; Jansen, D. A digital low frequency transceiver for biomedical implants with enclosed titanium housing. In Proceedings of the 2012 International Conference on Signals and Electronic Systems (ICSES 2012), Wrocław, Poland, 18–21 September 2012; pp. 1–4. [CrossRef]
22. Omicron Lab. Available online: https://www.omicron-lab.com/fileadmin/assets/bode/Documents/Bode_Technical_Data_V2.7.pdf (accessed on 12 December 2016).
23. Thyssen Krupp Schulte. Titan—Werkstoff mit Perspektive. Avaliable online: https://www.yumpu.com/de/document/view/34917172/titan-werkstoff-mit-perspektive-thyssenkrupp-schulte-gmbh (accessed on 1 August 2018).

MDPI
St. Alban-Anlage 66
4052 Basel
Switzerland
Tel. +41 61 683 77 34
Fax +41 61 302 89 18
www.mdpi.com

Materials Editorial Office
E-mail: materials@mdpi.com
www.mdpi.com/journal/materials

www.ingramcontent.com/pod-product-compliance
Lightning Source LLC
Chambersburg PA
CBHW061840210926
51850CB00006B/934
9783039289875